Flying the Frontiers

NACA and NASA Experimental Aircraft

Arthur Pearcy

Naval Institute Press
Annapolis, Maryland

Dedication

"We shall not cease from exploration
And the end of all our exploring
Will be to arrive where we started
And to know the place for the first time"

T. S. Eliot, 'Little Gidding'

This book is dedicated to all those aerodynamic explorers, past and present, who have served the NACA and the NASA faithfully since 1915, researching horizons unknown for the sake of mankind and the future.

The National Aeronautics and Space Administration maintains an internal history programme for two principal reasons: Firstly, sponsorship of research in NASA-related history is one way in which the NASA responds to the provision of the National Aeronautics and Space Act of 1958 which requires NASA to "provide for the widest practicable and appropriate dissemination of information concerning its activities and the results thereof." Secondly, thoughtful study of NASA's history can help agency managers accomplish the missions assigned to the agency. Understanding NASA's past aids in understanding its present situation and illuminates possible future directions. The opinions and conclusions set forth in this book are those of the author; no official of the agency necessarily endorses those opinions or conclusions.

First published in the UK in 1993 by
Airlife Publishing Ltd.

Published and distributed in the United States of America and Canada by the Naval Institute Press, 118 Maryland Avenue, U.S. Naval Academy, Annapolis, Maryland 21402-5035

ISBN 1-55750-258-7

Library of Congress Catalog Card No. 92-85550

Printed in Singapore by Kyodo Printing Co. (S'pore) Pte. Ltd.

Contents

Foreword by Ron Gerdes   6

Acknowledgements   7

Author's Note and Preface   8

Introduction   9

Chapter One   The Birth of the NACA   12

World War I — revival of interest in Langley's old laboratory. Walcott determined to establish agency for advancement of aviation. Act passed in 1915 creating NACA. Committee formed and approved. Survey of universities, schools, etc. Excavation of first laboratory at Langley, Virginia July 1917. Construction of airfield by US Army. Recommended move to Bolling Field. Dedication of NACA Langley on 11 June 1920. Law establishing the NACA — copy of document.

Chapter Two   Virgin Days at Langley   16

Construction of first wind tunnel 1919. First flight test programme with Jenny aircraft. Roll of engineering test pilot recognised. Comparison of wing-model test and full-scale flight test. Fleet additions during 1921. Log book records, more additions. Interest by US military services. Flight test of US Army and Navy types. Purchase of first NACA aircraft. Collier Trophy for NACA in 1929. More aircraft including autogyro. Second engine research laboratory added 1925. John Victory fights for finance. Hoover signs order to abolish NACA in 1932. House of Democrats kill any merger plans. Aircraft borrowed. Relationship with US Army and Navy. Sperry M-1 Messenger arrives, used 1924/29. Rotary wing research. Flight research laboratory opened 1932. NACA pilots fly autogyro. Rigid blade concept. Interest in helicopters.

Chapter Three   World War II — Priorities and Expansion   24

Langley laboratory enlarged, new installations, personnel moved to Ames and Cleveland. New prototype aircraft tested for both Army and Navy. Bell P-39 and cooperation with manufacturers. More World War II types tested. High speed flight and jet propulsion. Bell XP-59 Airacomet. Dropping of instrumented bodies from B-29. Flying wind tunnel and transonic research. Helicopter research. Sikorsky R-4B for tri-service use. Increase in NACA wartime personnel. Gas turbine technology.

Chapter Four   West to California   32

Preparedness for war. New laboratory at Moffett Field. Challenge from Congress. Construction at Moffett, named Ames in 1940. Consideraton of Sunnyvale site, advantages, which were many. History of base. Ice research transferred from Langley. De-icing systems. Smith De France appointed engineer-in-charge. Seth Anderson describes early days at Ames. New hangar built. Evaluation of new military aircraft. High-speed flight research. Ames Flight Research Section. Lockheed P-80 flight test. Wind tunnel research on jet aircraft engine installation. Models tested in 16ft tunnel. A look into the supersonic age.

Chapter Five   Further Wartime Expansion   40

Authorisation for Aircraft Engine Research Laboratory at Cleveland. New facility built. Transfer of personnel from Langley. First aircraft arrives. Fuels, lubricants, tunnels and test laboratories. Many engine types tested. Whittle gas turbine engine performance investigated. Tests into performance of Allison engine. Supercharger and jet propulsion. Workload of static test laboratory. Sub-sonic ram-jet test-bed research. Further expansion for guided missiles. Wallops Island developed. Pilotless aircraft research division moved to Wallops. Rocket propelled device developed. Model research. Helicopter research at Langley. Many new helicopter manufacturers. Purchase of helicopters by NACA. Convertiplane arrives. Hingeless rotor helicopter.

Chapter Six   The Road to Sonic Flight   48

German advanced aerodynamics. Supersonic sweep. Robert T. Jones. Model testing on Mustang. 'Operation Paperclip'. Transonic flight research with F-86 Sabre. John Stack. Bell research team and XS-1. High speed aircraft for US Navy. Contract to Douglas for D-558. Description of contract. Construction of Bell XS-1. Langley contingent to Muroc. NACA team established. Contractor's flight testing of XS-1. USAAF test pilots. Chuck Yeager and sonic boom. Robert J. Collier Trophy award to John Stack. Programme details of XS-1 and D-558. Transonic flight problems.

Chapter Seven   Shapes for the Future   58

High-speed wind tunnel tests and programme. Test models dropped. Ames investigates buffeting. Variable stability aircraft. Flight test hazards. Century series of fighters. Hypersonic high-flying research aircraft investigation. Richard T. Whitcomb at Langley. Area rule applied. Muroc becomes Edwards Air Force Base. NACA flight operations. Douglas D-558-2 Skyrocket research flights. Research aircraft programme expands. Bell X-5. Douglas X-3. Convair XF-92A. Northrop X-4. Second generation of Bell X-1s. X-3 Stiletto. Summary of X-series and NACA research aircraft. NACA test pilot flight record.

Chapter Eight   Advanced Research Aircraft   74

Research aircraft programme expanded. High altitude and Mach 4 to Mach 10. Becker group and tentative new design for Mach 7 research aircraft. North American X-15 contract awarded. Rocket engine selected. Flight test and introduction of the century series of fighter and interceptor aircraft. Mysterious crashes with F-100. The arrival of NASA. Introduction of the X-15. Biomedical aspects. Century series aircraft operated.

Chapter Nine   Transition to NASA   84

Transonic area rule. Variable stability aircraft. Competition for technical manpower. Boundary layer control at Ames. VTOL and STOL. Douglas C-47 workhorse. Simulator trials. Laboratory projects related to space flight. US space programme outlined, and planning of same. Orbits and trajectories plus lifting vehicles. The NACA ceases to exist.

Chapter Ten   Ames and V/STOL Research   94

Ames acquires more land and extends. Space Sciences Research Laboratory. Growing fleet of research aircraft. Coronado flying laboratory. Ryan VZ-3 Vertiplane. Light aircraft technology. Starlifter flying astronomical laboratory. STOL research. Survey of archaeological sites etc. Studies on Mach 3 bomber. Competition for contract and XB-70A. V/STOL studies. Flight test of various types. Convertiplanes. Bell-XV-3 and flight test programme. XV-15 tilting testbed. Transfer of helicopter research from Langley to Ames. List of research convertiplanes and tilt rotor vehicles.

Chapter Eleven   Blackbirds, Valkyrie, to the Shuttle   104

X-15 research flights and records. X-15A-2 programme. X-20 Dyna-Soar. Douglas F5D-1 Skylancer at Edwards. Space Task Group and trials with F-104. Rogallo-type 'Parawing' request, construction and flight trials. Flying Bedsteads for lunar landing vehicle. Astronauts fly LLVs. North American YF-12As to NASA. YF-12C joins flight test programme. Langley and SST research. Edwards and NASA involvement. Variety of aircraft utilised. General Purpose Airborne Simulator Jetstar. XB-70A Valkyrie. Loss of a Valkyrie and NASA pilots. F-4 Phantom with NASA. Robert D. Reed and lifting bodies, the family and flight programme. The ugly lifting bodies. Space Shuttle discussed. Martin Marietta X-24B.

Chapter Twelve   The New NASA   122

Sputnik and US space research programme. Advanced Research Projects Agency formed. The act establishing NASA and a new organisation. Goddard Space Flight Center and its aircraft fleet. Johnson Space Center and mission control. WB-57F aircraft used for research. Lockheed NP-3A and NC-130B aircraft introduced. Earth Survey programme commences. Jet Propulsion Laboratory and its fleet. Kennedy Space Center and Marshall Space Flight Center, supporting units plus aircraft fleet. Pregnant Guppy. Update of research centers.

Chapter Thirteen   Diversity at the NASA Centers   134

Establishment of Launch Operations Directorate. Super-critical wing and F-8U Crusader, flight test of same. Transonic Aircraft Technology programme and F-111A. Advanced Fighter Technology Integration. Remote piloted vehicles. F-15 RPVs. Fly-by-wire on F-8C Crusader. Integrated Propulsion Control System and F-111E. Aviation safety. Lockheed F-104N fleet. New fighters used. Contracts awarded for Space Shuttle. Air launch of *Enterprise* from Boeing 747 and further flights. Flight Research Center named Dryden. Tilt Rotor Aircraft Office established at Ames. Bell XV-15 programme, Langley introduces Boeing 737. Cooperation with United Kingdom at RAE Bedford. Ames aircraft on international overseas expeditions. Variety of advanced research aircraft at Ames. Chinook flight simulator, Langley to Ames. Rotor Systems Research Aircraft.

Chapter Fourteen   Earth Resources Aircraft Project   146

Ames investigates terrestrial and astronomical phenomena. Research aircraft acquired including U-2. Elaborate cover story and background to spyplane. Lockheed involvement with U-2 at Ames plus camera configuration and operations. Overseas surveys. High altitude sampling programme activity with Earth Resources aircraft. Deployments and updates. Douglas DC-8 acquired by Ames. The problem of CFCs — chlorofluoro-carbons. Global ocean flux study and P-3A Orion. TR-1A to Ames. List of Earth Resources Aircraft.

Chapter Fifteen   Past, Present and Future   160

Seventy-five years of aeronautical research. Types employed by NASA and survey. Drag research with F-15 at Dryden. Highly Manoeuvrable Aircraft Technology RPVs at Dryden. Mini-sniffer. Aircraft safety and crash worthiness design experiments. Satellite communications. Gulfstream shuttle trainers at Johnson. Coronado testbed for shuttle landing systems. Acronyms. F-18 Hornet to Dryden. Stratofortress veteran. *Pegasus* air-launched booster. Second Boeing 747 shuttle carrier acquired. Douglas C-47 workhorse. History of F-104 operations. Rotary-wing research. V-22 *Osprey*. British Kestrel and Harrier. Boeing 707 *Dash-80* Wallops research aircraft. Introduction of X-29 and X-31. National Aerospace Plane — NASP. Fuel efficiency. Programme for the future. List of F-104 test aircraft.

Appendix One   The NACA/NASA Aircraft Markings   172
                    and Numbering System

Appendix Two   Aircraft — Langley Research Center   180

Appendix Three   Aircraft — Ames Research   187

Appendix Four   Aircraft — Lewis Research Center   191

Appendix Five   Aircraft — Dryden Flight Research Facility   193

Appendix Six   Aircraft — Johnson Space Center   197

Appendix Seven   Aircraft — Wallops Flight Facility   199

                    Bibliography   200

Foreword

This book is about a subject that is very close to my heart — a history of NACA/NASA aircraft spanning three-quarters of a century. I was privileged both to observe and be part of that history for a 33-year period.

NACA-Ames Aeronautical Laboratory employed me as a summer-time engineering student while attending the University of California at Berkeley in 1955 and 1956. Upon graduation the following year I hired on at Ames with high hopes of being selected for an engineering test pilot position. I had to be content with plotting data and report writing until the summer of 1961 — then airborne at last. Those three decades, from the day I plotted my first data point as an engineering aide until my final flight as an engineering test pilot in NASA 719 (British Aerospace AV-8C BuNo. 158387) one week prior to my retirement in early 1988, were certainly filled with challenge and rewarding adventure.

Arthur Pearcy and his lovely wife Audrey have been close personal friends for several years now, and I have come to admire and respect his talents as an aviation writer and historian. The saga of NACA/NASA aeronautical research, frequently overshadowed in the media by space programme events, needed to be told. When Arthur first approached me with his decision to write a history of NACA/NASA aircraft research, I was somewhat awestruck at the thought of such a monumental undertaking — maybe second only to placing two men on the moon. However, as the reader will soon discover, he was well qualified and fully up to the task, having recently authored *A History of US Coast Guard Aviation* in addition to numerous historical publications pertaining to the development and deployment of the venerable DC-3 and its variants.

Armed with endless energy and an uncanny knack for digging out the facts, Arthur Pearcy systematically leads us from the days of World War I, when America found itself shamefully behind in aircraft production and research facilities, to its present place of world leadership in aerospace research and development. This is the story of how NASA and its predecessor, the National Advisory Committee for Aeronautics, played a key role in that rise to leadership. It is not just a story about airplanes and pilots, but about men and women dedicated in an effort to keep the United States at the forefront of aeronautical research.

Ron Gerdes

Acknowledgements

On 26 July 1983, fellow author and aviation historian Robert Dorr introduced me to Robert L. Burns, a NASA employee from the Goddard Space Flight Centre. This resulted in an immediate exchange of correspondence on the history and unique numbering system adopted for NACA/NASA aircraft. Bob revealed that he had no less than twenty-four volumes of mostly eight by ten prints plus a large collection of colour slides. One really need say no more, for from our first contact, and our eventual meeting at his home in Silver Spring, Maryland, Bob has been responsible for providing all the massive listing of aircraft and information gleaned over many hours of meticulous research into the NACA/NASA archives at many locations. Credit must go to Bob Burns for the Appendices of aircraft in this volume. This is the very first time they have all appeared in print.

My next contact some years ago — when the thought of this volume was in its infancy — was my good friend and NASA archivist Lee Saegesser, who has hosted me in the archives during research visits to the NASA HQ over the years. Lee has guided me to the source of so much valuable information and photos. Lee is housed in the NASA History Division whose Director, Dr Sylvia D. Fries, is well known. The NASA Public Affairs Officer, Mary L. Sandy, located in the headquarters building in Independence Avenue, Washington DC, has been a tower of strength to the project, supplying books and information, plus contacts at various NASA centres.

Ames in California has produced some good solid friends and contacts including our good friend Ron Gerdes, whose encouragement and enthusiasm for the book project has been second to none. Lynn Albaugh head of the Ames Photo Library guided us through the massive microfilm records and patiently offered her experience. Don James in the Public Affairs Division processed the photos and became involved. Seth Anderson a veteran NACA/NASA employee consented to use some of his World War II historical experiences.

At Langley we were introduced by Mary Sandy to Richard Layman who has, on our behalf, dug into the archives and produced the impossible — well, almost. Dick is responsible for all the photos illustrating Langley. A. G. Price together with Harry Verstynen, the latter Head of Aircraft Operations Branch, have together answered all my queries on aircraft. Elizabeth Guthridge Redors kindly provided graphics in the form of the beautiful NACA logo produced on the title page.

Joyce B. Milliner, Public Affairs Officer for many years at the Wallops Flight Facility, and her staff have provided photos and information. Sunil Gupta made available a box of colour slides he took of Wallops aircraft plus others, all adding to the large collection. Down in dusty Dryden, Cheryl Gumm in the Office of History HQ Air Force Flight Test Centre (AFSC) plus friend Don Haley, Public Affairs Officer with the NASA, have between them produced some excellent photos covering the experimental era of both NACA and NASA at Edwards AFB. At Lewis, located at Cleveland on Lake Erie, the NASA Director of External Affairs, Americo F. Forestieri assisted by Sherrie Campbell have assisted in every way possible to ensure we have coverage of this important NASA research centre. Finally, at the huge complex at the Johnson Space Centre in Houston, Texas, initial thanks to James E. Hannigan with whom I corresponded with some years ago, and more recently to Jeffrey Carr in the Public Affairs Office. Nina M. Lawrence, Director of the Legal Division of the NASA guided us on the use of official logos etc.

This volume was well supported by being able to read and study NASA-sponsored books written over the years by experts in the field of aerodynamic research and aeronautical history. Some of these excellent books are out of print, but others are available from the Superintendent of Documents, US Government Printing Office, Washington DC, 20402. (See Bibliography). These authors include the late author and friend David A. Anderton, Alex Roland, George W. Gray and Frederick B. Gustafson. Special thanks go to the following for granting permission to quote from their NASA volumes: Dr Richard P. Hallion (*On the Frontier*); Dr James R. Hansen (*Engineer in Charge*); Elizabeth A. Muenger (*Searching the Horizon*); and Roger E. Bilstein (*Orders of Magnitude*). Publisher and author and good friend Jay Miller of Aerofac publishers, who wrote that magnificent volume *The X-Planes* not only granted permission to use extracts from the book, but hosted Audrey and I at his home in Arlington, Texas, enabling me to research through his large photo library and book collection. His delightful wife Susan catered for our domestic needs. Here in the United Kingdom, friend and author Chris Pocock who produced *Dragon Lady* — the history of the U-2 spyplane — assisted and gave permission to extract from this Airlife product. Others including *British Aviation Review*, the monthly journal of the British Aviation Research Group; *Aviation Letter*; *Aviation Week & Space Technology* and many more. Nigel Macknight, Editor/Publisher of *Space Flight News*, a monthly NASA-orientated UK publication, supported the project.

Historians and private collectors who assisted in many ways include Peter M. Bowers; Edward J. Davies; Frank A. Hudson; William T. Larkins and Ralph Peterson, the latter driving many miles in order either to visit NASA establishments or museums containing NACA or NASA aircraft. Brian Pickering of Military Aircraft Photographs, and Brian Stainer of Aviation Photo News, both offered slides and photos from their massive collection. The Superintendent of the Royal Aerospace Establishment at Bedford, Mike Dobson, most graciously allowed me to procure photographs of NASA's Lockheed XH-51N helicopter, which the RAE flight-tested. Our good solid friend Boardman C. Reed vividly detailed a flight he had with Ron Gerdes on Thursday 28 February 1980, in the Lockheed L-300-50A-01 NASA 714 N714NA from Ames. How this guy gets such exotic treatment I shall never know.

Dave Menard, a friend of long standing, introduced the book project to Nick Apple, Public Affairs Specialist at the huge USAF Museum at Wright-Patterson, Dayton, Ohio, so providing photo requirements. The US aerospace manufacturers have been more than co-operative and generous with both information and photographs. Eric Schulzinger, Chief Photographer in the PR Department at Lockheed has shared some of his wonderful photos with us. Harry Gann at the Douglas Aircraft Company, Long Beach, has dug into the archives, whilst at Grumman both Lois Lovisolo, Corporate Historian and Herman J. Schonenberg in the History Centre, answered all my many requests. At the Northrop Corporation, Department 173/CC decided to remain anonymous but have been of tremendous assistance. Terry Arnold and Dick Tipton with Bell Helicopter Textron have constantly updated me on the antics of the tilt-wing family, especially the successful XV-15. Likewise friend Bill Tuttle, Public Relations with Sikorsky Aircraft, has updated me on the status of the X-wing, RSRA and others.

Lastly to Audrey my wife, partner and proof reader, plus Alastair Simpson of Airlife, now joint Chairman and Managing Director whose understanding, patience and guidance has made this interesting project so worthwhile. I apologise profusely and hang my head in shame if I have omitted anyone — not intentional I can assure you. However, I nearly did forget to thank my good friends Craig and Mike at Studioprint, Studio 13, in nearby Rushden, who do all my photocopying, reducing, enlarging, colour and even typesetting when required.

A final tribute to all those within the huge NASA organisation, for without their indulgence to the author, initially a stranger from a foreign land, this volume would never have been possible. Thank you one and all . . .

Arthur Pearcy Jr AMRAeS
Dakota, 76 High Street
Sharnbrook, Bedfordshire

Author's Note and Preface

The Concise Oxford Dictionary describes the word 'curiosity' as 'the need to know.' There is a saying that curiosity killed the cat. It was certainly curiosity which instilled my good friend Jay Miller to compile his excellent X-plane book. I feel sure it was also curiosity that inspired Bob Burns to investigate the aircraft used by the NACA and the NASA over the years. Personally, I have to admit that curiosity was at the source of my own investigation some years ago into the NACA/NASA aircraft numbering system. However, I am by no means the first who has attempted the feat of collating this information. In addition to Bob Burns from Goddard, there Seth Anderson and Ron Gerdes at Ames, plus private historians such as William T. Larkins and Peter Bowers — just to name two.

My own life, and indeed that of my wife and partner Audrey, has been steeped in aviation. My interest stemmed from a quite early age when my father — Arthur Snr — took me during the biplane period of the 1930s to see Alan Cobhams' air circus near my home town of Driffield, East Yorkshire. My first flight, as a prospective air cadet, was made in the summer of 1939 in a *Whitley* bomber from RAF Driffield, from where I made many subsequent flights as a cadet in No. 1233 Squadron Air Training Corps during and after World War II. During 1943 a secondment to the Royal Observer Corps, still at Driffield, introduced me to radar in the form of a Canadian gun-laying — GL — set used to plot the Luftwaffe inbound over Flamborough Head on a bombing run to the busy port of Hull on the River Humber.

In the early 1950s I became a keen member of the Brough Branch of the Royal Aeronautical Society, and on Monday 21 December 1953 I attended a '50th Anniversary of Powered Flight' dinner at the Guildhall in Hull. A memorable occasion. It was to be 30 July 1964 before I received my parchment as a full member of this famous society founded in 1866.

During March 1954 I was seconded by the British Air Ministry to the US Air Forces (Europe) under Project Native Son, to be employed in the busy operations with the 7531st Air Base Squadron, 7500th Air Base Group, part of the huge US Third Air Force. This was at Bovingdon, Hertfordshire, and we were equipped with a dozen sturdy Douglas C-47 Skytrain or 'Gooney Bird' transports. My daily diary records that on Tuesday 2 August 1955, a Lockheed T-33A-1-LO 54-1536 flew in from the 417th Fighter Bomb Squadron based at Hahn, Germany with North American F-86 *Sabre* aircraft. The pilot was none other than Major C. E. Yeager 16076A, who had flown in to talk to Third Air Force personnel on his Bell X-1 experiences at Edwards AFB. He returned to

Germany two days later, and I had the opportunity to talk with Chuck and process his USAF Form DD175 flight plan. Another visitor who passed through at this time was Lieutenant Colonel John Stapp, an aviation physician who carried out the high-speed sled rides at Edwards AFB.

After four and a half wonderful years with the USAFE, during which I met a control tower operator named Audrey, we paired up as a team and joined the staff of the Royal Aircraft Establishment at Farnborough, Hampshire — the home of UK research and development, steeped in history since the early days of flying. In 1958 we transferred to a picturesque air base in North Wales to be trained as part of a team who flew target drones, or RPV's as they are now known. We operated a pitch control wobble stick and helped to land these unique but successful vehicles which we provided as targets for the missile range. Days of no telemetry and drones with fixed attitudes only, no brakes but a huge tough deck hook. In March 1982 I was transferred south to RAE Bedford, leaving Audrey to work hard flying the drones for *Harrier* pilots destined for war service in the Falklands conflict. On final retirement Audrey joined me and I finally retired from Air Traffic Control duties in September 1986. We liked the area so settled for the thatched-cottage village of Sharnbrook.

We commenced frequent visits to the USA in the Spring of 1967, and it was some years later, whilst visiting Eastern Air Line friends in Miami, Florida, that I met up with friend Frank Borman who in December 1968, made the first manned orbits of the Moon in Apollo 8 — LTA-B. Some years later whilst in sunny California we were invited to visit NASA at Ames and met Ron Gerdes for the first time. Shortly after he was here in Sharnbrook with us on detachment from NASA to fly our two-seat *Harrier* aircraft. Ron was following in the footsteps of veteran NASA pilots such as Fred Drinkwater and Neil Armstrong who visited the Royal Aircraft — now Aerospace — Establishment at Bedford to fly the many unique British research aircraft.

During a US Coast Guard Aviation banquet held in Elizabeth City, North Carolina, we made friends with Commander Bruce E. Melnick — one of the group of fifteen astronaut candidates selected by the NASA in June 1987. Bruce is a Mission Specialist selected to be on the crew of STS-41 with the Shuttle in October 1990.

I was recently reminded of a quote from rocket pioneer Dr Robert H. Goddard which characterizes NACA and NASA aeronautic research. 'The dream of yesterday is the hope of today and the reality of tomorrow.'

Introduction

On 8 December 1903, Samuel Pierpoint Langley's *Aerodrome*, a manned full-size 48ft span powered aircraft, crashed after fouling its launcher on the Potomac River. Its pilot, Charles M. Manly, was rescued. After the accident the US government withdrew its financial assistance. Nine days later — at Kill Devil Hills, near Kitty Hawk, North Carolina — the Wright brothers achieved the world's first powered, sustained and controlled flight by an aircraft with their *Flyer I*. On 26 May the following year, the Wright brothers commenced flights in their aircraft *Flyer II*, using the weight-and-derrick method of launching for the first time on 7 September. On 9 November Wilbur Wright achieved his first flight of over five minutes duration. During June 1905 the Wrights completed the world's first wholly practical aircraft, the *Flyer III*. On 4 October Orville made his first-ever flight of over half-an-hour's duration, but after 16 October neither brother flew again until 6 May 1908. At this time no one else in the world had succeeded in emulating the first flights by the Wright brothers.

A chronology of the achievements of the National Advisory Committee for Aeronautics or the National Aeronautics and Space Administration between 3 March 1915 and the present day, covering seventy-five years of aeronatical research, would require numerous volumes in order to do it full justice. In this one volume an attempt to has been made to put an emphasis on the many aircraft employed since the first Curtiss JN-4H *Jenny* of 1919, including many of the research contributions and at the same time sketching a broad outline covering the activities of the two agencies involved. Much and many have been omitted for obvious reasons. Aeronautical research and development has taken great strides between the time when the *Jenny* was employed at Langley and today, when high-technology research craft such as the Grumman X-29 and others are braving the elements in preparation for the future.

Despite the Wright brothers' production of the first successful aeroplane, the United States had shamefully allowed itself to slip far behind the major European powers in the development of both aircraft and aeronautical research facilities. When World War I erupted in 1914, it was reported that France had 1,400 aircraft, Germany 1,000, Russia 800, Great Britain 400 and the United States only twenty-three. These other powers had seen the value of aeronautical research laboratories and facilities as far back as early 1866. In that year the Royal Aeronautical Society was formed, to stimulate research and experiments and to interchange information. Soon after 1870 two members of the society, Herbert Wenham and Horatio Philips, invented wind tunnels. On 14 May 1909 Samuel Franklyn Cody made the first flight of over one mile in the United Kingdom in *British Army Aeroplane No. 1* at Laffan's Plain, Hampshire.

Laffan's Plain was on the doorstep of the Royal Aircraft Factory, which between 1906 and 1916 designed thirty different types and actually built approximately 500 aircraft.

The National Advisory Committee for Aeronautics seen in session at their semi-annual meeting on 18 April 1929 in the NACA headquarters in Washington DC. Dr Joseph S. Ames, the chairman, is at the head of the table. Third figure on the left is Dr Orville Wright. (NACA 90-3727)

This site has long since incorporated the 1,400 acres which became the Royal Aircraft Establishment, Farnborough. The factory had been forbidden to design and construct production aircraft; its Superintendent Mervyn O'Gorman was dismissed, and a British government edict dictated that henceforth Farnborough should confine itself to 'the research and advisory aspects of aeronautics.' It has done that ever since, and over the years has co-operated closely on many projects involving both NACA and NASA. This has involved an exchange of not only information but research and design engineers, flight test pilots and even aircraft.

France had major installations; Sustare Eiffel's privately-owned wind tunnels were at the foot of the Eiffel Tower and at Auteuil. There was an Army aeronautical laboratory at Chalais-Meudon, and the Institut Aerotechnique de st. at Cyr. Germany had research laboratories at Göttingen University and at the technical colleges at Aachen and Berlin. The German government operated a laboratory at Adlershof, and the aircraft industry was well equipped with research facilities. Both Italy and Russia had aeronautical laboratories long before the United States.

On 20/21 May 1927, Captain Charles Augustus Lindbergh made the first solo non-stop crossing of the Atlantic in the single-engined Ryan monoplane *Spirit of St Louis*, flying from Long Island, New York to Paris. This history-making flight drew world-wide attention to the potential of the aeroplane and gave an impetus to aviation that no other single feat since the Wright brothers' first flight has ever matched. There were many other developments during that first decade which pointed the way towards the future of aviation. Record flights by the score showed the path to the future routine accomplishments of both civil and military aviation.

With the difficult problems facing the aircraft designer in the early post — World War I years, it is remarkable that any aviation progress was made at all. But it was. The strutted and wire-braced biplane had high drag, and a low lift-drag ratio. It had poor propellor performance, and engines possessed low horsepower and doubtful reliability. There was a complete lack of any means to control the landing speed and approach angle, a lack of knowledge of gusts and manoeuvring loads, together with stability and handling characteristics that varied between acceptable and highly dangerous.

The list of technological innovations of this first decade of flight is more than impressive. It includes the development of reliable, air-cooled engines; cantiliver design; the use of metal in structures; the concept of a tri-motored aircraft; the experimental use of superchargers; the trend towards the monoplane; and the development of blind flying equipment, plus many more.

This first decade at Langley was a startling growth for the aircraft. By 1919 it had a Curtiss *Jenny* and flight investigation into the characteristics of the type, including the lift and drag, began. The NACA full-time staff grew to eleven persons, four of which were professional. The first designers worked with a lack of data and filled the gaps with either their own experience or the experience of others. It was a decade of growing development, of luck and some solutions to the manifold problems of aircraft design. The aim at Langley was to reduce the luck and replace it with scientific data, carefully developed, and to point the way to improved aircraft design concepts.

After Lindbergh's historic crossing of the Atlantic there was an upsurge of a new wave of popular interest in aviation. There was a revolutionary change in the appearance and performance of aircraft and great strides were taken in commercial aviation. Transcontinental & Western Air — TWA — inaugurated the first coast-to-coast direct air service in 1930, between New York and Los Angeles. The Douglas DC-1 and the Boeing 247, progenitors of long lines of transports of the future, made their first flights in 1933. On 17 December 1935, on the anniversary of the Wright brothers'

first flight in 1903, there appeared the ubiquitous Douglas DC-3, the aircraft which was to revolutionize air transport throughout the world. Both NACA and later NASA were to employ the type as a workhorse at most of their establishments over the years. Pan American Airways flew air mail across the Pacific between San Francisco and Manila, and in 1936 the first passengers were carried on this route. During the following year both Pan American and Imperial Airways made survey flights across the Atlantic. With two-thirds of the world's surface being water there was great progress with flying-boat design.

During the second decade Boeing's Model 299, the prototype of its B-17 *Flying Fortress*, made its first flight — in 1935. In Britain, both the Hawker *Hurricane* and Supermarine *Spitfire* prototypes were being tested, and a top-secret report on radio detection and ranging — radar — was presented to the British Air Defence Research Committee. Three wars erupted during this period. During 1931 the Japanese Imperialists began its operations in China, Italy declared war on Abyssinia in 1935, and the Spanish Civil War commenced a year later in 1936. These conflicts led to an increased appreciation of airpower and the tragic Spanish Civil War drew other nations to the fighting, giving them an opportunity to test and develop new weapons and concepts.

The year 1928 had seen the first rocket-powered glider flight in Germany, and in the United Kingdom a fundamental paper on jet propulsion was published by Frank Whittle. Nine years later the first Whittle jet engine was being tested. The first volume of a nine-volume enclyclopedia was published by the Russians on interplanetary flight. The following year, the German Verein für Raumschiffahrt established a test site in Berlin, and the German Army Ordnance Corps established its rocket-weapon programme and moved into a test station at Kummersdorf. In 1937 the organization opened its rocket development establishment at Pennemunde on the Baltic. Static tests of a Heinkel He 112 were made in mid-1935 and the aircraft, fitted with an auxiliary rocket engine, first flew early in 1937. It was the foreunner of later German rocket-powered fighters. In Russia, three rocket test centres had been established near Moscow, Leningrad and Kazan.

Most commercial and military aircraft designs involved a basic machine, often a strut- and wire-braced biplane built of wood or steel tubing covered with fabric. Landing gears were fixed, engines — if aircooled — were uncowled. The propeller was a fixed-pitch type. A design revolution then brought into being the new monoplanes powered by sleek cowled engines and with retractable undercarriage and wing flaps.

During September and October 1936 George W. Lewis from the NACA headquarters visited Germany and Russia. On his return he briefed the Langley staff his principal impressions 'were of major expansions, especially in Germany'. Several large new centres for aeronautical research were under construction, and Lewis was even more impressed with the huge new staff, many times larger than that of the NACA and populated by a larger proportion of advanced-degree holders. He had little or nothing to say, however, about new aerodynamic or propulsion concepts or any new research results.

World War II dominated the third decade of Langley's research work. Before the conflict was concluded officially in September 1945, the shape of aircraft had once more changed. German engineers exploited advanced fighter and bomber designs using swept-back wings. Jet propulsion made great strides during the war. On 27 August 1939 the first turbojet-powered aircraft was flown in Germany, and before the end of the war both Germany and Britain had operational jet fighters. Rocket development was accelerated by the demands of war. The first German ballistic missile, the V-2, became operational in 1944, and guided missile warfare started in August 1943 with the Germans using radio-controlled rocket-powered glide bombs against shipping.

The first of NACA's rocket research aircraft, the Bell XS-1, had been conceived and designed during 1944. It made its first powered flight in December 1946, and in October 1947 US Air Force Captain Charles E. Yeager flew it past the speed of sound and pioneered the way into the age of supersonic flight. The remarkable series of X-planes grew into a stable of diverse types to probe and analyze new problem areas. From the barely supersonic performance of the original XS-1, the research series blasted first past Mach 2 and then Mach 3. in June 1951, the Bell X-5 first flew: it was characterized by its ability to change the sweep angle of its wing in flight.

The Korean conflict was the first involving the United Nations, during which the helicopter became an outstanding success; the war greatly furthered the helicopter's development and also led to greatly increased knowledge of its shortcomings. First tentative steps toward vertical take-off and landing (VTOL) aircraft were taken, especially in the United Kingdom with the Short SC.1 research aircraft and later the Hawker Siddeley P.1127 *Kestrel*. This latter was the forerunner of the two-nation *Harrier*, in use today for research tasks by NASA. New performance standards were set by the 'Century' series of fighters, commencing with the F-100 *Super Sabre*. All were evaluated by NACA and NASA since they posed new stability and control problems such as roll coupling and pitch-up difficulties — which were to plague both aircraft designers and the NACA research scientists.

Aeronautics & Space Adminsitration, Scott Crossfield and Joseph Walker.

In June 1963 President John F. Kennedy announced that the United States was going to develop a supersonic transport aircraft. France and the United Kingdom had formed a consortium which was hard at work building an SST; *Concorde* was to fly in 1969 and is still in airline service with British Airways and Air France. Since 1963 the major US aircraft manufacturers have devoted much time and many dollars to a SST research, and this has involved NASA.

The forerunner of the Lockheed SR-71 *Blackbird*, the A-11, was announced during 1964. There were three advanced models of this Mach 3 aircraft designated YF-12A, which were used by NASA for high-speed high-altitude research and breaking many records.

Today the US Air Force has retired its Lockheed SR-71 fleet, and during early 1990 the first two of three of these unique *Blackbird* aircraft were delivered to NASA at Dryden, pending a high-speed research programme to commence later in the year. Also in 1964, two aircraft which owed much of their conceptual design and development to Langley made their first flights. These were the General Dynamics F-111, a variable-sweep fighter, and the XC-142A VTOL transport built by a consortium of Ling-Temco-Vought, Ryan and Hiller. The latter type was not proceeded with but provided an enormous volume of data for future use.

North American Aviation was awarded the contract for the huge XB-70 bomber, known as the *Valkyrie*, an awesome aircraft intended to fly at three times the speed of sound. Then the unexpected happened. On 4 October 1957, a 'beep-beep-beep' signalled that the first man-made artificial satellite of the earth was in orbit, and that it was Russian. The signal from *Sputnik* had a mocking tone to frustrated US research scientists and engineers. The affront was repeated less than one month later by *Sputnik II* and a passenger, the dog Laika. This was the second event which affected the course of the NACA Langley history in the past seventy-five years. This crisis was indeed traumatic: *Sputnik* not only triggered the demise of NACA and the birth of NASA, but it also triggered what future historians might well call the 'space technology revolution.'

In 1960 the Apollo project was officially announced and Colonel Allan B. Shephard Jr. became the United States' first astronaut with a sub-orbital flight as part of Project Mercury in 1961. Four pilots who flew the North American Aviation X-15 research aircraft won the 1961 Collier Trophy: Major Robert White of the US Air Force; Commander Forrest Peterson, US Navy; and two civilian pilots from the National

The Bell X-1 46-062 was nicknamed 'Glamorous Glennis' by Capt Charles E. "Chuck" Yeager USAF as a tribute to his wife. On the 50th X-1 flight on 14 October 1947, Yeager made the first supersonic flight by a manned aircraft in 46-062 reaching a speed of Mach 1.06 700 mph at 45,000 ft.
(Office of History — USAF Edwards)

In mid-1977 the Space Shuttle *Enterprise* flew for the first time. Earlier, astronauts from NASA lifted the first samplings of the lunar surface in the start of scientific exploration of the Moon, an extension of research begun at Langley and other establishments in the organization, and carried on throughout the seventy-five-year history of NACA and NASA. On 20 April 1989, President George Bush established the US National Space Council. 'I sign this Executive Order with one objective in mind — to keep America first in space, and it's only a matter of time before the world salutes the first men and women on their way outward into the solar system. All of us want them to be Americans.'

In 1990, the seventy-fifth anniversary of its founding as the National Advisory Committee for Aeronautics, the National Aeronautics and Space Administration is a robust and diverse agency. Experiencing and meeting continuing challenges in a diversified environment of air and space that it has helped to create.

CHAPTER ONE
The Birth of the NACA

For its beginnings, the story of the NACA — later NASA — goes back to the days of World War I when aviation was in its infancy. During 1913 the new Secretary of the Smithsonian Institution, Charles D. Walcott, had tried to revive US government interest in the subject by reopening Samuel Pierpont Langley's old laboratory and appointing a committee to supervise it and conduct research. The proposed project had to be abandoned because the US Controller of the Treasury ruled that public funds and government employees could not be used to operate an agency that had not been established by law. Walcott was determined to get some such agency legally established. He had no personal knowledge of aeronautics, his own particular field of science being geology. However he was proud of Langley's long record of research and, as Langley's successor as secretary at the Smithsonian, Walcott felt a direct responsibility for the further advance of aviation.

He shared his ideas and enthusiasm with several Congressmen and other government officials, seeking their espousal of the cause. President Wilson was not receptive to the idea, since a conflict had started in Europe and he felt that to launch a project of this kind could reflect on US neutrality. However, Congress took a favourable attitude and late in the winter of 1915 it passed an act — attached as a rider to the US Navy Appropriations Bill which the President signed on 3 March — creating a government organization to be known as the National Advisory Committee for Aeronautics.

A Tidewater Virginia map showing the location of Langley Field and dates back from the late 1930s. The bridge at Newport News over the five-mile-wide James River was built in the late 1920s. During World War I, US highways 17 and 60 were just primitive dirt roads. (NACA 36942)

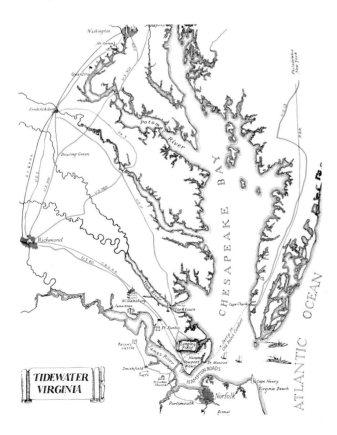

This act empowered the President of the United States to appoint the Committee, but specified that in making his appointments he should give representation to certain groups that had a natural interest in aeronautics. He was thus directed to select two members from the US Army; two from the US Navy; one each from the Smithsonian Institution, the National Bureau of Standards and the US Weather Bureau. In addition to these government members, it was felt that the public at large, in particular the engineering and scientific professions, should be represented. So the act provided for five additional members from the ranks of those acquainted with the needs of aeronautical science, either civil or military, or skilled in aeronautical engineering or its allied sciences.

The new agency, the NACA, was thus conceived as a broadly representative group consisting of a joint body of civilian and military personnel. Scientists and engineers have always constituted a large majority of the membership. The original committee, appointed by President Wilson in April 1915, included Brigadier General G. P. Scriven and Lieutenant Colonel Samuel Reber from the US Army, and Captain Mark L. Bristol and Naval Constructor H. C. Richardson representing the US Navy. These were technical men associated with the aeronautical branch of the two services. Dr Walcott was appointed to represent the Smithsonian, Charles F. Marvin the US Weather Bureau and S. W. Stratton the Bureau of Standards. To represent the public the President nominated Joseph S. Ames from the physics department of John Hopkins University, W. F. Durand from the engineering department of Northwestern University, and Byron R. Newton, who at that time was Assistant Secretary of the US Treasury.

Of the twelve members, four were connected with military aeronautics, seven others were directly engaged in some branch of science or engineering and only one could be regarded as a layman. But the keen personal interest of Byron Newton in science and his awareness of the future potential of aviation made him a particularly appropriate spokesman for the non-technical citizen.

The Committee elected Brigadier General Scriven as its first chairman, and when he resigned in 1916 the Stanford engineer, Dr Durand, was elected. Ever since then, by unwritten rule, the NACA chairman has always been a civilian scientist. John R. Freeman, the distinguished civil and mechanical engineer from Providence, Rhode Island, followed Durand, but he resigned within a year and was succeeded by Dr Walcott. This geologist of the Smithsonian, who had been the indefatigable prime mover from the beginning, served as chairman until his death in 1927. He was succeeded by Joseph S. Ames who occupied the NACA chairmanship for twelve years. General 'Hap' Arnold of the US Army Air Forces has called him 'the great architect of aeronautical science' and today the Ames Research Center in California proudly carries his name. Failing health caused Ames to resign in 1939, a few weeks after the commencement of World War II, and Vannevar Bush — an electrical engineer and president of the Carnegie Institution of Washington — succeeded him as chairman. After only two years in the post Dr Bush resigned in order to take the leadership of the Office of Scientific Research & Development, and in August 1941 the NACA elected Jerome C. Hunskar. He was the first aeronautical engineer to occupy the chairmanship, being head of the mechanical and engineering departments at the Massachusetts Institute of Technology.

During 1929, Congress amended its act of establishment to increase the NACA committee by three members, bringing its total membership to fifteen. In 1938 an additional act of Congress provided that two of the fifteen should represent the civil aeronautics activities of the Commerce Department. The committee now included representatives from the three executive branches of the US government with a prime interest in aeronautics — the Departments of War, US Navy and Commerce together with three scientific research agencies of the US government — the Smithsonian, the US Weather Bureau, the Bureau of Standards — and from the civilian ranks of scientists and engineers. This joint form of organization was unique at the time it was constituted, and has since been a model for subsequent organizations.

Congress specified that the function of the NACA committee should be 'to supervise and direct the scientific study of the problems of flight with a view to their practical solution,' and also to 'direct and conduct research and experiment in aeronautics.' During 1915 a survey of American universities, engineering schools, engineering societies and industrial establishments was conducted to see what facilities and personnel were available. The findings were far from encouraging. Only two educational institutions had courses of instruction in aeronautics. The committee did find some laboratories which could do work in the aeronautical field, but comparatively little was being done. 'The interest of the colleges is more one of curiosity than that of considering the problem a true engineering one,' said the report. After this appraisal, the committee decided that a well-equipped laboratory would be essential to its own work — a project not to be undertaken precipitously, however, but to be the result of gradual development.

Langley Field

In the heat of July 1917, excavation began on the first research laboratory to be built for the National Advisory Committee for Aeronautics at Langley Field, Virginia. The site had been authorized for the NACA's experimental aeronautical establishment just one month before, in June. A contract for the construction had been awarded to the J. G. White Engineering Corporation of New York. The estimated cost of the new laboratory was $80,900.

Lacking the money to purchase and develop a site for its laboratory, the NACA had circulated the idea of a joint civil-military experimental station. Inter-service rivalry, however, defeated a proposal to combine the aeronautical researches of the NACA, the US Weather Bureau, and the aviation section of the US Army and US Navy. General Scriven advised his colleagues on the NACA committee to support the request for $50,000 from Congress, to be included as aport of the FY 1916 US Navy budget, to build the laboratory. Finally on 29 August 1916, Congress appropriated the $87,000 requested by the NACA, of which $53,580 was to be earmarked for construction of a lab.

During 1915 Congress had directed the War Department to survey for a military reservation to house an experimental facility complete with airfield. The NACA co-operated with this Army Air Service project and it was General Scriven, also head of army aviation, who appointed a team of officers to investigate locations for the new research establishment. After considering fifteen sites — six in Maryland, four in Virginia and one each in West Virginia, Tennessee, Ohio, Illinois and Missouri — the NACA committee were informed of the army's choice; 1,650 acres in Elizabeth City County, Virginia, just north of the town of Hampton.

Following an enquiry to the Surgeon General concerning health conditions in the Hampton area, and an inspection of the site by one of the NACA sub-committees, it was recommended that the site be obtained for use of the US Government at an early date. Hampton, close to Chesapeake Bay, stood in relative proximity to Washington, D.C. and to shipbuilding and repair industries at Newport News, Norfolk,

and Portsmouth. A temperate but changeable climate and the location alongside a tidal river permitted experimental flying above both land and water under nearly all conditions which aircraft of the military could expect when in service. For a combined research facility sponsored by the US War and Navy departments, this site was ideal.

The new military establishment was named after Samuel Pierpoint Langley, former secretary of the Smithsonian Institution. The new NACA facility was later named Langley Memorial Aeronautical Laboratory, soon shortened to the familiar cryptic *Langley*. Earlier the NACA had suggested that the US Army flying fields be named after aviation pioneers. Following this suggestion the War Department named its new property Langley Field.

Construction of the airfield commenced on 17 July 1917, hampered by the great confusion following the US declaration of war on Germany. It was also hampered by heat, wet weather and marshy terrain which soaked up the soil as quickly as it was dumped. This Virginia tide-water region was worked on by gangs of all races and conditions. Forty-six workers died of influenza between September 1918 and January 1919, the epidemic being so severe that the undertaker who had the government contract for burial was unable to secure sufficient coffins to take immediate care of the bodies.

The chaos of World War I finally forced the US Army to abandon its plan to make Langley Field its aeronautical research and development centre. The construction contractors found it difficult to employ sufficient labour or obtain materials when needed. There were too many bosses and too much division of responsibilities, which exacerbated the confusion. The War Department complained that the constructors put up structures without consultation with or authority from government officials. Contractors, on the other hand, complained of work-order cancellation, red tape, and improper use of their equipment and supplies by military personnel.

Paul B. King, depicted in his test flying gear, was one of the very early civilian flight test pilots employed at the Langley Memorial Aeronautical Laboratory — LMAL. He was born in 1892, was the son of a US senator from Utah, had no degrees, taking classes at the University of Utah. He served the NACA at Langley from 1922 to 1927. (NACA 1118)

As mentioned earlier, the US Army, under heavy wartime pressures, dropped the plan to share the installation with NACA and reassigned aircraft research and development to their engineering division at McCook Field, near Dayton, Ohio. However Langley Field remained a large base, and military influence remained strong. Isolation, mosquito bites, influenza, inadequate housing and poor relations with the military began to plague the NACA personnel. On 9 July 1919, John DeKlyn, the engineer in charge of building and construction, complained to Chairman Joseph Ames that 'Langley Field can never be an efficient or satisfactory place for the NACA to carry on research work.' John Victory, the NACA executive secretary in Washington, D.C. agreed and recommended that the laboratory be moved to Bolling Field, near Washington, which was then under construction.

Congress was reluctant to change the location. This, and the cutting of the NACA postwar budget requests, for the committee to make the best of a bad situation. Visitors to Langley Field found that the corner occupied by NACA was comparatively modest. It consisted of an atmospheric wind tunnel, a dynamometer laboratory, the administration building and a small warehouse. There was a staff of eleven NACA personnel.

The formal dedication of the Langley Memorial Aeronautical Laboratory on 11 June 1920 guaranteed that the NACA would remain at Hampton. Ceremonies included an aerial exhibition highlighted by a 25-aircraft formation led by Brigadier General William 'Billy' Mitchell. Addresses by prominent military and civilian officials congratulated the NACA, extending its best wishes. Tours and a demonstration of the wind tunnel were included. Morale improved. There was plenty of room for growth, but there were still many tough battles left for the National Advisory Committee of Aeronautics — NACA — to fight.

Thomas Carroll, born 1890, was one of the earliest of the National Advisory Committee for Aeronautics civilian flight test pilots. He served at Langley from 1920 to 1929 after studying law at Georgetown. (NACA 439)

The early flight test research team at the NACA Langley was usually made up of a test pilot plus a flight test engineer. Depicted in the rear cockpit with Thomas Carroll in the front is John W. Crowley Jr. born 1899. He obtained a Bachelor of Science degree in 1920 as a mechanical engineer and served at Langley from 1921 to 1947 when he was transferred to the NACA headquarters in Washington DC. (NACA 298)

A well-known and highly-respected of the many NACA characters at Langley was John Stack, a hard-charging, persuasive man whose attitude toward unproven technology was usually — 'Let's try the damn thing and see if we can make it work.' John is seen studying models of the Douglas D-558 research aircraft. (NACA 48989)

Hugh L. Dryden, Director of Research and Head of the National Advisory Committee — NACA — on the left, is greeted by the engineer-in-charge at Langley, Henry J. E. Reid. In the centre is John F. Victory, executive secretary of the NACA since 1915. The transport in the background is a Douglas DC-3 which had brought the visitors for a visit to Langley. (NACA)

L TO RIGHT: DR. DRYDEN, HEAD OF NACA; DR. VICTORY, DR. REID

Law Establishing the National Advisory Committee for Aeronautics

(Public Law 271, 63rd Congress, approved 3 March 1915)

An Advisory Committee for Aeronautics is hereby established, and the President is authorized to appoint not to exceed twelve members, to consist of two members from the War Department, from the office in charge of military aerobatics; two members from the Navy Department, from the office in charge of naval aeronautics; a representative each of the Smithsonian Institution, of the United States Weather Bureau, and of the United States Bureau of Standards; together with not more than five additional persons who shall be acquainted with the needs of aeronautical science, either civil or military, or skilled in aeronautical engineering or its allied sciences: Provided, That the members of the Advisory Committee for Aeronautics, as such, shall serve without compensation: Provided further, That it shall be the duty of the Advisory Committee for Aeronautics to supervise and direct the scientific study of the problems of flight, with a view to their practical solution, and to determine the problems which should be experimentally attacked, and to discuss their solution and their application to practical questions. In the event of a laboratory or laboratories, either in whole or in part, being placed under the direction of the committee, the committee may direct and conduct research and experiment in such laboratory or laboratories. And provided further, That rules and regulations for the conduct of the work of the committee shall be formulated by the committee and approved by the President.

That the sum of $5,000 a year, or so much thereof as may be necessary, for five years is hereby appropriated, out of any money in the Treasury not otherwise appropriated, to be immediately available, for experimental work and investigations undertaken by the committee, clerical expenses and supplies, and necessary expenses of members of the committee in going to, returning from, and while attending, meetings of the committee: Provided, That an annual report to the Congress shall be submitted through the President, including an itemized statement of expenditures.

CHAPTER TWO
Virgin Days at Langley

During mid-1919, with construction of the first wind tunnel under way at Langley, the first NACA flight test with full-scale aircraft was authorized. The purpose of these tests was to compare in-flight data with wind tunnel data for the same aircraft to show the degree of correlation, and to determine a method of extrapolating wind-tunnel data to full-scale results.

The first flight-test programme used two Curtiss JN-4H *Jenny* training biplanes in a comprehensive investigation of aircraft lift and drag. It became the forerunner of a myriad of detailed investigations that would later lead to the development of a series of research aircraft to explore the unknowns of subsonic and supersonic flight.

This first programme with the two *Jennies* also produced a second, very important, result. The NACA Technical Report which described the tests noted that there was a need to develop a special type of research *pilot*. For the first time the role of the engineering test pilot was recognised.

The first experimental work at the new Langley laboratory commenced in 1919, involving two US Army Air Service Curtiss JN-4H Jenny aircraft borrowed from the flight line at Langley Field. Records show the first arrived with the NACA on 11 September 1919. (NACA 67-H-1242)

Three faithful Curtiss JN-4H *Jenny* aircraft served in a variety of tests over the early years. Amongst other things they pioneered in-flight investigations of pressure distribution, so that aircraft designers could calculate the air loads acting on the wings and tail of the aircraft. In the first programme beginning in 1920, the NACA technicians installed 110 pressure orifices in the horizontal tail of the wood-and-fabric Curtiss *Jenny*, hooked to a battery of liquid-in-glass mano-meters which could be photographed in flight.

During early January 1921, research was begun to compare the characteristics of wings in model tests and full-scale flight tests, so that designers could be furnished with complete and accurate data on which to base their performance estimates. In the same year, new instruments were developed and tested in flight to measure control position and stick forces exerted by the pilot. This was done to understand and improve handling characteristics and thus increase flight safety. More refined and miniaturized instruments used for the same basic purposes find continued employment in tests of high-speed jet aircraft and research vehicles today.

A major portion of the flight test work at Langley covered pressure distribution investigations. From measurements of loads in normal flight, the work was expanded to study the

effects of accelerated flight or manoeuvres. At that time there was virtually no data available to designers on the distribution of the loads on the wings of aircraft in accelerated flight. Later work extended the pressure distribution measurement to the nose of a non-rigid airship, initially under steady flight conditions and then during manoeuvres over a range of airspeeds and atmospheric conditions.

A small fleet of five aircraft shouldered the work load of flight test work during 1921. These included the three Curtiss JN-4H *Jennies*, a single Lewis & Vought VE-7 (A5667) which arived at Langley in April 1921, and a Thomas-Morse MB-3. The three *Jennies* logged 110 hours of flight time in 260 flights during the year.

Aircraft log-book records indicate that 26 April 1922, was a day like any other at Langley. Test pilot Thomas Carroll and scientist W. G. Brown were testing a fog landing device in Curtiss JN-4H *Jenny* No. 2. It was a mission aimed at eliminating a hazard that today still exists. The log sheet indicated that the *Jenny* needed four quarts of oil. Test pilot Kearns flew *Jenny* No. 3 whilst Arnold flew the Lewis & Vought VE-7. It is interesting to note that although the official numbering of the NACA aircraft fleet was not officially authorized until 1928, the Langley-based aircraft in many cases were numbered.

This modified Model T Ford, painted bright red, was used to start aircraft engines, and known as the 'Huck Starter' during World War I. It is depicted at Langley, complete with the NACA emblem and attached to the engine of Vought VE-7 NACA 4 in the 1920s. (NACA 67-H-1244)

On 5 September 1922, a British SE.5A — followed by a de Havilland DH-4 two days later — joined the Langley flight test fleet, thus raising the fleet total to seven aircraft. In addition, four more were being refitted for either test programmes or support work. These included a Fokker D.VII (6328), a Nieuport 23, a SPAD VII (AS7142) and a de Havilland DH-9. During 1922 the US Navy Bureau of Aeronautics asked NACA to undertake a comparative study of the stability, controllability and manoeuvrability of four aircraft. These four included the Lewis & Vought VE-7, the Thomas-Morse MB-3, the British SE.5A (which was one of the most widely used pursuit aircraft used in World War I) and the famous Fokker D.VII, mainstay of the German Imperial Air Service during the same conflict. Also during 1922, research began at Langley for the first systematic series of take-off and landing performance measurements.

The Curtiss JN-4H *Jenny* was used as a test vehicle during 1922 for an extensive investigation of manoeuvrability. The aim was to find a satisfactory definition of the word, in an aerodynamic sense, and to establish ways of measuring it. A muscular pilot might judge an aircraft light on the controls and manoeuvrable, whereas a lesser man may judge the same aircraft heavy and sluggish. Langley instrumented the *Jenny* to measure its angular velocity following control input, as a first approach to defining just what 'manoeuvrability' was.

It was natural that the calibre of the flight-test work being done by NACA at Langley began to attract attention from the military services. In 1923, the US Navy Bureau of Aeronautics came to Langley with a request that the laboratory run a series of flight tests in low-speed regime on the new Curtiss TS-1 float scout aircraft. The US Navy was particularly interested in accurate determination of its stalling, plus take-off and landing speeds. The TS-1 (A6249) arrived at Langley during November 1923, departing finally on 4 December 1928.

The US Army Air Corps was concerned with similar problems. In 1924 the service requested NACA to study the acceleration, control positions, angle of attack, ground run and airspeed during the take-off and landing of most of the aircraft then in their inventory. The types included the Curtiss JN-6H (AS44946) which arrived on 4 September 1923; the Lewis & Vought VE-7; the de Havilland DH-4B; the Fokker XCO-4 — the prototype of the C.IV two-seat biplane then in service with several countries; the SE.5A, the SPAD VII; the Thomas-Morse MB-3 which arrived in June 1923; the Martin MB-2, a biplane bomber; and the Sperry M-1 *Messenger*. The NACA Langley flight-line thus sported eleven test aircraft, some now numbered in the unofficial NACA inventory. During 1924 a total of 297 hours of flying were logged in 918 flights.

By the end of 1923 the US armed services had transferred seventeen aircraft to the NACA at Langley, of which thirteen were a temporary assignment from the US Army Air Service, including this British designed SE.5A AS8049 which became registered NACA 5 serving from 15 September 1922 to 10 September 1926. (Peter M. Bowers)

In the same year the US Army Air Corps requested flight research investigation of the pressure distribution over the wing of a Lewis & Vought VE-7 tandem trainer and the service transferred one of the type to Langley for the programme. The VE-7 soldiered on through other work after those tests were completed, and including ice formation research and a landmark programme using seven different propeller designs aimed at determining the effects of different propeller parameters on performance.

Those tests (along with tests with a series of no less than six interchangeable wings, each with a different aerofoil section, on a Sperry M-1 *Messenger* biplane) became the first of many NACA comparative tests where a systematic approach was used to develop a better installation or to design a better component. Sophistication had come to both flight testing and wind tunnel testing. By mid-1924 NACA was able to make complete pressure distribution surveys, either in the wind tunnel or in flight, during the same day. Formerly, such

tests had required a series of runs over a time period as long as two months.

Later in 1924 NACA reported a further refinement in flight testing techniques. Recording instruments had been developed, the committee revealed, to make a continuous record of pressure distribution, accelerations and other parameters during flight tests of aircraft.

The flight-test programme continued to grow during 1925, and at one time there were no less than nineteen aircraft involved in various phases of test work at Langley. The SPAD VII (AS7142) departed on 24 February whilst a Douglas DT-2 floatplane (A6425) from the US Navy had arrived at Langley on 14 April 1923. It departed on 15 June 1925. During this year the laboratory flew a total of 245 hours in 625 flights.

Men and Machines

NACA never owned many aircraft. Congressional suspicion of the NACA need to do so, modest budgets and the increasing availability of military aircraft for loan restricted the actual number. During 1924 NACA ordered its first aircraft, a Boeing PW-9 pursuit machine built with especially strong tail surfaces and fuselage for use in a systematic investigation of pressure distribution. It was delivered to Langley on 5 January 1927, finished in a US Navy silver colour scheme without stars on the wings but with tail stripes. It departed after tests on 15 October 1928.

George Lewis subsequently testified before a congressional subcommittee that the purchase of aircraft was necessary because the military services could not provide an aircraft of the special construction required for the research. It was revealed that NACA had requested an appropriation for the purchase of the Boeing PW-9 after the order had been placed with the manufacturer.

During 1928 NACA purchased a Fairchild FC-2W2 five-passenger high-wing monoplane with an enclosed cabin and detachable wings for testing a family of aerofoils. The aircraft became 'NACA 1', the first in a unique numbering system which identified the NACA fleet and which today is still in use by the NASA organization. A NACA badge symbol was introduced about this time.

What interested the NACA most about the famous Fokker D.VII was the thick, internally braced cantilever wings of the German fighter. Depicted is NACA 7 — 6328 — at Langley on 6 December 1922. It was used by the laboratory until 1923. (Peter M. Bowers)

On 6 December 1922, there was a fatal mid-air accident involving the Fokker D.VII (6328) numbered '7' and the Martin MB-2 bomber then at Langley. Records reveal that the Fokker D.VII survived since it remained with NACA until 1923. In the early years the NACA flight research team was usually made up of a test pilot and an engineer. Some of the early civilian test pilots included Thomas Carroll, William McAvoy and Paul King. Paul was the son of a US senator from Utah. The work of talented mechanics and other technical employees was instrumental to the Agency's success.

During the late 1920s the NACA announced a major innovation which resulted in the agency's first Robert J. Collier Trophy, presented annually by the National Aeronautic Association for the year's most outstanding contribution to American aviation. In 1929 the Collier trophy went to NACA for the design of a low-drag cowling. This trophy was first awarded in 1911, the first winner being Glenn H. Curtiss for developing the 'hydroaeroplane'. By the late 1920's the Collier Trophy was recognised as the most prized of all aeronautical honours to be awarded in the USA. The winner received the award from the President of the United States.

The de Havilland DH-4 was the only aircraft built under licence in the USA to serve in combat during World War I. It was powered by the US Liberty engine. Depicted is the DH-4B NACA 25 used at Langley between December 1927 and 1 January 1930. It carries the NACA tail badge, the number '25' on the fuselage and was ex-US Army Air Service AS31839. (Peter M. Bowers)

Early ice formation research on aircraft was conducted at Langley with this Vought VE-7 A5950 — NACA 14 — delivered on 22 June 1928. It was one of two Vought VE-7 trainers transferred from the US Navy, the other being NACA 4 ex-A5669 delivered in April 1921. (Peter M. Bowers)

Most American aircraft of the post-World War I decade used air-cooled radial engines, with their cylinders exposed to the air stream to maximize cooling. But the exposed cylinders also caused high drag. Because of this the US Army had adopted several aircraft with liquid-cooled engines, in which the cylinders were arranged in a line parallel to the crankshaft. This reduced the frontal area of the aircraft and also allowed an aerodynamically contoured covering, or 'nacelle', over the nose of the aircraft. But the liquid-cooled designs carried weight penalties in terms of cooling chambers around the cylinders, gallons of coolant, pumps and radiator. The US Navy decided not to use such a design because the added maintenance requirements cut into the limited space aboard aircraft carriers. Moreover, the jarring contact of aircraft with carrier decks created all sorts of cracked joints and leaks in liquid-cooled engines. The issue was simplified with air-cooled engines, although their inherent drag meant reduced performance. In 1926 the US Navy Bureau of

Aeronautics approached NACA to see if a circular cowling could be devised in such a way to reduce the drag of exposed cylinders without creating too much of a cooling problem.

Significant work on cowled radial engines had taken place in the United Kingdom. However the NACA technicians and aerodynamicists had the advantage of a new propeller research tunnel completed at Langley in 1927. With a diameter of 20 feet, it was possible to run tests on a full-sized aircraft. Following many tests, a NACA technical note by Fred E. Weick in November 1928 announced convincing results. At the same time, Langley acquired a Curtiss P-1A *Hawk* biplane fighter (25-411) from the US Army Air Service and fitted a cowling around its blunt radial engine. The results were exhilarating. With little additional weight, the speed of the *Hawk* went from 118 to 137 mph — an increase of 16 per cent. The virtues of the NACA cowling received public acclaim the following year, when Frank Hawks, a highly publicized stunt flier and air racer, added the NACA cowling to a Lockheed *Air Express* (7955). In February 1929 he established a new Los Angeles to New York non-stop record of 18 hours 13 minutes. The NACA cowl increased the aircraft's maximum speed from 157 to 177 mph. Lockheed Aircraft sent the Agency a telegram: 'Record impossible without new cowling. All credit due NACA for painstaking and accurate research.' By using the cowling, NACA estimated savings to the aircraft industry of over $5 million — more than all the money appropriated for NACA from its inception through 1928.

President Herbert Hoover presented the Collier Trophy to Joseph Ames, chairman of the Agency in 1929. Three years later, as part of his plan to increase efficiency in government, President Hoover was to sign an executive order to abolish the National Advisory Committee for Aeronautics.

In the meantime NACA purchased more aircraft. On 21 July 1931 a Pitcairn PCA-2 autogyro arrived at Langley, becoming registered 'NACA 44' and in use until 13 September 1933. Other types joining the increasing fleet included two Fairchild 22's, one delivered on 15 May 1933, becoming 'NACA 60'; a Stinson SR-8E *Reliant* delivered on the last day of July 1936 becoming 'NACA 94'; a Ryan ST Sport aircraft which arrived on 8 June 1938 to be registered 'NACA 96'; a Piper *Cub*, ex NC26899, became 'NACA 98' whilst two Lockheed 12 twin-engined transports became 'NACA 99'. Though all were ostensibly purchased for research purposes, several were also used as general transports.

An engine research laboratory with a dynamometer, to measure output and other performance data on aircraft engines, had been in full use since 1919. By 1925 a second had been added. Both were kept fully employed, as were the powerplant engineers. Early work on superchargers, investigated at Langley in 1924, led to consideration of supercharging to boost engine power for high altitude bombers, and to obtain a good rate of climb for interceptors. This extensive engine research later became the nucleus of the Lewis Flight Propulsion Laboratory at Cleveland. Ohio. One specific study made at Langley was to determine the adaptability of supercharging to an air-cooled engine and its effect on the flight performance of the engine.

A Fight for Existence

The early years of NACA had proved a period of very slow growth. At first there was no paid organisation at the committee headquarters in Washington, D.C., other than a clerk who kept the records and took care of the correspondence. The first employee, John F. Victory, engaged in June 1915, was appointed Assistant Secretary to the committee in 1917, became Secretary in 1927, and in these two posts had been responsible for the business administration. John Victory and George W. Lewis were the two individuals who first took firm control of the routine affairs of the organisation. More than anyone else at the NACA headquarters, these men left their lasting, if contrasting, impressions on Langley.

George W. Lewis was appointed executive officer of the Agency in November 1919 and director of aeronautical research in July 1924. This position he held until 1947. Lewis was a 1910 master's graduate in mechanical engineering from Cornell University, and he possessed considerably more technical competence than John Victory.

Financial problems plagued the NACA from its inception. After the $1,200 salary for John F. Victory was subtracted from the first year's budget in June 1915, the committee was left with only $3,800. The NACA had wisely kept its budget requests modest, and until 1930 none exceeded a million dollars a year. Obtaining appropriations was always a tricky business. The construction of Langley was funded through legislative contrivance as part of the US Navy appropriations bill. After the Bureau of the Budget was created in 1922, the Agency had to fight the same battles for money and live by the same budget cycle as other branches of the federal government. Legislation regulated how NACA spent its money and transferred its funds from one account to another.

On 9 December 1932, came the most serious direct threat to NACA since its inception. As part of the plan to reduce expenditure and increase efficiency in government by eliminating or consolidating unnecessary or overlapping federal offices, President Hoover signed an executive order to abolish the Agency — just as he had wanted to do a few years earlier when he was secretary of commerce. The original resolution that had become the Air Commerce Act of 1926 contained a provision insisted on by Secretary Hoover calling for the transfer of NACA to the Department of Commerce. Though the provision was eventually removed from the bill, Hoover continued to believe in its wisdom. As a lame-duck president, he finally acted on that belief. A war of words ensued in various editorials of the day, one accusing NACA of ceasing to be a research body and had become 'an advertising club, a rest home, a comfortable refuge for the two who have controlled it.'

The Agency responded to Hoover's order by eliciting support from its most influential friends. Chairman Joseph Ames appointed a dozen prominent personnel in military and civil aviation, including the chief of the US Army Air Corps, the chief of the US Navy Bureau of Aeronautics, Orville Wright and others to a Special Committee on the proposed consolidation of the NACA with the Bureau of Standards. As might be expected, they expressed strong opposition. Charles Lindbergh wrote a letter supporting the committee's report.

In January 1933 the House of Democrats voted unanimously to kill the mergers Hoover had proposed and left readjustment of the federal establishment to the new Roosevelt administration. NACA's continuous existence from 1915 to 1958 as an independent organisation of the federal government testifies not only to real merits in its record of research but also to skill in the art of survival.

The majority of aircraft flight-tested at Langley throughout its history came on loan from either the US Army or the US Navy. When the first experiments involving aircraft commenced in the spring of 1919 the two Curtiss JN-4H *Jennies* were borrowed from the flight line at Langley Field, where they were being used to train pilots and observers in gunnery, aerial photography, bombing and communications. It was not until 1920 that the Agency employed its first test pilot, so the first pilots were from the military. However, NACA still required routine assistance from the Langley Field base operations and flight operations. By the end of 1923 the military had transferred no less than seventeen aircraft to the Langley laboratory. Thirteen were on temporary assignment from the US Army including five *Jennies*, a Thomas-Morse MB-3 pursuit aircraft, the British SE.5A, the captured German Fokker D-VII, a French SPAD VII and two de Havillands — a DH.4 and DH.9. Four were transferred from the US Navy: two Lewis & Vought VE-7 trainers, a Douglas DT-2 torpedo aircraft and a Curtiss TS-1 floatplane.

As with the hundreds of other aircraft which were to serve at Langley over the years, the Agency conducted comprehensive aerodynamic investigations and research with some and used others as test beds for various innovations such as superchargers and high-speed cowlings. Over the years the NACA laboratory personnel also made brief evaluations of a considerable number of aircraft temporarily based with the military at Langley Field.

The decision to lend any military aircraft to Langley was often informal and personal. The US Navy aircraft utilized during the 1920 to 1930 period came through the good offices of Lieutenant Commander Walter S. Diehl, who was officer-in-charge at the Bureau of Aeronautics in Washington, D.C. He often approached his superiors at the BuAer with news that the Agency required to borrow a certain aircraft type for investigation at Langley. Walter Diehl met regularly with George Lewis and the NACA headquarters staff — his office was also in the Navy Building, and he frequently visited Langley. It was a most useful liaison on the friendliest of terms.

The huge full-scale wind tunnel at Langley opened during 1931. It measured 434 × 222 feet and was 90 feet high. The first aircraft to be tested was this Vought O3U-1 Corsair. In January 1975 the wind tunnel was closed for the first time, for a one-year rehabilitation. (NACA 75-H-238)

Early NACA markings were as varied as the types of aircraft used. Depicted is the Doyle 0-2 high-wing monoplane with the NACA emblem on the fin and the registration 'NACA 34' on the top of the wing. It was known as the Doyle 0-2 Oriole and used at Langley during 1929. (Peter M. Bowers)

A large number of the aircraft borrowed by NACA came to Langley direct from the manufacturer's production line. The US Navy aircraft were often experimental types flown in from the air stations at Anacostia or Norfolk. Though the US Army sent NACA many aircraft from Bolling Field, near Washington, D.C., and its engineering division at McCook Field, near Dayton, Ohio, most aircraft loans came from the local flight line at Langley Field — typically from US Army operational squadrons.

The Langley laboratory could retain most borrowed aircraft for only a specified period, usually several weeks. Some it retained for an undetermined period of research or on permanent transfer. On the majority it could make modifications and install special equipment as long as the aircraft was restored to its original configuration prior to return. On a few it could make no changes whatsoever or conversely, could make whatever permanent modifications it saw fit. This latter category mostly involved older aircraft for which the military had no further use.

One of the earliest test programmes requested by a branch of the military to be undertaken by Langley involved the loan of a Sperry M-1 *Messenger*, a small biplane which the US Army had procured to replace motorcycles for certain liaison uses. The engineering division of the US Army Air Service at McCook Field provided the *Messenger* (S68473) which was delivered to Langley on 19 January 1924. It was to be February 1929 before NACA closed the file covering job orders for nearly six years of occasional *Messenger* free-flight and wind tunnel testing. The testing involved determining the biplane's lift and drag characteristics when equipped with each of six interchangeable sets of wings. It was a very long and involved programme which produced numerous test reports and mounds of paper work. It also caused some friction between Langley and the US Army Air Service.

On 24 May 1926, the Agency held its first joint conference with representatives of the aircraft manufacturers and operators at Langley. This was the first of what was to become a

The Atlantic Model 7 version of the Fokker F.VIIA/3M trimotor was designated C-2A with the US Army Air Service. Its military serial was 28-123 and it is seen at Langley during 1929 fitted with experimental NACA low drag cowlings. Chief test pilot Tom Carroll was involved in the flight test programme and is among the group of NACA personnel seen with the aircraft. (NACA 3249)

recurring event and a great NACA tradition. It included an inspection tour, and provided the guests with an opportunity to criticize current research and to suggest new avenues they believed to be promising. The second of these conferences, held the following year, was expanded to include representatives of educational institutions that taught aeronautical engineering, and trade journals which played an important part in the dissemination of aeronautical information. This interchange of information between industry and the NACA, always one of the major factors in directing the course of the Agencies' research, has been maintained over the years since the first formal joint conference in 1926.

As recorded earlier, Langley purchased a Pitcairn PCA-2 autogyro which was delivered on 21 July 1931 becoming 'NACA 44'. This marked the beginning of rotary-wing research. The PCA-2's rotor was tested in the full-scale wind tunnel for correlation between tunnel and flight tests, and a model of the rotor was tested in the propeller research tunnel to determine scale effects. A camera was mounted on the hub of the rotor to photograph the blade behaviour during flight. Flight tests included some measurements during severe manoeuvres, with results still applicable to fast helicopters today. The Pitcairn autogyro had a fixed wing surface to carry some of the weight of the aircraft in normal forward flight. The flight tests at Langley included tests in which the incidence of the wing was varied, so that it carried a different proportion of the aircraft weight in each of a series of trials. These experiments highlighted some of the problems which today face designers of high-speed helicopters, who wish to unload the rotor by using a fixed or variable wing surface to generate additional lift. This was the first major project involving the small rotary-wing research group.

The flight research laboratory was officially opened in 1932. It was a separate area, with aircraft hangar space, a repair shop, plus office space of the staff. Aircraft delivered for tests during 1932 included a Curtiss O2C-1 *Helldiver* (BuNo. 8455), a Boeing F4B-2 (BuNo. 8628), a Boeing P-12C, Curtiss P-6E *Hawk* and a Fleet N2Y-1. More were to follow.

By 1937 the biplane had virtually disappeared. Both military and commercial aircraft were internally braced, unstrutted monoplanes with sleek cowled engines, retractable undercarriage, and wing flaps. The design revolution of the early 1930s had been largely sparked by the developments at the NACA.

Rotating Wing Aircraft

During 1928, Harold Pitcairn — who later formed the Pitcairn-Cievra Autogyro Company of America — purchased a Cievra C.8 autogyro for his own investigations in the USA. On 13 May 1929 Pitcairn flew his Cieva C.8 on its first cross-country flight from Philadelphia, via Washington, D.C., to the NACA at Langley. This flight was made to coincide with NACA's fourth annual engineering research conference. Just over a year later, in July 1930, a Curtiss-Bleeker helicopter (373N) visited Langley. Designated the Curtiss-Bleeker SX5-1, it was the most unconventional machine ever built by Curtiss. Maitland B. Bleeker designed it and persuaded Curtiss to built it and provide the finance. At that time, one of the major problems of single-rotor helicopters was how to overcome the tendency of the fuselage to rotate in a direction opposite to the directly-driven rotor. Bleeker sought to correct this by having each rotor blade driven by its own propeller through shafts and gears from a single engine in the fuselage. Each blade, which was more like an aeroplane wing than the traditional narrow autogyro or helicopter rotor blade, also had its own trimming surface called a Stabovator and could almost be regarded as a separate captive aeroplane. The SX5-1 was built at Garden City late in 1929 and was powered by a 420hp Pratt & Whitney Wasp engine. The project was terminated when the drive gear broke.

Prior to NACA receiving its Pitcairn PCA-2 autogyro in 1931, one of the Agency's two research pilots was indoctrinated into the mysteries of autogyro flight at the Pitcairn plant in Philadelphia. Jim Ray of the Pitcairn-Cievra Autogyro Co. and Bill McAvoy, the NACA pilot, delivered the PCA-2 to Langley on 15 July 1931. It was later purchased by the Agency on 21 July, becoming 'NACA 44'. Melvin N. Gough, NACA research test pilot, quickly went solo on the PCA-2 at Langley; a few years later he trained a nucleus of military pilots on the type. Gough had started his NACA career in the Propeller Research Tunnel. After flight training and serving as a pilot in the US Navy Reserve in the 1920s he joined the flight test section at Langley. He soon became one of the most accomplished experimental test pilots in the USA and later became Langley's chief test pilot.

A commercial Fairchild Model 22, two-seat high-wing monoplane, was procured as the XR2K-1 BuNo. 9998, being delivered to the NACA at Langley as NACA 82 on 16 September 1935 and used as a flight test bed. During October 1940, it was fitted with an experimental flap and spoiler installation. It survived until 1946. (NACA 22160)

Boeing Model 266, a P-26A 33-56, seen during flap development tests, mounted in the full-scale tunnel at Langley. It was delivered to the NACA on 26 June 1934. It was known as the Peashooter and was the first US Army fighter to be constructed entirely of metal and to employ the low-wing configuration. The wings, through, were externally braced, and the landing gear was fixed. (NACA 9819)

A Stinson SR8E Reliant became NACA 94 after delivery to Langley on 31 July 1930. Thirty-five of the gull wing SR8E aircraft were bult. The NACA used the aircraft for a preliminary study of control requirements for large transport aircraft. It survived until 1947 at Langley. (NACA 12264)

This Fairchild Model 22 designated C-7A was delivered to Langley for wing research in the early 1930s becoming NACA 47. On 10 June 1936 it was modified into a low-wing configuration and re-designated J-2. (Peter M. Bowers)

Over the next few years the Langley log books showed an increase in the number of autogyro types and hours flown. During early 1933 a Pitcairn PAA-1 autogyro was delivered to Langley. Meanwhile the PCA-2 autogyro averaged two flights per week during its three years with NACA. In addition to the rotor hub mounted camera mentioned earlier, other items of flight research instrumentation used on the PCA-2 included a trailing-bomb airspeed and flight-path-angle recorder.

Whilst private and commercial use of autogyros was not thriving military interest in the utility aspects of a vehicle capable of safe flights at low speeds picked up. Consequently NACA research became closely allied to the needs of the US Army Air Corps. In fairly quick succession, the Langley laboratory had a series of Kellett autogyros. A Kellett KD-1 was delivered on 2 April 1935: a Kellett YG-2 autogyro (35-279), ex X14776, arrived at Langley on 17 December 1935. On 3 January 1936, Kellett YG-1 (35-279) went to the Agency followed by a YG-1A and YGF-1B. The records reveal that a Pitcairn XOP-2 and a Wilford XOZ-1 autogyro seaplane — the latter belonging to the US Navy — were also tested at Langley.

The blade motion and bouncing tests with the Kellett KD-1 provided informally-reported experimental blade dynamic twist and flapping-motion results used later in theory-data studies, and also the first manifestation of a new rash of dynamics problems. The NACA pilot test assessments of the Kellett YG-1 provided recommendations on manoeuvre limitations and on redesign for better serviceability, and also the recommendations for service test by the US Army Air Corps on experimental types. The larger Pitcairn YG-2 autogyro was found by the NACA to have such heavy control forces and violent fluctuations as to make the aircraft unsuitable for service use. Related measurements were taken and provided as a basis for corrective development work.

Parachuting from a rotating-wing aircraft is quite unusual, and fortunately, flight accidents of any kind are fairly rare in research organizations such as NACA. For these and other reasons, the date — 30 March 1936 — when Bill McAvoy and engineer John B. Wheatley, left an autogyro by parachute is still discussed as a part of the Agency's rotary-wing history. The autogyro involved was the Kellett YG-2 (35-279), ex X14776.

At 3,000 ft there was apparently 'a noise like a steam locomotive,' and the control stick — which was a heavy one because the test flight involved use of the stick-force recorder — abruptly whirled around the cockpit. The pilot-to-engineer intercom was torn loose and a short 'after-you' period ensued with Bill McAvoy being driven back and up as the rotating stick, which was directly connected to the damaged blade system, ate into the seat pan beneath him. After both occupants left the autogyro, the pilot, with parachute open, watched what appeared to be a delayed-opening descent by John Wheatley, the flight test engineer, whose parachute opened just in time to put him in an apple tree. It seems his hand had initially gripped too much of his complex flight gear, hence no release action.

It is understood that the fabric on the outer part of one blade broke open as a result of internal air-column centrifugal loads generated by reason of inadequate venting at the tip.

This Kellett YG-1 autogiro 35-278 was delivered to Langley on 3 January 1936. This unique type could make extremely short take-offs and landings, but it was not capable of vertical flight or hovering like a helicopter. It remained at Langley until 26 May 1936. (NACA 11981)

The autogyro continued to fly itself on a generally even course without any applied control, and the autorotating rotor kept turning fast enough to provide lift in spite of high blade drag from the torn and flapping fabric.

During April and May 1936, US Army Air Corps pilots Lieutenant Gregory and Lieutenant Nichols received instruction in the flying and maintenance of autogyros at the NACA laboratory at Langley. Thereafter the military was able to conduct its own rotating-wing acceptance testing, so NACA was able to restore emphasis on test and research.

By the autumn of 1938, international events required that the USA's military emphasis return to preparedness, which meant that fighters and bombers with better performance than anyone else's were required. Dr Lewis, head of NACA at that time, still wished he could afford attention to rotating-wing aircraft and in fact entered a proposal for fundamental research work on the subject to the War Department in connection with the Dorsey Bill. This Bill authorized but did not immediately provide $2 million for rotary-wing aircraft development. It kept a few projects active and eventually led to American attempts at military helicopter development.

Over the next few years there was relatively little experimental rotating-wing work, but extremely important basic groundwork was done. Some work was carried out on the sometimes catastrophic phenomenon of ground resonance. During 1939 and again in 1940, flight measurements of blade bending, control position and the like were made with Wilford's XOZ-1 Gyroplane. This machine, incidentally, was

During the summer of 1939 the NACA at Langley tested an experimental cowling and cooling system on this Northrop A-17A attack aircraft 36-184. This aircraft was built by Douglas at El Segundo and was tested at Langley between 13 February and 11 June 1939. (NACA 19673)

number 117 on the list of aircraft tested or used by NACA at Langley. These flights were the Agency's first with the 'hingeless' or rigid-blade concept. However by the late 1930s there were many diverse reasons for the restoration of interest in the helicopter as opposed to the autogyro. It is not surprising that when NACA was able to resume a small but intensive attack on the problems of rotating-wing aircraft in the early 1940s, it was the helicopter and not the autogyro which was studied.

Of all the many events that have affected the course of the history of Langley, only two have caused a major trauma. The first was World War II, which had a very profound effect on the Agency.

This Lockheed Model 12, NACA 97 c/n 1268 ex-NC17396 was acquired by Langley on 26 January 1939. The NACA fitted an extra central fin to attempt to improve the directional stability. It flight tested the first heated leading edge de-icers. Also used as an executive transport, being transferred to Ames on 4 October 1940. (NACA 20252)

CHAPTER THREE
World War II — Priorities and Expansion

Prior to World War II, NACA at Langley and its parent headquarters organization were rather obscure operations. It was well known in government circles that there were US congressmen who did not even know that the Agency existed. The conflict altered this status dramatically. First the Langley laboratory was enlarged — in 1938 the total staff totalled only 426, increasing to 524 in 1939, but by 1945, in order to meet the increase in workload, it had risen to 3,200. In 1935 employees were attached to six different research divisions, working in one of a dozen buildings on a few acres surrounded by US Army property. Ten years later, employees worked in eighteen divisions located either in the old East area or in the large new West area, separated not only from activities of the Langley Field military base but also, by a few miles, from other NACA operations.

With the setting-up of new installations, a large number of Langley personnel moved. There was the Ames Aeronautical Laboratory set up at Moffett Field, California, and the Aircraft Engine Research Laboratory in Cleveland, Ohio. The personnel who remained were less uniform, and for the

such a way as to produce high drag. The Agency demonstrated that with modifications, the top speed of this prototype could be increased by 31 mph to 281 mph — an improvement of more than 10 per cent.

This test was the first in a long series of clean-up programmes performed at Langley for the US Army Air Corps and the US Navy Bureau of Aeronautics. The Brewster XF2A-1 set two precedents. It was the first aircraft to use NACA's new 230-series aerofoils. All high-performance US military aircraft built up until the end of World War II, with the exception of the North American P-51 *Mustang*, employed an aerofoil from this efficient series. They were highly effective because of their high maximum lift and low minimum drag. Secondly, Langley did such an outstanding job in reducing the drag of the *Buffalo* that both the US Army Air Corps and the US Navy were soon sending all their prototypes to the laboratory for drag reduction. Between April 1938 and November 1940 Langley put eighteen different military prototypes through tests in the Full-Scale wind tunnel.

Prototype Aircraft Tested by Langley 1938–1940

RA No.	Date	Type	Serial No.	Name
603	Jun 1938	Brewster XF2A-1	BuNo.0451	Buffalo
606	Jun 1938	Grumman F3F2	BuNo.0967	
607	Jun 1938	Grumman XF4F-2	BuNo.0383	Wildcat
633	Aug 1938	Vought-Sikorsky SB2U-1	BuNo.0726	Vindicator
635	Aug 1938	Curtiss XP-37	AAC 37-375	
636	Aug 1938	Curtiss P-36A	AAC 38-1	Mohawk
637	Aug 1938	Curtiss XP-40	AAC 38-10	Kittyhawk
646	Dec 1938	Northrop XBT-2	BuNo.0627	
647	Dec 1938	Curtiss YP-37	AAC 38-474	
672	Jun 1939	Seversky XP-41	AAC 36-430	
674	Jun 1939	Bell XP-39	AAC 38-326	Airacobra
695	Sep 1939	Curtiss XP-42	AAC 38-2	
698	Sep 1939	Grumman XF4F-3	BuNo.0383	Wildcat
709	Nov 1939	Curtiss XP-46	AAC 40-3053	
739	May 1940	Republic XP-47	AAC 40-3051	Thunderbolt
746	Sep 1940	Chance-Vought XF4U-1	BuNo.1443	Corsair
796	Oct 1940	Brewster AF2A-2	BuNo.0451	Buffalo
797	Oct 1940	Curtiss XSO3C-1	BuNo.1385	Seamew
811	Nov 1940	Consolidated XB-32	AAC 41-141	Dominator

RA No. = Research Authorization number.

first time a large number of women worked with NACA — many of them doing a man's job. Langley's fiscal posture changed dramatically. Between 1940 and 1945 laboratory expenditure amounted to approximately $33 million, more than twice what had been in the first twenty years of Langley's history combined — approximately $14 million.

An experimental US Navy fighter, the Brewster XF2A-1 (BuNo.0451) and named *Buffalo*) was delivered to Langley during April 1938. The US Navy was unhappy with the 250 mph flight-test performance, and the Bureau of Aeronautics wanted the NACA staff at Langley to sort out any bugs. The aircraft was immediately mounted in the 30 × 60 ft Full-Scale wind tunnel. At the end of five busy days of tunnel tests, the team concluded that the Brewster company had overlooked the aerodynamic importance of several small but highly significant details of the design. The landing gear, engine exhaust stacks, machine-gun installations and gunsight all projected outside the smooth basic contour of the aircraft in

A summary of the tests was published in November 1940 as a NACA Advanced Confidential Report, to be circulated only to industry and the military. 'The drag of many of the aircraft tested was decreased 30 to 40 per cent by removal or refairing of inefficiently designed components. In one case the drag was halved by this process. Emphasis on correct detail design appears at present to provide greater immediate possibilities for increased high speeds than improved design of the basic elements.' — stated the report.

In the case of the Seversky XP-41 (36-430) which arrived at Langley on 10 May 1939, for example, Langley studied the drag of the fighter in eighteen different configurations. The Agency did its best to help industry realize dramatic increases of speed in production aircraft. The fruits of such efforts can be seen in the results of work on the Bell XP-39 *Airacobra* (38-326) which arrived on 6 June 1939, eleventh in the series tested. Bell's chief engineer, Robert J. Woods, was a former Langley employee working in the variable-density tunnel, and had designed the unconventional fighter. It had its engine amidships, at the centre of gravity, and a cannon in the nose and was theoretically capable of 400mph. Despite approval by the US Army Air Corps of its performance, General Henry 'Hap' Arnold felt the speed of the new fighter could be increased by cleaning up the drag. On 9 June 1939, he formally requested that NACA carry out immediate full-scale tunnel testing on the XP-39. Bell later incorporated changes recommended by NACA which improved speed by about 16 per cent. During 1941 the US delivered nearly 700 Bell P-39 *Airacobras* to the United Kingdom and the Soviet Union under Lend-Lease agreement. Reverting to the tests with the prototype XP-39, it was discovered at Langley that by cuffing the propeller at the point where it met the hub, streamlining the internal cooling ducts in the wings, lowering the cabin by six inches, decreasing the size of the wheels so that they would be completely housed within the wing and removing the turbo-supercharger and certain air intakes, the speed of the XP-39 fighter for a given altitude and engine power could be increased significantly. On the basis of an airframe of the same weight as that of the prototype but fitted with a more powerful 1350hp engine with a geared supercharger, it was estimated that the top speed attainable with the XP-39 might be as high as 429mph at 20,000ft. The head test engineer at the Full-Scale wind tunnel did not know precisely how much additional air would be required to cool the larger engine, but he did believe that even if this increase was very large, it would not prohibit the aircraft from obtaining at least 410mph.

This Curtiss Model 75S XP-42 fighter 38-4 was one of nineteen aircraft involved in the Langley drag reduction programme between April 1938 and November 1940. It is depicted in flight in the vicinity of Langley with an experimental engine cowling design. (NACA 43212)

Bell incorporated the changes recommended, including the installation of an engine which could be equipped with a gear-driven supercharger. However, the new power unit produced only 1,090hp, which was 60hp less than that of the engine which had taken the unarmed XP-39 to 390mph at Wright Field in the spring of 1939. The US Army Air Corps then resumed flight trials. The less powerful fighter, redesignated XP-39B, weighed some 300lbs more than the original, and without the turbo-supercharger it reached a maximum speed of 375mph at 15,000ft in the first trials. Both the US Army Air Corps and Bell expressed satisfaction with the NACA test results. In January 1940 Bell was told to finish the production of the first sereis of YP-39s without turbo-superchargers. The Bureau of Aeronautics called the NACA report on the XP-39B the 'worst condemnation of turbo-supercharging to date.' Soon after this Lawrence D. Bell, president of the Bell Aircraft Company, informed NACA's George Lewis that:

'As a result of the wind tunnel tests at Langley Field, we are getting extraordinarily satisfactory results. From all indications the XP-39 will do over 400mph (even) with

1,150hp. All of the changes were improvements and we have eliminated a million and one problems by the removal of the turbo-supercharger. The cooling system is the most efficient thing we have seen. The inlet ducts on the radiator are closed up to three per cent and the engine is still cooling. I want to convey to you personally and your entire organisation our very deep appreciation of your assistance in obtaining these very satisfactory results.'

The US Army Air Corps left the problems of increasing the speed of the XP39 to over 400mph to Langley. On 6 February 1940, General Arnold's office advised the Agency to make any modification its engineers thought necessary 'which do not involve structural change to the airplane.' The response was that 'the entire investigation should be carried out in flight' at Langley. At first, this appeared possible: during a telephone conversation with George Lewis on the morning of 28 February, General Arnold said that if NACA felt the best way to increase the speed of the *Airacobra* to over 400mph was to make flight tests with the aircraft at Langley, Langley 'should do that and, if necessary, get a pilot from Wright Field.'

However, the US Army Air Corps, Bell and the Agency soon agreed that 'these tests could be better conducted first in the Full-Scale wind tunnel.' In early March the XP-39B was flown to Langley from Bolling Field, where it had undergone performance tests, and was again mounted in the Full-Scale wind tunnel. Within a few weeks the team had finished another systematic drag investigation, this time concentrating

During the early 1930s Langley took delivery of a single Martin XBM-1 tandem two-seat biplane BuNo.9212 for flight test duties. During 1939 it was fitted with an experimental heated wing for icing research, a programme which was continued at Ames in the 1940s. (NACA 18577)

on internal flow problems. Little more could be recommended to improve the airframe, however, because within the poorly designed ducts were structural members for the wings which could not be altered without some basic reconstruction of the aircraft. A flight test programme for the XP-39B Airacobra (38-326) followed at Wright Field.

George Lewis at NACA headquarters notified Langley to keep the research authorisation No.674 covering drag cleanup of the XP-39 open 'in order to provide for the possibility of additional tests being requested by the Air Corps.' For the next several months, Langley sent representatives to both Wright Field and the Bell factory in Buffalo to ensure that the major modifications called for by the wind tunnel analysis were being carried out properly. During September 1940 the first YP-39 (40-027), having incorporated most of the suggestions called for by the Agency, reached a top speed of 368mph at 13,300ft.

There were to be more tests at Langley involving the *Airacobra*. On 5 January 1943 a Bell P-39D-1-BE (41-28378) arrived. The maximum speed of the production P-39D was only 368mph, which was due to the US Army Air Corps adding a new and bigger power plant and heavier armour plate. The first unarmed P-39 prototype had flown at 390mph, faster than any subsequent *Airacobra*, but ten miles per hour slower than Bell had advertised.

The drag reduction programme required precisely the kind of systematic wind tunnel work by Langley did best. Indeed, it cleaned up the drag problems of most US military aircraft which fought in World War II. Most of these were aircraft designs which came off the drawing board before or early in the war, with many drawing on basic NACA data for their design. Secretary of the US Navy Frank Knox said in 1943: 'The Navy's famous fighters — the *Corsair*, *Wildcat* and *Hellcat* — are possible only because they are based on fundamentals developed by NACA. All of them use NACA wing sections, NACA cooling methods, NACA high-lift devices. The great sea victories that have broken Japan's expanding grip in the Pacific would not have been possible without the contributions of the NACA.'

Commencing in 1940 Langley assisted North American to test fly its prototype XP-51 Mustang, the first aircraft to employ the NACA laminar-flow airfoil. On 27 December 1941 the prototype 41-38 ex-RAF AG348, was delivered to Langley for flight testing remaining until 14 December 1942. (NACA 34304)

The US Navy flew the prototype Brewster XF2A-1 Buffalo BuNo.0451 to Langley on 25 October 1940, for tests in the full-scale tunnel for drag reduction studies. After modification its top speed was increased by 31 mph giving more than a two per cent improvement in performance. It stayed at Langley until 9 June 1941. (NACA 15337)

As World War II progressed there was an increasing flow of military aircraft in and out of the NACA laboratory at Langley. The 200th test aircraft to undergo trials was the Republic XP-47F *Thunderbolt* (41-5938) which arrived on 22 February, departing on 15 October 1943. The 201st aircraft was a captured Japanese *Zero* fighter (4593) which arrived on 5 March 1943, remaining for six days. From the NACA aerofoil catalogue, the US aircraft industry produced the wings of the best aircraft of their era — including the Douglas DC-3 transport and the Boeing B-17 *Flying Fortress* bomber, as well as a number of postwar general — aviation aircraft.

Both transports and bombers were included in the NACA flight test programme at Langley. During September 1937 Langley performed both stalling and icing studies with a DC-3 Mainliner NC16070 passenger transport belonging to United Air Lines. In order to warn the pilot of an approaching stall, the NACA engineers installed sharp leading edges on the section of the wing between the engine and fuselage. These sharp edges disturbed the airflow sufficiently to cause a tail buffeting which could be felt by the pilot on his control column. In order to simulate the effects of ice formation on the DC-3's performance, the engineers cemented pieces of sponge rubber to the forward part of the wings where ice was thought to form most often and then measured the resulting changes in the transport's climb, cruise, and stalling speeds. It is worth recording that this DC-3 — NC16070, delivered to United on 25 November 1936 — is still flying in the United States.

At one time during July 1944, a total of 78 different models of aircraft were being investigated by NACA, most of them at Langley. Spin tests were made in the Langley free-spinning tunnel on 120 different aircraft models. The atmospheric wind tunnel crews tested thirty-six US Army and US Navy aircraft in detailed studies of stability, control and performance. From these tests came a wealth of data, first for the correction of existing problems and secondly for the design and manufacturing handbooks.

During 1940 the British gave North American Aviation just 120 days to produce a fighter prototype which met their requirements. It was the Agency at Langley who assisted the company to test-fly its prototype of the famous P-51 *Mustang*, the first aircraft to employ the NACA laminar-flow aerofoil. On 27 December 1941, North American XP-51 No.41-38 arrived at Langley, remaining for a year on and off and finally departing on 14 December 1942. By 1941 the first British fighter types arrived at Langley for testing. These included a Hawker Hurricane and a Supermarine Spitfire (R7347) on 24 November, followed by a second Spitfire (W3119) on 24 December.

Flight-test research work on a wide variety of aircraft inevitably built a large body of experience in the flying and handling qualities of various flying machines. Early pioneering work at Langley in this field had given test pilots a new appreciation of flying qualities, and the huge World War II test programme sharpened that appreciation. As performances increased, so naturally did some of the flight test problems. Again using the systematic approach, Langley test pilots and flight test engineers developed quantifiable handling and flying parameters for the variety of types they flew, and further defined them in terms of wind-tunnel measurements.

After nineteen aircraft had been systematically flight-tested, Langley test engineers prepared a summary report. The report included suggestions for minimum criteria to define an aircraft from the point of view of its handling characteristics. This report became the foundation of the extensive work to be carried out later by NACA, the military services and the aircraft industry. Also it was an impetus to the writing of a military specification on handling qualities — the first such to be written up in the United States.

Other work at Langley during the busy wartime period included an extensive study of wing planform shapes and their effects on the stalling characteristics of an aircraft. Variations in taper and thickness ratio, sweepback and twist were investigated in the wind tunnels. Aircraft loads in manoeuvring flight, still something of a mystery, were studied in test flights, in the wind tunnels, and by theory. Changes in stability and control due to engine effects, another misunderstood flight phenomenon, were delineated in flight test and in the tunnels at Langley. The famous NACA cowl was refined further for a higher speed range, whilst a special flush riveting technique was developed to reduce the parasite drag of aircraft.

Some of the aircraft used in World War II could not have been more poorly designed for landings on water. Fuselage intakes on the underside, bomb-bay doors or wheel wells all scooped up water and served to somersault the troubled aircraft, which often sank inverted. With more over-water combat and ferry flights, and the number of aircrews lost through ditching in the water, there was considerable interest in finding a way of getting the aircraft to ditch safely. Consequently, the hydrodynamics test facilities at Langley were suddenly turned to a high-priority programme of testing scale models in simulated landings on water and recording their behaviour on movie film. One answer was to develop some kind of ditching flap that would counter the effects of the intakes, bomb-bays and wheel wells. Langley produced such a flap, but it was never used since the production changes were regarded as far too extensive.

High Speed Flight and Jet Propulsion

As World War II progressed, aircraft speeds increased and some experienced compressibility problems. These effects emphasised the need for understanding this new characteristic of high-speed flight. It was one thing to fix a problem of high-speed flight temporarily; it could be done empirically, through tests in the Langley wind tunnels, or by carefully controlled and instrumented flight testing.

Despite the wartime work load, Langley research personnel had been thinking about some of the problems of high-speed flight. During 1939 the Airflow Research Division had a further look at the basic concepts of jet propulsion, a long-known principle that had briefly come to light in a 1923 Technical Report published by NACA. In this area, the Agency's research scientists were not working alone: their counterparts in other countries were looking at and working on the problems of jet propulsion. The Germans were very close to flying an experimental jet-propelled aircraft; the British had written a specification for the first jet-powered machine, whilst the Italians were flying a rudimentary jet-propulsion scheme in a test-bed aircraft.

The US military services were keen to get the best performance in range and speed from their aircraft for strategic reasons. Jet propulsion in 1939 and the early war years was not their concern, so reports on its possibilities remained in the files. But not for long. On 5 September 1941, to take advantage of early British work on gas turbine power plants for aircraft, the US Army Air Force requested the Bell Aircraft Corporation to undertake development of a jet fighter design. Preliminary drawings of such an aircraft, the Bell Model 27, were submitted before the end of September and were approved. Power was to be from two General Electric type 1-A turbojets, developed from the Whittle-type engine developed in the United Kingdom. The first US jet-propelled aircraft, the XP-59 *Airacomet*, first flew on the first day of October 1942. One of the thirteen YP-59A aircraft built was extensively tested in the Langley Full-Scale wind tunnel.

Work in a further area of high-speed flight research commenced during 1941, when a group began to work in the eight-foot high-speed wind tunnel to develop propeller designs which could be used to propel an aircraft at the then unheard-of speed of 500 mph. The NACA personnel involved at Langley formed the nucleus of those who carried out later work on high-speed flow which was to win the agency two more of the coveted Collier Trophies. Wind-tunnel methods guaranteed a way of unearthing high-speed problems, but it was only one of the methods NACA traditionally used to obtain design data. Flight tests had to supplement the wind tunnel, plus a variety of other kinds of tests in special facilities such as the free-flight tunnel. These had to be integrated into a test programme before the scientists believed the data was good enough to provide a design base.

At speeds of 500 mph. designers would be working near the fringe of the transonic region and the speed of sound. That speed had been defined as a problem some years earlier, when a British scientist, W. F. Hilton, had said that sonic speed 'loomed like a barrier' against the further development of flight. The words 'sound barrier' passed quickly into the literature and folklore of flight. At that time there was no available way to discover whether it was a barrier or a smoke screen.

During July 1943, NACA directed Langley to investigate major impediments to safe and efficient flight at high speeds by creating a new Compressibility Research Division at the laboratory. This new division incorporated all the ground-based high-speed research sections then at Langley, including the eight-foot and 16-foot high-speed tunnel sections and the model supersonic tunnel section, together with a small section involved with the study of fundamental gas dynamics. Five months later George Lewis formed a special four-man panel

The contributions of women in aeronautical research began well before World War II, expanding to great proportions after 1941. The US Navy's largest flying-boat, the Martin JRM-1 was ordered on 23 August 1938. Models were constructed and tested in the wind tunnels at Langley. Photo depicts NACA personnel, including one female, seen installing flaps and wiring on the wing of a one-twelfth model of the JRM-1 in the dynamic model shop. (NACA 46826)

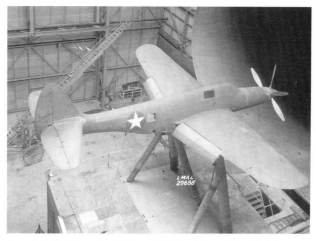

Rare photo depicting a nine-tenths scale NACA wind tunnel model of the Republic XP-69 pursuit fighter in the Langley facility during late 1942. It was a sleek and graceful design with a large wing and a powerful engine.
(NACA LMAL-29688)

to co-ordinate NACA high-speed research. There were many problems which prevented fruitful research until practical solutions were found. Why did strange things happen in Langley's own tunnels as airflows approached Mach 1, the velocity of sound? Why could one get Mach 1 in an empty high-speed tunnel but not in one with a model installed in the test section? There were many more, and they plagued aerodynamicists.

The technical merits of building a transonic research aircraft were successfully sold to the management at Langley in the spring of 1942. By the early summer of 1943, research engineers had finished the preliminary design of a small turbojet-powered aircraft capable of flying safely to high subsonic speeds, from Mach 0.8 to 1.0. Military personnel who learned about the proposed transonic research aircraft had solid reasons to be cautiously interested in it. The Chief of Aeronautical Research of the US Air Service Material Command at Wright Field, felt, too, that the speed of sound was only a wind tunnel and mental barrier. However, one military person who actively advocated support for the Langley concept was Captain Walter Diehl of the US Navy. A longtime friend and close working associate of NACA, Diehl argued in late 1943 in the course of many meetings with the chief of BuAer's structures branch that a transonic research aircraft was the only way to convince people that the sound barrier was 'just a steep hill.' According to the NACA spokesman, the purpose of the aircraft would be to collect the aerodynamic data needed for transonic flight which could not be obtained in any wind tunnel. When the military services became aware of the project there were many differences of opinion — some of which were quite outspoken.

The high speeds reached by diving aircraft and missiles during World War II took the Agency by surprise, for there were no wind tunnels in the Langley system capable of exploring transonic aerodynamics. One of the stopgap

The 201st aircraft to be listed on the large Langley inventory was none other than this captured Japaneze Zero fighter No. 4593 delivered on 5 March 1943, departing six days later after evaluation by NACA flight test pilots. (NACA 32130)

During 1946 the NACA at Langley modified this Stinson L-5E-VW 44-17939 with a unique five-bladed propeller system to demonstrate with a quiet aircraft could be developed. It was demonstrated with great success during the annual NACA inspection at Langley in 1946. (NACA 53020)

methods developed by Langley after 1944 to acquire meaningful data near Mach 1 involved dropping instrumental bodies for free falls from high altitude. The first aircraft used by the NACA to drop its instrumented bodies was a Boeing B-29B (44-83927) *Superfortress*. Earlier the Agency had abandoned the drop-body technique, but in December 1943 the falling-body idea was revived in response to a proposal for a joint American-British effort on the transonic research problem made by William S. Farren, Director of the Royal Aircraft Establishment located at Farnborough, Hampshire. The RAE in the United Kingdom had experimented with dropping weighted bodies from high altitudes.

Langley commenced new falling-body tests with the Boeing B-29 borrowed from the US Army Air Force, which it equipped with the US Navy's most advanced SCR-584 tracking radar unit. The test missile was carried to 30,000 ft and then released. Ground observers received radio signals relayed from instruments inside the body, which measured the forces on it as it fell at velocities sometimes equalling or exceeding that of sound. The NACA research engineers considered the data reliable enough to estimate the drag and power requirements of a transonic aircraft.

During 1944 Robert R. Gilruth, a NACA engineer, devised an alternative method of transonic research. A flying wind-tunnel had one great advantage over gound tunnels: it would not have walls to constrict and distort the airstream. In the first application of the 'wing-flow' technique the young Gilruth mounted a small aerofoil model perpendicular to the upper surface of a North American P-51D *Mustang* wing. The model was placed vertically just above the *Mustang's* wing, making sure, in order to generate uniform flow for valid testing, that it rested in that part of the supersonic flow region where the induced velocity was most constant. A NACA test pilot then flew the P-51 to the desired altitude and put it into a safe diving attitude, which took its speed to about Mach 0.81. As the speed of the aircraft in its dive increased, airflow around the wing-mounted model passed from subsonic through the transonic region to supersonic velocities of the order of Mach 1.4. A small balance mechanism fitted within the fighters' gun compartment and tiny instruments built into the mount of the model recorded the resulting forces and airflow angles.

Approximately one year after commencing to use the drop-body and wing-flow techniques extensively, Langley began to develop a third stop-gap method of acquiring transonic data: rocket-model testing. From June 1941 to the time it finally approved this form of testing in early 1945, Langley had worked on practically every guided missile project operated by the US military services, including the development of

glide, shrouded, and buzz bombs, gliding torpedos, pilotless aircraft or drones and various types of interceptor missiles. Together with its support of the military's top-secret guided-missile projects, Langley began an ingenious programme of more basic aerodynamic tests. This programme moved from Langley to a new NACA research establishment located on the remote Atlantic coast beaches at Wallops Island, some distance from Hampton off the eastern shore of Virginia.

Conferences and meetings with the military services suggested that it was impossible for the US Army Air Force to co-operate in a high-speed aircraft programme, so it was up to NACA to attempt a new approach: it should try to persuade one of the services, or each of them individually, to procure its own transonic research aircraft.

Helicopter Research

Because of the wartime necessity for concentrating efforts on production aircraft flight-testing, the first few years of World War II witnessed only sparse man-hours applied to rotating-wing research. Soon, however, there grew an interest in special missions such as anti-submarine warfare and the rescue of ditched pilots. Research by way of flight test is often viewed as painful and slow but essential. Research test flying with one of the first production helicopters, the Sikorsky YR-4B *Hoverfly*, did not happen without special attention to maintenance, careful piloting technique, specially devised instrumentation and other non-routine efforts. A Sikorsky YR-4B arrived at Langley on 23 January 1944 and a series of useful results — chiefly involving check of or correlation with the forward-flight rotor discussed — was obtained with a modest number of flights for each step; they were also available form the flight trials sooner than from the Full-Scale wind-tunnel work. A special airspeed boom was fitted for the flight test, and stopped-action rotor-blade photos were taken in the Langley wind tunnel.

All services in the US including the US Coast Guard, had evaluated the R-4B and placed orders from the manufacturer. Igor Sikorsky, its designer, was no stranger to the NACA organisation. As early as 1939, at the annual engineering meeting, the entire audience rose to its feet to pay tribute to Sikorsky after he had delivered the contents of his paper on 'Commercial & Military Uses of Rotating-Wing Aircraft.' His eloquent statements of belief in the potential of the helicopter were all the more impressive to the audience by reason of his reputation as a solid citizen in the fixed-wing aircraft manu-facturing business. He was also already working towards his successful VS-300 helicopter.

Langley experienced problems in obtaining parts for its Sikorsky YR-4B. Some smart alec in the stores channels, responsible for obtaining spare parts, ordered the parts required for the counterpart Sikorsky supplied for the Full Scale wind tunnel, instead of the flight test *Hoverfly* and it was discovered the parts arrived a lot quicker. It is rumoured that upon seeing an order from the NACA at Langley for a new clutch for the YR-4B in the wind tunnel, which apparently had only recently had a new clutch, Colonel Frank Gregory's voice could be heard from his office at Wright Field without the benefit of the telephone. Gregory was a veteran helicopter pilot with the US Army Air Force in charge of the division handling the Sikorsky R-4B programme.

Even during 1944 the helicopter was such a novel vehicle that crowds would gather whenever the Sikorsky R-4B hovered by the NACA hangar. These crowds tended to form a tight circle just outside the rotor blade tips, and on one occasion an open vehicle full of personnel halted just behind the tail rotor. In the interest of safety, rules precluding such close inspection by onlookers during hovering had to be made and enforced. Those directly involved with the helicopter programme kept at a safe distance at all times. There were the odd eccentrics who recalled the contents of the 'NACA Technical Note No.4' published in 1920 which stated: 'the

The Bell L-39 BuNo.90060 was a purely experimental version of the P-63 Kingcobra fighter, and designed to study the low-speed handling characteristics of swept wings. It arrived at Langley on 22 August 1946 being transferred to Lewis at Cleveland on 12 December 1949. (NACA 54988)

This North American NA-147 B-45A-1-NA Tornado was delivered to Langley on 1 October 1948. This was the first four-jet bomber to fly in the USA, appearing on the drawing board during 1944/45. It first flew at Edwards AFB on 17 March 1947 and was powered by 4,000lb Allison-built General Electric J35-A-11 engines. Tornado 47-21 was lost in a fatal crash on 15 August 1952 killing Herbert H. Hoover, Head of Flight Operations who had flown at Langley since 1940. (NACA LMAL-63471)

gravest charge against the helicopter is its lack of means of making a safe descent when the engine has stopped.' For most, the thrill of seeing a helicopter perform a 360-degree turn over a designated point on the ground is one they would not forget.

Development efforts with the first large load-lifter helicopter model designed and built at Langley with the designation XH-17 provided both a need and an opportunity to demonstrate the value of dynamic models involving the problems of ground resonance, blade flutter, etc. Its low cost (estimated at approximately $15,000) was in sharp contrast to general precedent. The model was designed and used to provide information applicable to the full-scale helicopter and general results showing the effect of wide changes in properties such as control stiffness and distribution of weight in the blades.

Forty-one Sikorsky YR-4B helicopters were built, many of these involved in development, including 43-28225 for full-scale tunnel tests at Langley. Photo shows NACA technicians setting up equipment for stop-action blade photos. This early helicopter served at Langley between July 1944 and October 1948. (NACA 40416)

The Full-Scale wind tunnel tests on the Sikorsky R-4B were extensive. Comparative hovering tests were completed on six sets of rotor blades of varying construction and geometry on the YR-4B wind tunnel rig. Since these rotors were not a systematic series, the effects of plan form and twist could not be clearly brought out; the effect of surface condition and contour accuracy was, however, unmistakable. World War II helicopter research at Langley appears to have been restricted to the Sikorsky YR-4B, which finally departed on 20 October 1948. By that time Langley had a Sikorsky HO3S-1 (BuNo. 122520) which arrived on 24 February 1948 for flight testing. During 1947 a helicopter test tower was built and put into operation, handling rotors up to about 50ft in diameter mounted at a height of 40ft. Fuel was supplied by an overhead feed line, and there was a two-layer, loosely-hung safety net which served its purpose on several occasions.

Wartime Personnel

The World War II peak figures for NACA personnel reached a total of 6,804 with Langley employing 3,253. Almost every kind of worker was needed by the agency, ranging from highly-trained technical staff to truck drivers and messengers. Scouts were sent out to strategic cities and colleges in the US, and every candidate for a job was interviewed. Prior to World War II, women had served only in office positions, but by 1943 there were hundreds employed in sheet-metal, riveting, polishing, operating test machines and instruments and serving in many other fields and facilities.

Schoolboys proved to be extremely useful in the model shop were aircraft models were made for wind-tunnel studies. The NACA made special efforts to find talented boys, and the young model makers were to fill the gaps when World War II dried up the usual sources of supply for woodworkers and mechanics. At one time Langley employed no less than 500 in this category.

One of the biggest handicaps was the drain to selective military service: the US military draft board did not initially accept employment in aeronautical research as a reason for deferment. Finally in 1943 the military recognized the value and importance of the Agency's research work, and a plan was drawn up in which draft-eligible NACA employees who were essential to the heavy research operations could be retained in their jobs. Presidential approval was received in February 1944, and the scheme went a long way toward preventing the loss of the Agency's key personnel. To assist with the requirement for extra staff, the US Army Air Force agreed to release personnel in certain categories who had served overseas or were over twenty-six years old. Several hundred were obtained by this means, though unfortunately less than twenty turned out to be research engineers.

As early as August 1939 Congress voted favourably on a project to establish a new research laboratory at Moffett Field, California. This received final approval the following month, and early in 1940 the first grass sod was cut. Among the projected equipment listed in the original prospectus for the establishment at Moffett Field was an engine research laboratory. A special survey committee, which cast the deciding vote for the new site in California, had been asked to review the whole array of research installations throughout the United States. It was dismayed to find a paucity of first-class facilties for fundamental research on aircraft engines and recommended that a major research station devoted wholly to aircraft engine and allied problems be established.

The outcome of this recommendation was the building of the Flight Propulsion Research Laboratory, at Cleveland, Ohio. The plan for this new laboratory was reported by a special committee in January 1940, laid before the Bureau of the Budget in March, approved by the President and submitted to Congress in May and authorized by an act of Congress in June.

A bunch of Langley NACA flight test pilots pose in front of a Republic P-47 Thunderbolt during 1945. Pilot on the left of the group is Melvin N. Gough, born 1906, receiving a Bachelor of Science degree in mechanical engineering during 1926. He then joined the NACA becoming the chief test pilot at Langley and later transferred to the NASA. (NACA 42613)

During World War II requests were coming in to NACA for research on guided missiles as well as transonic and supersonic aircraft. Several new techniques were proposed for these high-speed studies. The work was highly secret, and a range for experimenting with rocket-propelled missiles required ample space. Wallops Island, an uninhabited stretch and sand dunes and marsh on the Atlantic side of the Virginia Eastern Shore Peninsula was selected during the winter of 1944/45 as the site for these experiments.

Despite the success of American warplanes, two of the major aeronautical trends of the era escaped the attention of NACA. The agency endured much criticism in the early postwar period for its apparent backwardness in the development of jet propulsion, and in the area of high-speed research. History suggests that there may have been a lapse of sorts, although not as total as many of NACA's critics believed. When the first US jet aircraft, the Bell XP-59A *Airacomet*, made its maiden flight on 1 October 1942, at Muroc — a remote area of the Californian desert — there was no official NACA observer present. It is rumoured that the Agency did not even known the aircraft existed. After World War II, the failure of the USA to develop jet engines, high-speed swept-wing aircraft and supersonic designs was most unfortunately blamed on the NACA. The many critics argued that the Agency, as America's premier aeronautical research establisment, had somehow allowed leadership to slip to the British and the Germans during the late 1930s and during World War II.

In retrospect, NACA's record seems mixed. There were some areas, such as gas turbine technology, in which the US clearly lagged although the NACA research engineers had begun to investigate jet propulsion concepts. There were other areas, such as swept-wing designs and supersonic aircraft, in which the Agency had made important steps. It was very unfortunate that the lack of advanced propulsion systems — including jet engines — made such investigations academic exercises. During this period, when NACA found it difficult to progress into the future due to lack of co-operation from the US military services, it is difficult to deny that the Agency trailed behind the rapid strides made in Europe.

During 1946 a new Induction Aerodynamics Laboratory was introduced at Langley. In this facility, research engineers investigated the aerodynamics of subsonic and supersonic internal flows, concentrating on solving such basic problems as the optimum method of inducing air and supplying it to high-speed conventional and jet engines.

Despite the critics there is no doubt that the National Advisory Committee for Aeronautics was making progress and aiming to climb and cross new frontiers.

This Lockheed P-80A-5-LO Shooting Star 44-85382 single-seat jet-fighter arrived at Langley with the NACA on 15 November 1946. It became NACA 112, serving at the centre until 8 January 1959, when it was salvaged. It is seen parked at Langley showing the USAF buzz number PN-352. (NACA 50653)

The east hangar at Langley in 1947, showing a Douglas C-54D Skymaster 42-72713, and two Boeing Superforts including TB-29 44-69700 and B-29G-6-BW 45-21808 registered NACA 124. A Republic P-47 Thunderbolt and a Lockheed F-80A 44-85352, later NACA 112, are seen parked centre right. In the background is the 19ft pressure and the full-scale tunnels. (NACA 52313)

CHAPTER FOUR
West to California

Early in 1936, some three and a half years before the preliminary tremors in Europe awoke the free world to the gathering clouds of World War II, the NACA committee in Washington, D.C., completed a careful review of its resources in connection with the United States' preparedness for national defence. The matter was brought up at a meeting of the executive committee on 3 March, when the chairman presented an eye-opening account of recent aeronautical research developments abroad. The NACA office in Paris had reported several laboratories recently established, new wind tunnels of improved design under construction and new high-performance aircraft of advanced design. George W. Lewis NACA's Director of Aeronautical Research was to visit Germany and Russia during September and October 1936 and on his return addressed the Langley staff on his findings.

The NACA appointed a special committee on aeronautical research facilities which had its first report ready before the end of the month. It pointed out that the general trend in aeronautics was toward larger and faster aircraft, and the report warned that new and intensive research in the United Kingdom, France, and Italy 'will within a few years, enable European countries to surpass the United States in the technical development of aircraft.' It was therefore recommended that, 'in the interests of national defence and of American commercial aviation, NACA should promptly take steps to improve its research facilites to meet the existing needs of aviation in the United States.'

The first aircraft to arrive at Ames was this North American O-47A-1 37-323 on 5 September 1940. It was heavily involved in the icing research programme originated at Langley. Note the wing section fitted to the starboard wing, suitably braced. This test-bed was used by the NACA at Ames until 13 March 1946. (NACA AAL-1435)

There was no time after 1936 that the committee did not have a group working on the problem of strengthening its preparedness for war. One of these groups, working on long-range planning, pointed out that the area at Langley set aside for laboratory use was congested, questioned the policy of concentrating all equipment at one locality and suggested that a study be made of the advisability of establishing a second research centre in another part of the USA. These discussions were under way in 1938. During October, Chairman Dr Joseph S. Ames, appointed a committee to appraise the situation, and in December it submitted a report recommending three integrated projects. The first involved the enlarging of the existing laboratory at Langley: the second, establishment of a new laboratory at Moffett Field, California; whilst the third embodied a plan of more effective co-ordination of research in the aircraft industry.

During the last week of 1938 these recommendations, supported by a memorandum containing reasons for the proposed expansion, were delivered to the chairman. Dr Ames sent the report to the White House with a letter to President Roosevelt calling attention to its contents.

Despite Presidential approval, there were sharp challenges from Congress. The estimated needs were disputed and the idea of setting up a second research centre was particularly attacked. The proposal for a second laboratory was flatly rejected by both House and Senate committees. It was not until August 1939, after a hue and cry boosted by newspapers, aeronautical scientists and the military and pressure generated by further reports from Europe, that Congress finally voted favourably on the project. The act specified that the new research station should be constructed at a place to be chosen by NACA from amongst the sites brought to its

attention within the next thirty days. After careful consideration and examination the fifty-four sites which were on the list, the committee finally returned to California for its choice. The deciding criteria were proximity to the aircraft manufacturers on the west coast, location on an active aviation field at Moffett, an adequate 100 acres of space, accessibility of ample electric power — and climatic conditions favouring all-year-round flying. The decision to build at Moffett Field was reported on 14 September 1939, and construction began early in 1940.

The list of projected equipment for the new station included an engine research laboratory. There was already a small one at Langley, and the idea was that each establishment would study aircraft powerplants along with propellers, wings, and other components. But a new idea began to take form, and it was finally recommended that a new major research establishment, devoted wholly to aircraft engine and allied problems, be established. The outcome was the Flight Propulsion Research Laboratory at Cleveland, Ohio, which was authorised by Congress in June 1940.

Construction began at Moffett Field in the early part of 1940, when a small group from Langley arrived to supervise operations. The design of wind tunnels and other structures for the new laboratory had been started at Langley, so blueprints were ready to guide the erection of the first building. Ground for it was broken on a windy day in February, and thereafter — right up to VJ-Day — there was never a time when a major piece of research equipment was not under construction. In March the foundations were also being laid for the flight research building.

The question of naming the new laboratory was high on the agenda when the NACA committee met in Washington, D.C., Dr Joseph S. Ames, then in his seventy-sixth year, had been the Agency's chairman from 1927 to 1939, and had served continuously since President Wilson appointed him in 1915. It was wholly appropriate that NACA named the Moffett facility 'Ames Aeronautical Laboratory' in April 1940.

Lockheed Model 12A Electra NC17397 named 'Sky Baby' seen at Ames being delivered on 16 January 1942, possibly from Lockheed. It appears to be fitted with wing leading edge de-icing equipment. At this time Ames was heavily involved in a de-icing flight programme, with a Lockheed 12A NACA 97 ex-NC17396 included in the fleet. (NACA AAL-1774)

After the first building had been completed in August 1940, a programme of aerodynamic investigation of the problems of high-speed military aircraft was inaugurated. A technical service building to house the machine, model and sheet-metals shops and other service facilities was completed next, followed by the erection of the utilities building. Three wind tunnels were in operation before the end of 1941, and during the following year the science laboratory came into use. The largest installation at Ames, the 40 × 80ft Full-Scale wind tunnel was not completed until 1944. By the end of World War II there were five wind tunnels and eleven laboratory buildings in use and an enlarged flight-research hangar under construction. The staff then numbered 328 engaged in research, 412 in technical services and 84 in administration and clerical work.

Early History and Projects

The choice of Sunnyvale site in California was bound up with several important considerations. Being located on the Pacific coast would at least protect the new laboratory from the possibility of attack from Europe, although NACA had carefully considered the threat from Japan. The vulnerability of the site, however, was offset by other factors. It was located on an active military base and could therefore take advantage of existing airfield facilities; there was good flying weather throughout the year with a relatively low density of air traffic. There was also adequate and economical electric power — this had been a problem at Langley — and access to the aircraft industry located along the Pacific coastline from San Diego to Seattle. industrial centres and academic institutions of repute was excellent.

By 1939 almost half the United States aircraft industry was located on the west coast. Those research engineers from Langley who liaised with the aircraft companies on a regular basis had either to endure a four-day train ride or some thirty exhausting hours by aircraft. A further factor was the competitive relationship which had developed between the NACA and the Guggenheim Aeronautical Laboratory at the California Institute of Technology. Pasadena was easily accessible to the aircraft industry in California, but its facilities were often heavily stretched with industrial research requirements. The choice of Moffett Field answered the needs of industry and reasserted the Agency's research pre-eminence.

Moffett Field had been conceived as a base for the US Navy's rigid airships during 1931. At that time, the US Navy was building ZRS-4 and ZRS-5, later christened *Akron* and *Macon*. The former was to be assigned to a base at Lakenhurst, New Jersey, *Macon* being located at Sunnyvale. Local committees in California had donated 100 acres for the base, and an additional 750 acres was purchased by the government. The new facility was expected to provide employment both in its construction and maintenance, and the local business community envisaged steady orders for supplies and materials. The Pacific Gas & Electric Company proudly announced that the electricity load at Sunnyvale would be 2,200 horsepower.

A huge hangar 198 ft high and 1,133 ft long was constructed, covering eight acres. It was reputed that the huge capacity induced its own weather conditions, which could vary at both ends. On 4 April 1933, the *Akron* crashed into a stormy Atlantic off the New Jersey coast, leaving the *Macon* as the country's only rigid airstrip. Commanding the *Akron* when it crashed was Rear Admiral William A. Moffett, who had been influential in both the US Navy's commitment to rigid airships and in the selection of the Sunnyvale site for a west-coast base. In June 1933 the base was named Moffett Field in his honour. It was finally completed in January 1935, but the following month *Macon* crashed off the California coast at Point Sur, 150 kilometres south of the air base. This crash marked the end of the US Navy's involvement with rigid airships and the Sunnyvale base became surplus to requirements. During October 1935 it became a US Army Air Corps training site and it was coincidental that one of the first squadrons sent to Moffett Field was from Langley Field.

The US Army Air Corps was busy throughout 1939 constructing barracks and mess halls, whilst NACA were preparing to make Moffett Field the location for its second laboratory. The US Navy still felt there was a future for rigid airships and requested that the NACA buildings should be located outside the still-present mooring circles.

Since 1930 the NACA had been involved in icing research. A workhorse operated by Ames was this Curtiss C-46A-5-CU Commando 41-12293 delivered on 10 March 1943. It is seen in flight on 22 March 1946 with TEST on the fin and rudder and a wing section mounted on top of the fuselage. (NACA A-9556)

The land acquired by the Agency was completely un-developed as was Moffett Field. Adjoining the property were farms, fruit trees and dairy cattle; the pastures would be concrete in twenty-years. The Bayshore Highway led to the main gate at Moffett Field. The geography of the NACA site was well investigated, and the high water-table and the possibility of earthquakes given special consideration. Building specifications had to be adjusted for maximum protection. This southern end of San Francisco Bay today bears little resemblance to its appearance in 1939.

Back at Langley, planning began in autumn 1939 for the facilities needed at the new laboratory. The design group was headed by Smith De France, who — it was rightly assumed — would head the new establishment. However, he remained at Langley over the winter of 1939/40 designing the wind tunnels and building required at Moffett. At the end of August 1940 De France and twenty-two others arrived in California from Langley. The Agency had hired, through the usual civil service procedures, additional junior engineers and support staff. A few were Stanford University graduates, and by the end of the summer there were fifty-one people on the staff. There was one division — the Research Division: subordinate groups were defined as facilities became available. In addition to the three research sections there was a small flight research unit consisting of two test pilots, a research engineer and a handful of aircraft maintenance personnel.

Ice Research

The first aircraft to arrive at Ames — a North American 0-47A-1 (37-323) — appeared on 5 September 1940, to be used in an icing research programme. Test pilot William H. McAvoy was involved, having transferred from Langley along with Lawrence A. Clousing. NACA had been working on the problem of heat de-icing at Langley since 1930; a Martin XBM-1 bomber had been equipped with an experimental heated wing panel for icing research during 1939 at Langley and was flown by McAvoy and Clousing. Tests on the bomber were rudimentary but the results convinced that Agency of the need to purchase a small twin-engined Lockheed 12 transport for ice research work. This was 'NACA 97'. The wings of the Lockheed were rebuilt to incorporate an ice-protection system utilizing the engine exhaust as a means of de-icing. The project had the support of the US Army Air Corps and Lockheed, and Kelly Johnson and his staff at Burbank designed, built and installed the new wings. It had been decided earlier that the ice research project should be transferred to Ames, and whilst waiting for the Lockheed 12 to be modified the test pilots carried out minor icing flight-research tests with the O-47A; William McAvoy made the first flight on 16 November 1940. On 20 January 1941, the Lockheed 12 was delivered to Ames and two days later — with McAvoy and Clousing at the controls — the initial flight was made. From then on the programme proceeded at a good pace. On 13 May 1942, a Consolidated XB-24F-CO *Liberator* (41-11678) joined the programme.

By the end of 1942, the combined efforts of Ames, the aircraft industry and the military had produced prototype installations of heat de-icing systems for both the Boeing B-17 *Flying Fortress* and B-24 *Liberator*. The US Navy had also made an installation on a Consolidated PBY flying-boat. The NACA team involved in the de-icing project felt that a special base should be established in the north-central part of the USA for the purpose of evaluating prototype de-icing installations such as those incorporated in the B-17 and B-24. The idea was approved, and in the winter of 1942/43, a base was set up at Minneapolis, Minnesota, near the headquarters of Northwest Airlines. The New Ice Research Base was staffed by NACA personnel and the US Army Air Corps, but it also received help from both Northwest and United Air Lines. Other agencies also became involved. The US Weather Bureau assigned William Lewis to work directly with the Ames group, whilst the Royal Aircraft Establishment at Farnborough in the United Kingdom assigned a very able research specialist, J. K. Hardy, to work with the group. The RAE quite rightly felt that it was foolish to duplicate work done by NACA on icing research.

On 3 October 1943, a Curtiss C-46A-5-CU Commando cargo aircraft (41-12293) was delivered to Ames and was soon modified to incorporate the most complete ice-protection system yet provided for any aircraft. It was thoroughly instrumented and carried special recently-developed equipment for obtaining basic information on the character of icing cloud. Bill Lewis of the US Weather Bureau made major contributions during this phase of the programme. The C-46 became a flying research laboratory operated from the Ice Research Base at Minneapolis and flew far and wide searching for conditions. Airport habitués were often astonished to see the C-46 approaching through the murk when most other aircraft were grounded. It was flown by a number of pilots, but most often by Captain C. M. Christensen, Senior Pilot of United Air Lines who contributed greatly to the project.

In the early days at Langley, Lewis A. Rodert had taken over the heat de-icing project — then having a low priority. With the move to Ames he had lived with the project and deserved much credit for his important share in its success. In 1947 his good work was recognised when, at a White House ceremony, President Truman presented him with the Collier Trophy 'for his pioneering research and guidance in the development of a practical application of a thermal ice prevention system for aircraft.' It should be added that, earlier in 1943, the Institute of Aeronautical Sciences had given Ames test pilot William H. McAvoy the Octave Chanute Award for 'continuous service in the flight testing of experimental planes under hazardous conditions imposed by aeronautical research.' Certainly an important part of the rationale for this award came from the hazardous de-icing flights involving the Lockheed 12 'NACA 97' on which McAvoy had served as pilot.

In April 1942, the US Army Air Corps left Moffett Field and the US Navy returned. With responsibility for Pacific coastline security, the US Navy planned to patrol the coast using non-rigid airships or 'blimps', and the large Moffett Field hangar was the logical place to moor them. The transfer was accomplished routinely, and relations between NACA and the new residents became as cordial as had those with the US Army Air Corps. Apart from influencing the direction taken by early research work, World War II visibly affected Ames in other ways. Early in 1942 blackout shades became compulsory, and the establishment continued its mushrooming growth in both facilities and personnel. The original number of fifty-one personnel in September 1940 had grown to 341 by February 1943 and the wartime high of 844 was reached during August 1945.

Until 1944, Ames had been more successful than other laboratories in obtaining deferments from the military draft board. In the end most laboratory personnel were inducted into military service and continued in the same jobs. Those at Langley and Cleveland joined the US Army, but because Ames was a US Navy base its personnel joined the Navy. To help ease the manpower shortage, the US Navy assigned some 200 men — from machinist mates who had seen combat to engineering students from the V-12 college programmes — to the laboratory.

The Flight Research building went into operation during August 1940, only two weeks behind schedule. It housed not only the flight research engineering staff, who were about to resume the de-icing research programme started at Langley, but also doubled as an aircraft hangar and maintenance shop. As forecast, Smith De France, the assistant chief of aerodynamics at Langley, was named Engineer-in-charge at Ames. He was regarded by those who worked with him at Langley as a fine engineer, a hard worker and a thoroughly professional civil servant. In 1924 at Langley he was flying a newly-arrived engineer around the laboratory site in a Curtiss JN-4H *Jenny*. Whilst attempting a landing, the aircraft crashed into the marshy ground surrounding the field and the young engineer was killed: De France lost an eye. He never flew again, honouring a promise to his wife.

The Early Days of Flight Research

Editors note: Seth Anderson was an early flight research engineer and became the chief of the Flight & Systems Simulation Branch at Ames. He is currently Research Assistant for Interagency Programmes. This item was written by Seth and was published in one of the publications produced during 1989, the 50th Anniversary Year of Ames.

Ames Flight Testing had barely started when I entered the hangar on 7 July 1942, to join the Flight Research Section headed up by Larry Clousing. The Flight Research Building, the first permanent structure of the newly established Ames Aeronautical Laboratory, also served as temporary quarters for all administration functions including the offices of the Engineer-in-Charge, Personnel, Fiscal, Library, etc. Ames had no cafeteria — we ate on the US Navy side. I counted five aircraft in the hangar, three used for icing research; a North American O-47A, a Consolidated XB-24F-CO, and a Lockheed 12A. A Vought Sikorsky OS2U-2 *Kingfisher* (BuNo. 2189) and a Brewster F2A-3 *Buffalo* (BuNo. 01516) were being used in performance and handling qualities studies.

Clearly, World War II not only dominated research activities at AAL, but also one's lifestyle. I had hitchhiked to California from Illinois — it took a while, the wartime national speed limit being 35 mph. Getting to work was in itself an exciting challenge. Because of fear of invasion on the West Coast, there were no street lights; black out shades covered all the windows, and no street signs denoting locations of Ames or Moffett Field were evident. Crossing the four-lane Bayshore via Moffett Boulevard was like Russian Roulette since there were no stop lights to regulate traffic flow. Finding transportation was not easy. Cars were not produced during World War II and a bicycle could be purchased only if you had a signed form stating that use of the bike was essential to the war effort. I remember riding to work in the trunk of Willie Harper's coupe along with Steve Belsey. Fortunately, there were no stops from Palo Alto to Moffett, only open farmland. The carbon monoxide exhaust was debilitating.

The second prototype Douglas XSB2D-1 Destroyer BuNo.03552 seen at Ames on 30 August 1945. It arrived on 12 June 1944 followed by a production BTD-1 BuNo.04968 on 28 July. The Destroyer BuNo.03552 made a forced landing in a prune orchard on 10 January 1946, with flight test crew George Cooper and Welko Gasich on board. (NACA A-8668)

Flight research played a lead role early in AAL activities. The first programme assigned to Ames, and the first Ames research report published in September 1941, involved icing research using the O-47A aircraft. The military needed quick answers on operational problems and service aircraft showed up with clock-like regularity for testing. Because the aircraft were taken from squadron use, time was of the essence and work continued through Saturdays (no extra pay) to help return aircraft promptly. In addition to normal engineering duties, the researcher, in many cases, helped install research instrumentation and wiring in the test aircraft, working alongside women aircraft mechanics who skilfully contributed to prompt completion of the programme. In sharp contrast to today's leisurely pace of flight testing, only three months ensued from time of aircraft arrival, completion of flight tests, and return to operational use. In the time span of World War II, Ames flight tested approximately 56 aircraft.

Typical of most of these flight research programmes, the short time available didn't allow fixes to be made for aircraft deficiences. For example, after participating in flight tests of a Consolidated Vultee A-35A *Vengeance* (41-31174) attack dive bomber. I authored a report noting that the aircraft failed to meet current military flying quality standards in several areas. Shortly after an advanced copy was given to the manufacturer, the vice president of Vultee — who also was the project test pilot for the A-35A — appeared at Ames wanting to know how we could possibly find any shortcomings in his aircraft, which he personally developed and expected to sell to the Army Air Corps. I was summoned to the office of Smith J. De France, Engineer-in-Charge, expecting to suffer both in job longevity and credibility. I reviewed the factual evidence of the deficiencies from the flight data and pointed out how the aircraft could be improved with only minor modifications. In the end, both 'Smitty' and the Vultee representative were smiling. I returned to my office, remembering that a 'good' engineer must also be a diplomat.

Lawrence A. Clousing, flight test pilot, seen prior to test flying one of the two Lockheed XP-80A Shooting Star fighters, 44-83023. It arrived with the NACA at Ames on 19 September 1944. Larry Clousing transferred from Langley along with William H. McAvoy born 1896 and employed with the NACA since 1921. (NACA A-15412)

Many people may not be aware that Ames was the first to conduct flight research at Muroc Dry Lake (now Edwards AFB) in the latter years of World War II. This was before the High Speed Flight Station (now Ames-Dryden) was established in 1946, primarily with Langley personnel. Two high performance aircraft, the North American P-51B *Mustang* and the then 'secret' Lockheed YP-80 *Shooting Star*, were tested at Muroc to take advantage of the large unrestricted flight area and ample room for landing in the event of an emergency. As it turned out, the test area served its purpose well but with mixed results. the P-51B (43-12114) was used to correlate flight drag measurements with results obtained from a ⅓-scale P-51B model in the 16-foot wind tunnel.

Because the wind tunnel tests used an unpowered model, the flight tests were conducted without a propeller to eliminate slipstream effects. The P-51 was towed to altitude by a Northrop P-61 *Black Widow* aircraft by means of two long tow cables. At 28,000 feet, pilot Jim Nissen released the tow line and glided down, taking records to obtain drag performance. Things went well until the third flight when the tow cable prematurely released from the P-61 during climb out. It flew back and wrapped around the wing of the P-51, bending the airspeed boom. In part, as a consequence of erroneous airspeed readings, the pilot overshot the vast lake bed, ending up in a gravel pit. This totalled out the aircraft, but Jim crawled from the wreckage relatively unscathed. After he was taken to the base hospital to be X-rayed for possible broken bones, it was discovered that the X-ray machine lacked power because Jim had clipped the power transmission line in the ill-fated landing. Fortunately, enough data had been obtained to establish a reasonable drag correlation.

Tests of the YP-80 (44-83023), the first US jet-powered fighter, were equally exciting. The aircraft had been thoroughly instrumented at Ames for performance and handling qualities and for measurements over a yet unexplored speed and altitude range. In 1945, shortly after a series of successful checkout test flights made from the Muroc North Base, the value of testing over a large dry lake bed would once again become apparent. At 35,000 feet, measurements were being made to document strange directional oscillatory characteristics associated with an audible 'duct rumble' which started in sideslipping flight. During one large sideslip excursion, the engine flamed out and could not be restarted. The pilot, Larry Clousing, expected no problems in making a 'dead stick' landing on the dry lake bed. Because the

With the end of hostilities in 1945, Ames took advantage of a surplus of aircraft types to use as research vehicles. Seen on 31 May 1946, is Vought F4U-4 Corsair BuNo.97028 in original US Navy livery but with TEST on the fuselage. (NACA A-9972)

engine-driven hydraulic pump was inoperable, it was necessary to use the hand-pumped emergency hydraulic system to extend the landing gear. Alas, because of peculiarities in the poorly-designed hydraulic system, it was not possible to extend the gear with the emergency system and the aircraft was safely belly landed at Rogers dry lake. After repairs were made (one year later) the YP-80 flight research programme continued, helping to uncover and explain several mysteries of flight in the transonic speed range.

Although there were several harrowing flight experiences, there were no serious flight accidents during the World War II test period. As one of my duties as a flight test engineer, I was part of the flight crew responsible for overseeing test procedures and operating the on-board recording test equipment. In March 1944, I was involved in flight tests of a Martin B-26B-21-MA *Marauder* twin-engine medium bombardment-type aircraft. This aircraft had marginal performance during take-off when power was lost on one engine. Its notorious engine-out safety record had inspired the nickname 'Widow Maker' and many aircraft had been lost in combat and training.

A strato-cumulus overcast limited our ceiling to 7,000 feet as we headed north on a Saturday morning to begin low-speed tests. With flaps and gear down, the left engine throttled to idle and the right engine delivering full take-off power, airspeed was gradually reduced. Straight flight was maintained by use of considerable right rudder force and full nose-right rudder trim tab input. I called on the intercom that the data was being recorded. After reaching the lowest controllable airspeed, pilot Larry Clousing, endeavouring to resume normal flight, abruptly removed power from the right engine without returning the rudder trim tab to neutral from full nose-right. As a consequence, the aircraft yawed violently to the right and then back to the left as the pilot realized his mistake. Both the pilot and the co-pilot, Bill McAvoy, were attempting to regain directional control, sometimes with counteracting rudder and engine power inputs as the aircraft behaved like a carnival ride, rapidly losing altitude and departing toward an incipient spin. Speculation was rife whether control would be regained before available altitude was used up. I clipped on my chest pack parachute and amid shouts in the cockpit to the effect of 'I've got it!' the flight situation further deteriorated. I headed for the escape hatch, calling out to the pilots, 'I left the records on — I'm getting out.' 'Not until I give the orders, you don't!' said McAvoy.

This production model of the Lockheed P-80A-1-LO Shooting Star 44-85299 was delivered to Ames on 18 December 1946. It is seen in flight test markings as NACA 131 fitted with vortex generators installed ahead of ailerons to combat aileron buzz. It survived at Ames until 6 June 1955. (NACA A-16628)

Shortly afterwards, control was regained at 3,000 feet directly over the Southern Pacific Railroad tracks in the city of San Mateo.

Although the pace was hectic back then (we didn't get to use our annual leave), valuable information was obtained from evaluations of World War II aircraft which helped improve design of future aircraft. Looking back, I consider myself fortunate to have participated in the early days of flight research.

Flight Test Research

The ice research project represented a single and very important phase of the flight research programme at Ames. Flight research had always been an essential element of

NACA's research effort. Scientific techniques of modern flight research had first been established in the US by NACA, and the Agency remained a leader in advancing these techniques. NACA required its test pilots to have an engineering degree in addition to a minimum number of flying hours in their logbooks, together with the ability and motivation to perform precise preplanned manoeuvres previously worked out on the ground. The oral report of the test pilot generally formed only a minor part of the data obtained from a NACA test flight. The major part was in the form of data obtained with the help of special wire and tape recorders carried in the aircraft.

Faced with the growing threat of war, President Roosevelt in the early 1940s called for the production of 100,000 aircraft. Soon the aircraft industry was receiving orders from the government for new types of aircraft as well as production orders for existing types. After the US entered World War II in December 1941, a new family of aircraft began to appear. Most had more powerful engines and were capable of high speeds. These new factors generally introduced problems in the areas of handling qualities, stability and control, drag reduction and structural loadings. NACA adopted criteria for flight tests which were passed on to the military and proved extremely useful in evaluation of new aicraft types. Equally the US Navy (and sometimes aircraft companies) contributed the services of their test pilots to further the flight research at Ames. These activities expanded so fast during World War II that the hangar in the flight-research building became too

This Grumman F6F-3 Hellcat BuNo.42874, later NACA 158, delivered to Ames on 22 June 1945 was the original variable-stability aircraft developed by the NACA at Ames. It was used until 9 September 1960. (NACA A-20543)

small. A contract for a new much larger hangar with attached offices and maintenance and engineering shop went out to the contractors in March 1944.

Ames could, and did, evaluate the handling qualities of many new military aircraft, and when serious faults were discovered the service involved would request the aircraft company to make the necessary change to the aircraft design. At that stage, such a change could be both difficult and expensive to make. If it was possible to correct such faults while a new development was in the wind-tunnel model test stage, the situation would be vastly simplified. This was a problem which the Ames engineering and design staff attempted to solve, and eventually did solve. Responsibility was undertaken jointly by the 7 × 10 ft wind tunnel and the Flight Research Sections, and wind-tunnel tests were checked by flight tests of the actual aircraft.

This rigorous testing was carried out on numerous aircraft types, but the most extensive data confirming the technique were obtained on the US Navy's Lockheed PV-1 *Ventura* twin-engined patrol bomber. On 6 January 1944, PV-1 (BuNo.48871) was delivered to Ames for flight testing whilst a PV-1 model was being tested in the 7 × 10ft wind tunnel. The first attempt to interpret the PV-1 model test data in terms of flying qualities was made by Victor Stephens and George McCullough, whilst Noel Delany and William Kauffman reported confirmatory results obtained from flight tests of the actual PV-1 aircraft.

Prediction of performance factors, such as drag and speed, from such joint evaluations was quite a different problem. The test models were idealized with smoother, truer surfaces than the originals and lacked the gaps, excrescences and rivet heads found in the actual aircraft. Other influences in the wind tunnel were often difficult to evaluate, including the interference of the struts on which the model was mounted, the turbulence in the airstream and the subtle effects of the surrounding walls. The rather indeterminate effects of many of these factors called into question the accuracy with which the drag and speed of an aircraft could be determined from wind-tunnel tests. To obtain more information, Ames engineers undertook to make comparisons of the drag of an aircraft determined initially by model tests in a wind tunnel and then by measurement made in actual flight. The comparison was made for only two aircraft, the North American P-51 *Mustang* and the Lockheed P-80 *Shooting Star* — the latter being a P-80A-1-LO 44-85299, which became 'NACA 131' and was delivered to Ames on 18 December 1946. The Ames aircraft records reveal that North American P-51B *Mustang* 43-12111 was delivered on 11 August 1943, for handling-quality tests.

Used as a general utility transport, plus being used for gust research development, this Fairchild C-82A-30-FA Packet 44-23056 became NACA 107. It was delivered to Ames on 31 August 1947 remaining until 7 February 1961. This photo was taken in May 1957 at an Open House held at Moffett Field.
(William T. Larkins)

During early 1943, a spate of tail failures appeared during the operation of some new high-speed flight aircraft. These aircraft had been designed to dive at high speed and perform rolling pullouts and other violent manoeuvres, yet failures were occurring under what were considered safe conditions by the designers. The Ames Flight Research Section was assigned a Bell P-39 *Airacobra* to study the general effects of compressibility on air loads in addition to the specific tail-failure problem. A variety of special NACA instrumentation was installed, including instruments to record variables in the indicated airspeed, pressure altitude, normal acceleration, engine manifold pressure, engine rpm, approximate angle of attack of the thrust line and landing-gear position. Other parameters recorded were aileron, elevator, and rudder position, aileron and elevator forces, rolling, yawing, and pitching velocities, and the pressure distribution over extensive areas of the wings and tail surfaces.

The instrumented P-39 was used in several quite extensive programmes to determine handling qualities and airloads both in steady straight flight and in various manoeuvres. Valuable information was obtained which pointed the way to improved methods of design. Test pilot for most of these programmes was Lawrence Clousing, his principal ground engineers being William Turner and Melvin Sadoff. Tests showed that indeed the loads had been underestimated. A memorandum report written by Turner and Clousing calmly noted. 'Other miscellaneous failures of various degrees of severity occurred to both the elevator and stabiliser structure . . .' — 'other' indeed! Clousing exhibited remarkable fearlessness in all his flight assignments. He certainly did not do it for the flight pay, for all NACA test pilots received standard government civil service remuneration. Commercial pilots were paid treble the amount for far less dangerous jobs. Larry, an ex-US Navy pilot was reluctant to talk about his job.

Details of the many wartime and early post-war projects would fill many volumes. Amongst the aircraft tested were the Bell P-39 *Airacobra* and P-63 *Kingcobra*: Boeing B-17 *Flying Fortress*: Brewster F2A *Buffalo*: Consolidated B-24 *Liberator*: Curtiss C-46 *Commando*: Douglas SBD *Dauntless*: XBT2D-1 *Skyraiders*: A-20 *Havoc*: A-26 *Invader*: General Motors P-75 *Eagle*: Lockheed P-38 *Lightning*: P-80 *Shooting Star*: PV-1 *Ventura*: Martin B-26 *Marauder*: North American P-51 *Mustang*: B-25 *Mitchell*: Northrop P-61 *Black Widow*: Ryan FR-1 *Fireball*: Vought OS2U- *Kingfisher* and Vultee A-35 *Vengeance* dive-bomber. Flight studies of aileron buzz on Bell P-63 *Kingcobra* 42-68892 delivered to Ames on 17 February 1944, were conducted by John Spreiter and George Galster, with George Cooper — a new member of the Ames staff — as flight test pilot. Similar studies on the Lockheed P-80 *Shooting Star* were undertaken by a research team consisting of Harvey Brown, George Rathert and Lawrence Clousing as test pilot.

A remarkable aircraft, the P-80 was the United States' second jet aircraft and was produced in Kelly Johnson's 'skunk works' in approximately 140 days. It was subsonic, the fastest thing on wings, yet beset with an aileron buzz problem which seemed to grow in intensity as the speed increased. Exhibiting his characteristic fearlessness, Larry Clousing dived the aircraft to speeds higher than any man had ever reached before — to a Mach number of 0.866. The ailerons whipped violently but wing flutter did not occur. When the jet was later examined, the left aileron was found to have a buckled trailing edge. Lockheed P-80A-1-LO 'NACA 131' was later fitted with vortex generators installed ahead of the ailerons to deal with the aileron buzz problem.

At what was described as a 'somewhat tardy' dedication ceremony in 1944, Ames had five wind tunnels; the 7 × 10ft, the 16ft, the 40 × 80ft and the high-speed 1 × 3.5ft tunnel. The last two were new and represented the two diverging directions of research undertaken at Ames. The 40 × 80ft tunnel proved useful for testing full-scale aircraft during the last year of the war. One of the first aircraft tested in it was the unique Ryan XFR-1 *Fireball*, a promising new type powered by a conventional Wright reciprocating engine and propeller in front and a General Electric I-16 jet engine enclosed in the fuselage behind the pilot. US Navy flight tests at Patuxent River, Maryland, had shown the XFR-1 to be seriously lacking in certain stability and control characteristics, particularly in the carrier-approach conditions. To the delight of the Ryan company and the US Navy, testing in the 40 × 80ft tunnel and in-flight trials identified the flaws and led to modifications to correct them. The Ames aircraft records show that the first Ryan FR-1 *Fireball*, BuNo.39650, arrived on 17 February 1945; no less than seven of the type, including XFR-4 BuNo.39665, which did the wind tunnel tests — arriving on 11 June 1947 — were subsequently tested by Ames.

New problems appeared with the introduction of early jet-powered aircraft, and the Ames team operating the 7 × 10ft tunnel and other branches were directly involved. There were two major questions to be solved — Where should jet engines be located in an aircraft, and how should the huge volume of air that passed into and out of a jet engine be handled?

The world's first jet aircraft, the Heinkel He.178, had flown for the first time on 27 August 1939, powered by a turbojet producing 840lb of thrust. It was 1944 before NACA knew anything about this development. In April 1941, Gen. Hap Arnold, chief of the US Army Air Corps found to his amazement whilst on a tour that the British were not only planning to develop gas turbines but were actually within weeks of flight testing a turbojet-powered aircraft, the Gloster E.28/39, using an engine designed by Frank Whittle. General Arnold made arrangements to bring engine blueprints, and eventually a Whittle W.1X engine to the United States. On his return to the US, Arnold assigned the task of imitating the engine to the General Electric laboratory at West Lynn. Massachusetts. In the summer of 1941 the US Army Air Corps placed a top-secret order with Bell Aircraft Corporation for construction of an experimental jet aircraft. At the same period a Major Donald Keirn carried, manacled to his wrist, a set of design drawings for the Whittle engine from London to the US. Thirteen months later the Bell XP-59A powered by two Whittle W.1A 'Superchargers' — a name which was a disguise — flew at Muroc Dry Lake, California, complete with armament. Because of the tight security imposed on all jet propulsion developments, Langley and the rest of NACA, knew nothing of these events until by chance a NACA representative visiting the Bell plant on the west coast heard rumours in May 1942. This lack of information put Langley research engineers at a serious disadvantage.

A project undertaken in the 16ft wind tunnel was one of great urgency and importance. It was an attempt to find the cause very dangerous diving tendency displayed by the Lockheed P-38 *Lightning* fighter and cure it. At least one US Army Air Force pilot had been killed by 'tuck-under', as the phenomenon was often called. It was suspected that the problem had to do with compressibility since it seemed to occur only at high speed — at Mach numbers above 0.6. It was found that the problem was clearly due to the formation of shock waves on the wing and fuselage. In April 1943, flaps were installed on the lower surface of the wing at 33 per cent chord point. Ames had tested its first *Lightning* fighter when P-38F 41-7632 arrived on 30 December 1942. Nearly two years later a P-38J-15-LO 43-28519 was delivered, remaining until 15 March 1946.

Not long after the P-38 episode, the 16ft tunnel staff were hard at work once more. During early flights of the North American P-51 *Mustang*, a strange thumping noise was heard. Despite the advanced features of the P-51 design, the US Army Air Corps policy was to operate P-38 and P-47 only. The US government stipulated that two P-51s should be delivered for evaluation. Early orders were very modest since there was no 'Lend-Lease' at that time. Ames discovered that if the outer wing panels were removed, the P-51 fuselage with stub wings would just fit in the 16ft tunnel test section. The rumble appeared in the tests and its source was diagnosed to be the belly scoop on the under surface. The scoop was lowered, keeping it out of the boundary layer, and the cure delighted North Amercian.

Models of at least sixteen different aircraft were tested in the 16ft tunnel, these including the Lockheed P-80 *Shooting*

Star. During 1945 the US Army Air Force was making plans to develop a new family of jet bombers. They would be much faster than the Boeing B-29 *Superfortress*, and would require wings that would perform well at high subsonic Mach numbers. A series of six wings was included in a programme in the 16ft tunnel. The results were presented directly to the USAAF at a conference held at Wright Field in September 1945. Wing sweepback was also discussed at the meeting.

At the end of World War II, when the US scientific teams entered Germany, they discovered that the Luftwaffe had discovered the benefits of sweep. The teams included R. G. Robinson from Ames. Rumours of the German work on sweep had reached the United States a year or so earlier, but the significance of this work appears not to have been recognized until Robert Jones from Langley published his findings on sweeping the wings of an aircraft backward.

It was mid-1944 before the 40 × 80ft Full-Scale wind tunnel began operation. This tunnel and the Flight Research Sections comprised the Full-Scale and Flight Research Division. The 40 × 80ft operation and flight research proved to be an extremely compatible and important combination of activities working hand-in-glove with each other. The first flight test programme involved the El Segundo-built Douglas XSB2D-1, later BTD-1, aircraft called *Destroyer*. On 12 June 1944, XSB2D-1 BuNo.03552 arrived at Ames and on 28 July 1944 BTD-1 BuNo.04968 arrived for testing. It is stated that Ames spent more time on the BTD-1 than on any other single aircraft. Unfortunately soon after, the US Navy lost interest in the BTD-1 and it never went into quantity production. The 40 × 80ft tunnel also tested the Northrop N9M-2 flying-wing prototype, the Grumman XF7F-1, the General Motors P-75A and the Douglas A-26B *Invader*.

North American F-86A-1-NA Sabre 47-609 the fourth production single-seat jet fighter which served at Ames as NACA 135 between 10 April 1950 and 15 May 1956. It is seen parked at Ames fitted with an extra long pitot head, possibly carrying test instrumentation. (NACA A-15836)

The end of World War II on 14 August 1945, did not initially effect the operation of Ames Aeronautical Laboratory. Through sheer inertia, the work load carried on essentially unchanged into the next year. There was in operation or under construction an impressive array of modern aeronautical research facilities valued at approximately $21 million. There was a well-organized staff of 800 which included a large number of very skilled research personnel, a few of whom were truly outstanding. As 1945 ended Ames was ready to take the first faltering steps into the supersonic age.

CHAPTER FIVE
Further Wartime Expansion

On 26 June 1940, Congress authorised construction of NACA's Aircraft Engine Research Laboratory at a site near the municipal airport at Cleveland, Ohio. As with the Ames laboratory, the key personnel for this new facility were to be drawn from Langley. Amongst the projected equipment listed in the original 1939 prospectus for the new laboratory at Moffett Field, California, had been an engine research division. There was already a laboratory devoted to engine research at Langley, work which had been prominent in the NACA programme from the beginning. When the original Langley laboratory was dedicated in 1920. One of its three buildings was the engine dynamometer laboratory. At that time the manufacture of engines for aircraft was an infant industry, and research was elementary.

It was finally recommended that a major research establishment, devoted wholly to the aircraft engine and allied problems, be set up. The outcome of this recommendation was the building of the Engine Research Laboratory, later known as the Flight Propulsion Research Laboratory, at Cleveland. The plan was reported by a NACA special committee in January 1940, laid before the Bureau of the Budget in March, approved by the President and submitted to Congress in May and authorized by an act of Congress in June.

The initial programme adopted for aircraft engine research, like that pursued in aerodynamics, was directed at fundamental problems as a research staff was slowly employed and

The second aircraft to arrive at the NACA's new Aircraft Engine Research Laboratory at Cleveland, Ohio, in January 1943 was this Republic P-47G Thunderbolt 42-24929. It was used on radial air-cooled engine performance tests plus turbo-supercharger trials. It departed the NACA in June 1945 and is seen on ground test on 6 June 1945. (NACA C-10535)

specialized equipment developed and installed. By 1940 the power-plant division conducting research at Langley comprised a staff of seventy-five personnel. That was the year the decision was made by the Agency to establish a separate establishment in Cleveland. Personnel from Langley collaborated in planning and designing the new laboratory.

The new facility occupied an initial 200 acres adjoining the Cleveland Municipal Airport. This land had previously served as parking space for the famed pre-World War II Cleveland National Air Races, and in January 1941, except for an old farmhouse and a grandstand, was barren when the ground was broken for the erection of the first unit for NACA — the flight research building. Work was commenced on the engine-propeller research laboratory in February due to adverse weather and other circumstances, construction progress was severely delayed and it was not until January 1942 that the flight research building was finally finished. This was a spacious hangar which was immediately pressed into service as offices and for storage of materials. Until new specialized buildings were completed, the hangar served to house the entire laboratory.

Personnel from the power-plant research group at Langley were transferred as buildings were completed at Cleveland to inaugurate a variety of programmes of investigation and to form the nucleus of an extraordinary expansion for the future. The plans were originally for the study of the piston engine and any related problems, but the province of the new laboratory was soon extended to include turbines and jet propulsion. Whatever involved the propulsion of aircraft — be it engines, propellers, fuels, lubricants or even their components — had and still has a part in the programme of research.

By the spring of 1942 the engine-propeller research laboratory was sufficiently advanced for NACA engineers to mount an engine with propeller in the test chamber. Study of this test-piece, which began on 8 May, initiated the NACA research at Cleveland — research which has been active ever since. The aircraft record archives reveal that the first aircraft, a Martin B-26B *Marauder* bomber (41-17604) arrived at Cleveland in January 1943 for radial air-cooled engine performance tests. It remained until October 1943. Arriving at the same time was a Republic P-47G *Thunderbolt* (42-24929) also for radial air-cooled engine performance tests and turbo-supercharging trials. It finally departed in June 1945. Initially the test aircraft used at Cleveland did not carry NACA numbers since they were pure military service types.

The focus of all the research and test facilities located at Cleveland, and available to all divisions, is the engine research laboratory. It is equipped with facilities for studying practically every operational aspect of aircraft engines. The huge building is a consortium of many research laboratories. Test chambers and cells enable engines and their components to be operated at conditions of atmospheric temperature and pressure which would be encountered at high altitudes. The laboratories were involved initially in combustion, heat transfer, waste heat, engine parts, stresses and materials research: other interests included compressor and turbine studies.

Several other research facilities were housed in the huge engine building. The fuels and lubricants departments were located to the south, the administration and flight research areas to the north and the wind tunnels to the east. These included a tunnel for altitude tests, icing research and two supersonic tunnels. At a more remote location on the 200-acre site were located static test laboratories for jet propulsion, a high-pressure combustion laboratory and test locations for rocket research and other experimental stations.

At the end of World War II in 1945, NACA at Cleveland occupied fifteen large buildings and twenty smaller ones with construction still under way. The huge establishment employed 1,512 personnel in the laboratories; a further 600 were engaged in research tasks, whilst 400 were employed in the vital administration department. Cleveland, the youngest of NACA's main centres, quickly grew to dimensions which

certainly rivalled those of Langley and Ames in physical plant investment and research.

Because the laboratory at Cleveland was planned, built and put into operation during World War II, its initial programme was almost entirely conditioned by military needs. As the aerodynamicist experts at Langley and Ames were called on to investigate lift, drag, controllability, and other aerodynamic characteristics of specific military aircraft, the engine researchers at Cleveland were asked by the US Army Air Force and the US Navy to investigate specific engines and their installation in particular types.

Several of the standard makes of aero engines were studied in investigations of fuels, lubricants, fuel injection, water injection and other devices. In five instances the US Army Air Force asked the NACA to see what could be done to improve the overall performance of a particular engine. The engines in question included the Allison 12-cylinder liquid-cooled V-1710; the Wright 18-cylinder air-cooled R-3350; Pratt & Whitney's 18-cylinder air-cooled R-2800 and 28-cylinder air-cooled R-4360, and the Packard-built Rolls-Royce 12-cylinder liquid-cooled Merlin. Various tests were made on each of these engines. Efforts to improve the efficiency of supercharging, cooling, carburation, fuel injection, lubrication and other elements of operation were put in hand. Single cylinders and special crankcases were used in preliminary tests, whilst full-scale engines on torque stands and in dynamometer test-cells were used when overall performance had to be determined. Finally the complete engine installation was tested, either in actual flight or under simulated flight conditions in the altitude wind tunnel. Resulting from these studies, methods were found to operate the engines at outputs up to 150 per cent of their rated horsepower.

In 1943, the US Army Air Force called upon NACA to investigate performance characteristics of the Whittle gas turbine engine, which had been brought over from the United Kingdom in a highly secret mission. New equipment was constructed at Cleveland to provide test stands and rigs for jet engine units. Despite the expanding interest in jet propulsion, the summer of 1944 found the greater proportion of the

Bell P-59B-1-BE Airacomet 44-22650 seen at Cleveland on 29 May 1946 during water injection testing. It was delivered to the NACA in September 1945, remaining until January 1949, being used to test thrust performance, augmentation, etc. (NACA C-15037)

laboratory effort still focused on piston engines and these continued to be the main interest of the facility up until VJ-Day, although both turbo- and ram-jets were progressively claiming and receiving attention. Hardly anyone at Cleveland was aware that the situation would soon reverse.

Emergency Request

In the fall of 1942, the NACA chairman, Jerome C. Hunsaker, received an urgent letter from General Oliver P. Echols of the US Army Air Force submitting an emergency problem.

'It is understood that engine research facilities will soon be available at the Committee's laboratories at Cleveland, and it is requested that a study be made of the Allison V-1710-45 engine with a view to improving its power output from sea level to 20,000 feet. It is particularly requested that a temperature survey be made of the pistons and piston rings, showing the effects of different coolants, and to obtain the rating of the latest Allison with improved fuels.'

The Allison engine was rated at 1,650hp at sea level. It had been selected to power the Lockheed P-38 *Lightning*, the Bell P-39 *Airacobra*, the Curtiss P-40 *Warhawk* and the Bell P-63 *Kingcobra* — a new fighter which was still in the flight-test stage. This investigation involving the Allison engine made use of almost all the resources then at Cleveland. No single request from the US armed forces involved so many laboratories, so many talents and so much co-operation from so many scientists and engineers. Frequent consultations took place between the engineers and technical personnel from the manufacturers and the US Army Air Force.

A close look at this photo will reveal a huge intake system installed on top of this Consolidated B-24M-10-C 44-41986 Liberator. It was used at Cleveland between November 1945 and July 1949. Projects it was involved in included icing research and measurements, tests of axial flow and centrifugal jet engines. (NACA)

Results were many and varied. The speed of the engine was increased by the Allison company from 3,000 revolutions per minute to 3,200. By using different pistons, the engine was studied at various compression ratios. The capacity, pressure ratio and efficiency of the supercharger installation were increased. There were more recommendations involving high-octane fuels and cooling systems, all checked in flight by the NACA research test pilots using the modified engine in a Bell P-63 *Kingcobra*.

The aircraft record archives reveal that a Bell P-63A *Kingcobra* (42-68889) arrived at Cleveland in October 1943, remaining on test work until June 1945. The net result of the huge co-operative effort raised the output of the Allison at sea level to 2,250hp with corresponding increases for operation at altitude. As one NACA historian remarked: 'It was as though, without enlarging the 12 cylinders, fifty additional horses had been coralled into each of them and put to work.'

Basic research on supercharging was under way at Langley in the early 1920s, when only the occasional engine was supercharged and pilots rarely flew above 15,000ft. However early experiments on supercharging were so interesting to the US Navy that it authorized Lieutenant Apollo Soueck to attempt a new altitude record. His *Apache* aircraft was powered by a Wasp engine, to which he attached a supercharger borrowed from NACA at Langley. On 18 May 1929, he reached 39,140 feet. That was 722 feet higher than the existing record, but Soueck was not satisfied. A few weeks later, this time using a seaplane, he managed to reach 43,166 feet. Describing his feat to the public, the lieutenant explained:

'A supercharger was used to deliver air into the engine at a pressure above that existing at high altitudes. The supercharger, a vital part of our high-altitude equipment, was the result of long development work in the laboratories of the National Advisory Committee for Aeronautics.'

In 1939 a special subcommittee on supercharger compressors was appointed, its members drawn from the US Army Air Corps, the US Navy and the NACA. It decided to enlarge the work that had been under way for some years at Langley. On the eve of World War II an accelerated programme was launched, and the supercharger research staff at Langley, headed by Oscar W. Schey, was expanded. With the opening of Cleveland, Schey and his group transferred most of the wartime superchargers, both gear and turbo-driven, were centrifugal compressors. Finally, Cleveland and its research engineers increased the efficiency of supercharging from 69 per cent in 1939 to an overall efficiency of 85 per cent in 1945.

During May 1944, a Consolidated B-24D *Liberator* (42-40223) arrived at Cleveland for use as a flying laboratory. It was powered by Pratt & Whitney R-1830-75 Twin Wasp engines. One engine was used for fuel tests, including a superior fuel called triptane. The engine performance was evaluated whilst the aircraft flew on the other three engines. Performance with various fuel additives and mixtures was also tested. It was the most comprehensive study of fuel ever made by a single organisation. In the extensive series of tests, triptane was compared with thirteen other high-performance fuels. The main problem with triptane was its phenomenal cost. Sefright (op. cit. p.114) states '. . . promised to be a wonderful discovery until it was realised that production of it . . . would involve . . . the total chlorine production of the whole USA.' Triptane is 2-2-3 trimethyl butane. The B-24 *Liberator* continued its fuel research programme until March 1946.

---

* Quote from *Frontiers of Flight* by George W. Gray. 'Most of the wartime superchargers, both the gear-driven and the turbine-driven, were centrifugal compressors. The rotating element is a fanlike wheel with blades shaped to catch the air, speed it up, and throw it out radially at high velocity. This part which whirls is called the impeller, because it impels the slow-moving air to speeds that may reach hundreds of miles per hour in travelling a few inches. But a pump must do more than accelerate the flow. It must also compress the air to higher density, and the part of the supercharger that performs this function is the diffuser. In conventional design, the diffuser is a series of vanes affixed to the inside of the casing which houses the impeller. Thus, the impeller moves and by centrifugal force imparts velocity to the air; the diffuser stands still, and by interposing its vanes it dams up the outrushing air and imparts pressure.'

Early in 1944 the Agency was asked to investigate a problem with the engines on the Boeing B-29 *Superfortress*. Power was from four Wright R-3350 Cyclone engines each rated at 2,200hp. An overloaded bomber together with the pilot's effort to maximize the flow of cooling air through the engines, at a high cost in drag, caused problems. The cylinders overheated, exhaust valves softened and engines were severely damaged. Cleveland took delivery of a Boeing B-29 *Superfortress* (42-8357) in June 1944 and carried out extensive flight testing for two months on the cooling problems and engine performance. The nacelle of a B-29 complete with engine installation and propeller was mounted in the altitude wind tunnel. These combined ground and air tests produced three main recommendations: change the cowling and baffles to provide more efficient airflow; change the cylinder head to provide sufficient heat-conducting metal in the vicinity of the exhaust valve; and change the carburettor to provide more uniform mixing of the fuel-air charge.

Jet Propulsion

During the spring of 1943 General Electric had designed an improved jet engine known as the GEI-16 and by the summer experimental units were ready for testing. At this stage NACA was called upon to assist with the US Army Air Corps jet-propulsion programme. At Cleveland a new building appeared — a rectangular structure of steel frame and cement-asbestos board. Work on the jet-propulsion static test laboratory was begun in July 1943, completed in August, and in September its two test cells each held a GEI-16 unit. Of the 1,300 personnel at Cleveland, only eight who were to work on the project were given the necessary security clearance. A barbed-wire fence surrounded the new laboratory and a round-the-clock armed guard was assigned. During September 1945 Cleveland took delivery of a Bell P-59B *Airacomet* (44-22650) for jet thrust performance tests, thrust augmentation and water injection tests. It stayed until January 1949.

Within six months it became necessary to double the capacity of the static test laboratory. The engine research building was enlarged, with a huge west wing added to house research facilities for compressors and turbines. Supersonic wind tunnels and new laboratories for combustion work and research followed. Facilities originally planned for work on piston engines were converted to the research and study of jet propulsion, which soon became a major item involving the entire laboratory. The studies at Cleveland were closely allied to in-depth consideration of jet aircraft at both Langley and Ames. Far swifter and high flight required the successful mating of thermodynamics and aerodynamics.

For turbo-jet research under the conditions of actual flight, the Cleveland laboratory operated several interesting aircraft. These included a Boeing B-29A *Superfortress* (42-1808), which arrived in April 1946 and which was employed on a series of interesting flight-test tasks. It became a jet-engine test bed and investigated engine icing, and also a ram-jet test bed and launch vehicle. The roomy *Superfort* provided ample space for both instrumentation and personnel for direct study of any problem which developed at 40,000 feet or more. In-flight examinations were made of combustion, fuel-injection, cooling, lubrication, thrust augmentation and other problems of turbo-jet design and operation.

A unique sub-sonic ram-jet test-bed was a Northrop P-61 *Black Widow* (42-39754) which arrived at Cleveland in October 1945 and was used as a test-bed for three years, serving until October 1948. In October 1947 the Boeing B-29 was supplemented (and later replaced in October 1948) by a North American XF-82 *Twin Mustang* (44-83887) the second of two prototypes. Unfortunately this aircraft was damaged in a ground incident during July 1950. It was replaced by another *Twin Mustang*, this being the record breaking F-82B (44-65168) which was allocated the serial 'NACA 132' and

named *Betty Joe*. On 28 February 1947, this aircraft set the distance record for flights by flying from Honolulu to New York in 14 hours, 31 minutes, 50 seconds — a distance of 4,968 miles. The F-82B was used as a ram-jet and aerodynamic test missile launcher. Then came a need to increase the test altitude in order to give higher launch altitude and velocities, and a replacement aerodynamic missile and ram-jet launch vehicle was required to replace the *Twin Mustang*. In January 1955 a McDonnell F2H-2B *Banshee* (BuNo. 124942) arrived at Cleveland, which had by now been renamed the Lewis Research Center. The *Banshee* became 'NACA 209'.

Both the intermittent and steady-flow ram-jet were studied deeply at Lewis. In 1919 George W. Lewis had been appointed Director of Aeronautical Research with NACA at the headquarters in Washington, D.C. and in July 1948 Cleveland was renamed 'Lewis Flight Propulsion Laboratory' — today known as the 'Lewis Research Center'. George W. Lewis died on 12 July 1948. On 19 June 1957, a replacement for the McDonnell F2H-2B arrived in the form of a Martin B-57A *Canberra* (52-1418) which became 'NACA 218' in the Lewis fleet. In addition to being used for aerodynamic tests and as a rocket launch vehicle, the B-57A was employed on flight hardware performance tests. It was also chase aircraft for a Martin B-57B *Canberra* (52-1576) which had been in use at Lewis since May 1956. This latter was allocated 'NACA 637' — later NASA. The tasks allocated to the B-57B flying test-bed were numerous and included research into the use of hydrogen fuel in jet engines which was a real look forward into the future. Some of its other trials involved noise suppressor evaluation, solar cell calibration and high-radiation measurements. On 19 March 1973, this aircraft was delivered out to Langley becoming 'NASA 516'.

Initially the flight test used by the NACA did not carry NACA numbers or emblems, as they were US service types on temporary loan. This Lockheed F-94B-5-LO 51-5329 Starfire NASA 203 arrived at Cleveland during April 1956 remaining until December 1959 and used on aerodynamic noise studies. (NASA C-52344)

Wallops Island

As expansion of the NACA empire developed rapidly during World War II, the need to enlarge and improve the laboratory equipment on both sides of Langley and the requirements of both the US Army Air Force and the US Navy for intensified study of high-speed aerodynamics were increasing. Requests were also coming in to the Agency for research on guided missiles, as well as on transonic and supersonic aircraft. Several new techniques were proposed for these high-speed studies.

The work was highly secret, and a range for experimenting with rocket-propelled missiles required ample unrestricted space. Wallops Island, an uninhabited stretch of sand dunes and marsh land on the Atlantic side of the Virginia Eastern Shore Peninsula, was selected during the winter of 1944/45 as the site for these experiments and ably met both requirements.

This extension of the NACA Langley laboratory became known as the Pilotless Aircraft Research Station — later Division. Using both missiles and free-falling bodies for its experiments, this establishment today provides a valuable auxiliary to the high-speed tunnels and other on-the-ground equipment and facilities at Langley.

The Pilotless Aircraft Research Division — PARD — moved into Wallops Island during late June 1945, and on 18 October launched its first successful drag-research vehicle. This was a rocket-propelled model aircraft, designed to evaluate wing and fuselage shapes for the provision of basic design information at transonic and supersonic speeds. With advances in research, the test vehicles became more elaborate. In June 1946 Wallops launched a control-surface research vehicle which evaluated controllability in roll by deflecting the ailerons in a programmed manoeuvre. Wallops has long outgrown that original test site and today is sprawled over portions of the former US Naval air station in Chincoleague.

On 9 May 1945 Langley took delivery of a Boeing B-29B *Superfortress* (44-83927), followed by a TB-29 (44-69700) on the last day of July. With guns and ammunition compartments removed, the bomber carried only a limited fuel load on the task allotted. This was to climb to an altitude in excess of 40,000 feet over Chesapeake Bay and directly over the shore at Wallops Island. At a signal, synchronised by ground control on the radio-telephone, the flight test engineer pressed a switch and a long projectile-like object separated from its restraining rack, dropped free of the *Superfortress* and started its free-fall descent. Falling in a parabolic path, the object increased its velocity until it was soon moving at

the speed of sound. Forty-three seconds after it had been released, the projectile hit the ground at a speed of approximately 1,100mph and buried itself in the grass-covered sandy marsh.

On the ground, the research engineers had been carefully observing the performance of the projectile as it fell. They used a powerful telescope instrumented for the trial, so powerful that they could observe the B-29, and see the projectile's launch and follow its descent. Radar tracking was also used to record the velocity throughout the drop, whilst telescopic cameras mounted on the radar mirror took film of the trial. A radio receiver on the ground picked up signals from a transponder fitted in the projectile so that the research engineers had air pressure recordings, drag on the tail fins and acceleration during descent.

This series of high-altitude trials with the *Superfortress* involving the release of fully-instrumented and telemetered test bodies was a stratagem for studying airflow in the transonic region. Some wind tunnels gave reliable measurements for speeds approaching the speed of sound, others recorded events after the speed of sound had been passed, but in the region of transition between the two limits the airflow in a wind tunnel can become disturbed and measurements cease to have a scientific value.

The US Navy had donated the $500,000 radar ground installation, the US Army Air Force placed a B-29 at the disposal of the laboratory and NACA provided the technical support and the research personnel for a long series of trials and experiments involving measuring aerodynamic responses at high speeds. A major accomplishment at Langley during 1946 was a research missile capable of flight velocities above 1,100mph. This rocket-propelled device was developed at the Pilotless Aircraft Research Division at Wallops Island, and many varied wing forms were tested, ranging in sweep angle from 0° to 52°, and in aspect ratio from 1.5 to 2.7.

The Martin Model 272 was the British designed Canberra known in the US as the B-57. Depicted at Cleveland is B-57A 52-1418 delivered on 19 June 1957 having a research rocket vehicle loaded. It later became NASA 218 and used on a variety of tasks including flight hardware performance tests and as a chase aircraft for B-57B 52-1576. It replaced a McDonnell F2H-2B Banshee BuNo.124942. (NASA C-45903)

In April 1947 Wallops launched its first scaled-down aircraft, in a test for performance evaluation. It was a model of the Republic XF-91, a radical fighter design to be named the *Thundercepter*, which combined turbojet and rocket propulsion for performance at extreme altitudes. This single-seat supersonic interceptor was powered by a 5,200lb static thrust J-47-GE-3 turbojet and a 6,000lb static thrust XLR11-RM-9 rocket engine. It had an inverse-taper wing, with broader chord at the wingtips than at the roots. Two XF-91-RE aircraft were ordered, one being later tested with a V-tail assembly. The success of this test programme was followed by model flight tests at Wallops of most US Army Air Force — later US Air Force — and US Navy supersonic and subsonic aircraft designs.

Research and development at Wallops paralleled transonic tunnel studies and extended them into the supersonic speed range which could be reached with rocket-propelled models. Slowly the transonic region yielded to probing and analysis. Basic problems began to be defined.

During 1953 Wallops tested rocket-powered models of the new delta-winged Convair F-102 single-seat interceptor fighter, the *Delta Dagger*, before and after application of the area rule. The F-102 was a scaled-up version of the F-92 single-seat delta-wing supersonic interceptor with a projected top speed of Mach 1.5, but it was not built although a later XF-92A aircraft became the first powered delta-wing to fly. The F-102 went into quantity production. At Wallops Island on 24 August 1956, the PARD successfully launched a five-stage, solid-propellant rocket vehicle. It reached a speed of Mach 15, far into the hypersonic region and beginning to touch the high Mach numbers that would be encountered in ballistic missile re-entry bodies and in the return of men and women from space.

One of the more unusual types used by the NACA-NASA was this North American AJ-2 Savage BuNo.134069 NASA 230, delivered to Cleveland in January 1960. It was involved in zero gravity tests and facility behaviour of certain fluids. A second AJ-2 BuNo.130412 was used on loan from the US Navy. (NASA C-59878)

This Martin B-57B 52-1576 NASA 237 was delivered to Cleveland in May 1956 and involved in a number of research projects including use of hydrogen fuel in jet engines, noise suppressor evaluation, solar cell calibration and high altitude radiation measurements. It transferred to Langley on 19 March 1973 as NASA 516. (NASA C-62272)

Helicopter Developments

A Sikorsky HO3S-1 *Dragonfly* helicopter BuNo. 122520 was delivered from the US Navy to Langley on 24 February 1948, remaining with the laboratory until 6 September 1961. It was involved in general handling and stability studies, being fitted with special airspeed sensors below the nose and a hood permitting the flight test engineer/pilot in the rear to study the flight on instruments. It carried the NACA number 'NACA 201'. The research test pilots at Langley had noted several outstanding problems in helicopter handling, and continuous control action was required at cruise speeds to prevent the machine from diverging to attitudes from which recovery would be dangerous. Study of aircraft motion following control movement led to suggested flight-test procedures and criteria for examining these problems and assessing attempted improvements.

Work of this nature was encouraged by agencies who were quickly beginning to utilize helicopters in large numbers. Eventually a number of helicopter types were sampled at Langley to check the generality of the conclusions and uncover additional factors important to effective safe flying. On 13 February 1951, a US Navy Piasecki HRP-1 *Rescuer* tandem helicopter (BuNo. 11813) was delivered to Langley to be fitted with 'horse-collar' spoilers to permit exploration of the effects of changed directional-stability characteristics. NACA test pilot engineer Jack Reeder was involved with this programme of research. A Vertol H-25A *Retriever* from the US Air Force (51-16549) arrived on 24 February 1954, remaining until 26 January 1960. A second Vertol H-25 *Retriever* (51-16637), this time from the US Army, arrived at Langley on 14 September 1954 and was used for inflight studies of coupled frequencies. It was a tandem rotor helicopter and was fully instrumented, being fitted with airspeed-sensing devices.

On 9 February 1962 this Boeing-Vertol YHC-1A-BV prototype helicopter 58-5514 arrived at Langley, becoming NASA 533. It was converted into an inflight simulator, the spacious cabin occupied with bulky computer and recording equipment. During February 1972, it made the first fully automatic landings by a full-scale helicopter manned by NASA flight test crew. It was transferred to Lewis at Cleveland on 24 March 1975 for a series of crash tests. (NASA 72-H-194)

Flight test pilots James B. Whitten (left) and Jack P. Reeder, prepare to investigate the handling qualities of a US Navy Piasecki HRP-1 BuNo. 111813 which was acquired by the NACA at Langley on 13 February 1951. It was fitted with horsecollar spoilers to permit exploration of the effect of changes in directional stability characteristics. (NACA 71609)

In the early stages of helicopter research, most of the variations explored were obtained by modifying a particular machine. Electronics began to be used in 1953, with the rear pilot or test engineer moving controls which were directly connected to rheostats, whilst the safety pilot could take over if necessary.

By 1954 a NACA conference on helicopters which attracted over 200 people was held at Langley, the thirty-two papers divided among sessions on rotor aerodynamics, propulsion and parasite drag, including noise, stability and control, load stresses, vibration and flutter. Not all the authors present were connected with the helicopter industry although one was author and co-author of no less than five papers and presented two at the conference. New helicopter manufacturers were appearing on the scene in addition to the well-known Sikorsky and Kellett companies. These included Bell, Boeing-Vertol, Hiller, Kaman, Piasecki, McDonnell and McCulloch. Mention was made earlier of the huge XH-17 model of a load-lifting helicopter with a rotor diameter of 13ft which was built and tested at Langley. It is not generally known that in fact this dynamic experimental giant helicopter was built by Kellet-Hughes and that the single YH-17 (50-1842) completed its ground tests at Culver City, California in May 1952 prior to its first flight. A 42,000lb gross weight crane helicopter, its two modified General Electric turbojets supplied gas pressure through ducts leading up to the huge rotor shaft and out of the tips of the massive and long rotor blades. It was one of the largest helicopters built up to that date.

— work should replace rather than augment or parallel rotor research. It will be recalled that the 1954 helicopter conference at Langley included thirty-two papers on rotors. In contrast the NASA conference on V/STOL aircraft in 1960 produced only five papers on rotors from the twenty-six papers presented, and in 1966 at the V/STOL and STOL conference only three of the twenty-three papers were concerned with the rotor.

It soon became possible for NACA/NASA to purchase (rather than borrow) prototype or production aircraft for research, including helicopters. About this time, a co-operative arrangement between the armed services and the Ames laboratory was introduced which resulted in increased rotor research. A somewhat similar arrangement involved the Langley and Lewis laboratories. When the Ames laboratory was authorized in 1939, one of the objectives had been to provide facilities and personnel to work more closely with industry in development efforts. In keeping with that objective, the 40×80ft Full-Scale tunnel at Ames has been utilised for numerous measurements with both current and advanced rotating-wing test vehicles. The earliest of these tests, those involving the McDonnell XV-1 Model M.82 unloaded-rotor 'convertiplane' driving a rotor but having a conventional wing layout, were made in 1953/54. Due to its very high security classification, the existence of this unique vehicle was not widely known. The first of two XV-1s (53-4016) arrived at Ames during 1952 for tests and is today in the National Air & Space Museum in Washington, D.C.

This Vought F-8A Crusader BuNo.141354 NASA 666 was used at Cleveland for only a short while. It was delivered in May 1969, being used for programme support and chase for the NF-106B. On 14 July 1969 it was lost in a landing accident at Cleveland. (NASA C-69-1513)

This Convair F-102A Delta Dagger 56-998 NASA 617 was the replacement aircraft for the crashed F-8A. It was also used as a chase aircraft for the NF-106B NASA 607. It arrived in June 1970, and just four years later went into storage at Davis-Monthan, AFB, Arizona. (NASA C-71-2216)

The helicopter had been evaluated in action by the US Army Air forces in the jungles of Burma during the latter stages of World War II, this being the Sikorsky R-4B used on casualty evacuation from clearings. Helicopters had again proved their value as rescue vehicles in the United Nations conflict in Korea during the 1950s. It also fulfilled prophesies of the 1930s as to its life-saving abilities following hurricanes and floods. The US Coast Guard pioneered the Sikorsky R-4B which effectively produced techniques and the broad operating procedures for search and rescue — SAR — including use of the rescue hoist.

As the decade progressed there were views in some quarters that virtually all research work on the helicopter had been completed. Another view expressed was that, being in its infancy, non-rotor VTOL — vertical take-off and landing

On 9 February 1962, a Boeing-Vertol YHC-1A-BV prototype helicopter (58-5514) arrived at Langley, becoming 'NASA 533'. Production models were designated CH-47 and named *Chinook*. This twin-rotor helicopter became Langley's inflight simulator, with the spacious cabin occupied with bulky computer and recording equipment. It served Langley well until 24 March 1975, when it was transferred to Lewis at Cleveland for a series of crash tests.

On 17 September 1953, a US Army Bell H-13G (52-7834) had been delivered to Langley and converted as a hingeless-rotor helicopter, being actually what they termed a 'quick-and-dirty' modification of a two-blade see-saw-rotor type. It was used for flight studies until 25 July 1967, when Langley purchased from Lockheed a hingeless-rotor (also known as 'rigid-rotor') helicopter for flight research. More developments were to follow, including research into tilt-rotor technology.

CHAPTER SIX
The Road to Sonic Flight

Information on advanced aerodynamics found in defeated Germany was soon passed back to the United States by the teams which surveyed the scene. There were some unique and quite startling German aircraft, such as the small Messerschmitt Me 163 rocket-powered interceptor which had seen successful combat against the large formations of US Army Air Force bombers. There was the Junkers Ju 287 jet bomber with its forward-swept wings. The information, confirmed by photographs, immediately prompted the critics to question why US aircraft design appeared to lag behind that of Germany. It was the story of the turbojet all over again: the critics accused NACA of being responsible for US flight research falling precariously behind during World War II.

It is true to say that the outcome of wartime German research made an impact on post-war development of swept wings in the US which eventually led to high-performance jet bombers such as the Boeing B-47 *Stratojet* and the North American F-86 *Sabre* jet fighter. It must be remembered that early post-war aircraft such as the Republic F-84 *Thunderjet*, a fighter, and the B-45 *Tornado* jet bomber produced by North American had conventional straight wings. On the credit side, a great deal of independent progress had been made by US designers, including NACA scientists, by the time details of the German hardware were available at the end of World War II. Admittedly it was not as far advanced as the German research.

Seen flying over Muroc Dry Lake, part of the huge Edwards Air Force Base complex, on 8 June 1948 is Skystreak BuNo.37972 fitted with 50-gallon wingtip fuel tanks. (Douglas ES72429)

The concept of wings with subsonic sweep came to Robert T. Jones, a Langley aerodynamicist, in January 1945, and he eagerly discussed it with the US Army Air Force and his NACA colleagues during the following weeks. Jones was so confident that on 5 March 1945, he wrote to the NACA's director of research, George W. Lewis.

> 'I have recently made a theoretical analysis which indicates that a V-shaped wing travelling point foremost would be less affected by compressibility than other planforms.
>
> 'In fact, if the angle of the V is kept small relative to the Mach angle, the lift and centre of pressure remain the same at speeds both above and below the speed of sound'.

Only flight testing would provide the data to make or break Jones's theory. Langley personnel immediately went to work fabricating two small models to see what would happen. The NACA technicians mounted the first model on the starboard wing of a North American P-51H *Mustang* (44-64192). One 'bump model' represented a Grumman F9F-6 *Panther* and localized supersonic upper wing-surface airflow velocities could be generated around the model during dives from altitude. The instrumented model soon began to generate useful data. The wind tunnel model for tests was a truly diminutive article, crafted from sheet steel by Robert Jones and two fellow engineers. The supersonic tunnel had a nine-inch throat, so the model had a 1.5 inch wingspan in the shape of a delta. The promising test results were issued on 11 May

1945, and were released before the Allied investigation teams had the opportunity to interview captured German aerodynamicists on delta shapes and swept wing developments.

The presence of these gentlemen in the United States by courtesy of 'Operation Paperclip' helped to bestow the imprimatur of proof on swept-wing configurations. At Boeing, designers at work on a new jet bomber discarded sketches for a conventional aircraft with straight wings and built the B-47 *Stratojet* instead. With its long, swept wings the B-47 launched the company into production of a remarkably successful family of swept-wing bombers and jet airliners. At North American, a standard jet fighter design with straight wings went through a dramatic metamorphosis, eventually taking to the air as the famed F-86 *Sabre* — a swept-wing fighter which had an enviable combat record during the Korean conflict in the 1950s.

On 29 August 1949, North American F-86A-5-NA *Sabre* (48-291) arrived at Ames becoming 'NACA 116' and remaining with the laboratory until 11 January 1960. The aircraft was fitted with special instrumentation for transonic flight research, including spin tests. As many as ten more F-86 aircraft were used extensively at Ames, together with the US Navy version, the FJ- *Fury* series.

Unfortunately NACA was again the target for criticism from post-war Congressional and US Army Air Force committees. It may have been that the Agency was not as bold as it might have been, or that it was so involved in immediate wartime improvements that crucial areas of basic research received short measure. Historian Alex Rowland noted in his study of the NACA that its shortcomings 'should not be allowed to mask its real significant contributions to American aerial victory in World War II.' Moreover, NACA's postwar achievements in supersonic research and rapid transition into astronautics surely reflected a new vigour and momentum.

The Douglas Aircraft Company built a total of three D-558 Skystreak research aircraft. The second aircraft BuNo.37971 took the world's absolute record on 25 August 1947 reaching 650.796mph. All three aircraft were painted scarlet as depicted in this photo. (Douglas HG77-247C)

Rare colour photo of the Bell X-1E 46-063 named 'Little Joe' with a pair of dice insignia. Joe Walker made a total of twenty-one flights, remaining as the X-1E project pilot, making the first flight on 12 December 1955 which was an unpowered glide. (Office of History — USAF Edwards)

Three North American F-82 Twin Mustangs were used by the NACA at Cleveland between 1947 and 1957. Depicted is F-82B 44-65168 which was later NACA 132 and equipped with a variety of R&D flight test items. It was also used for ram-jet and aerodynamic test missile launching. (NASA C-88-11377)

This Boeing 737-130 c/n 19437 ex-N73700 NASA 515 is today on the Langley inventory. It was acquired on 17 May 1974 and is used as a flying laboratory. Initial use was in the NASA Terminal Configured Vehicle Programme in support of a joint US government effort to solve national airspace congestion and other air traffic control problems. (NASA WL79-48-7)

A unique subsonic ram-jet test-bed aircraft was this Northrop P-61B-15-NO 42-39754 Black Widow, a World War II night-fighter. It arrived at Cleveland during October 1945, being used as a research vehicle with the NACA until October 1948. (NASA C-76-1717)

Boeing P2B-2S (B-29-95-BW) BuNo.84029 (45-21787) registered NACA 137 seen parked at the Dryden Flight Research Centre, Edwards AFB, on 5 October 1980. The Superfortress is currently in Florida. (AP Photo Library Colour Slide)

Nose art on the Boeing P2B-2S Superfortress NACA 137 named 'Fertile Myrtle' and showing the tote board recording the launches of the Douglas D-558-2 Skyrocket research aircraft. (AP Photo Library Colour Slide)

This Short SC7 Skyvan 3 c/n SH.1844 NASA 430 ex-N30DA was acquired by the Wallops Flight Facility on 17 July 1979. It provides Wallop with an aerial recovery capability for sounding rocket payloads, and it is also used as an applications platform. It is depicted with recovery net extended flying over the coastline of Virginia. (NASA WI 1987 791-6)

Supersonic Flight

John Stack was appointed head of the compressibility research division at Langley in 1928, and was prominent in organising a large number of research projects. In 1933 he had conceived a high-speed aircraft, but this paper design had been merely an object for theoretical performance evaluation. With the coming of the compressibility crisis in 1940 and the growing recognition that there was some barrier preventing the acquisition of useful transonic data in existing wind tunnels. John Stack campaigned privately for NACA and military support for an actual aircraft for high-speed research. In 1944 there were research engineers such as Ezra Kotcher at Wright Field and some of Stack's colleagues at Langley who had competing ideas of the requirements of a high-speed research aircraft. Though many of the particulars of Stack's research aircraft concept would provide a solid foundation for the design of what became the Bell XS-1, the first supersonic aircraft, some of its particulars would not be accepted whilst others would undergo major compromise.

Prior to 1945, NACA studies of transonic and supersonic flows found practical application in propeller blades, cowling, ducting, wing-sections, turbo-jet compressors, turbines, burners and other critical components. A number of improvements came out of these efforts, but no supersonic aircraft. As far back as 1942 the compressibility research division at Langley was reviewing the available data. Macon C. Ellis Jr. and Clinton E. Brown jointly prepared a report entitled 'Analysis of Supersonic Ram-jet Performance and Wind-Tunnel Test of a possible Supersonic Ram-jet Airplane'. This was an investigation into the practical possibilities of supersonic flight.

On 15 March 1944, during two conferences held at Langley, representatives of the US Army Air Force Air Technical Service Command, the US Navy Bureau of Aeronautics and NACA expressed a desire to explore the little-known field of supersonic flight. The US Navy and the Agency further agreed to work together in preparing joint

specifications; a task posing problems as the US Navy wished to obtain an aircraft capable of being produced eventually for service use, whereas NACA wanted a pure research aircraft. The outcome was that the US Army Air Force contracted with the Bell Aircraft Corporation, and the US Navy with the Douglas Aircraft Company. These decisions were made early in 1945 when World War II was at its height of fury, and marked the beginning of a programme resulting in the construction of four high-speed research aircraft and the design of a fifth.

The Bell Aircraft Corporation called together a design team headed by Robert Stanley, who had been the pilot of the first US turbojet — the Bell XP-59 *Airacomet*. Under the project designation MX-524 the team commenced development of the 'Experimental Sonic-1' aircraft known as the XS-1. Apparently the designation MX-524 covered US Army Air Force high-speed projects in general. By the late spring of 1945, the XS-1 programme had been given the designation MX-653.

Stack was not at all satisfied with the Bell XS-1 project for the USAAF, so he contacted the US Navy who were ready to accept the NACA's advice and general guide lines. In September 1944, a Bureau of Aeronautics engineer and US Marine Corps officer, Abraham Hyatt, proposed that the US Navy procure a high-speed research aircraft capable of meeting both the military and NACA's research requirements. This design basically matched the conservative concept set out by Stack and his team. It would be a turbojet, not powered by a rocket; would take off from the ground and land under its own power; and it would have a maximum velocity not exceeding the speed of sound. These details made the design very different from the XS-1.

Douglas accepted the project and contract, and by the first week of 1945 the engineers were busy considering the design criteria for what would become the 'Douglas Model 558, High-Speed Test Airplane' known as the D-558. So the development of two different transonic research aircraft was in progress — the rocket-powered XS-1, and the jet-powered D-558. Though the research engineers at Langley would actively assist in the development and flight testing of both, they would prefer to assist the D-558 since it was nearer the specification they wanted.

The first two Bell X-1 research aircraft 46-062 and 46-063 were rarely seen together. Here they pose between research flights at Muroc during 1947. The Boeing B-29 launch aircraft 45-21800 is in the background. Bell X-1 46-063 is parked just in front of the loading pit ram. (Bell Aerospace Textron)

On 10 March 1945, the US Army Air Force notified NACA that it was awarding Bell a contract to develop the rocket-powered XS-1 research aircraft. It was planned for the aircraft to take off from the ground, rather than be air-launched and to climb through Mach 0.8 to the 1.0 speed region — the field NACA most wanted to know about. However, two month later after a design review meeting at Wright Field, Bell opted to change to air-launch. A specially-configured Boeing B-29 *Superfortress* would carry the XS-1 to an altitude of 30,000ft prior to release.

This change of plan further dampened any enthusiasm Langley had for the XS-1. The change to air launching meant that the small rocket aircraft could never be operated out of Langley, a busy airfield close to populated areas. If ever the XS-1 came loose accidentally, minus pilot, from the B-29 in flight, the resulting crash would probably cause many casualties. It was also felt that if the XS-1 was operated from another airfield, the Agency would not be able to retain control of the programme of test flights. However Langley complied with all requests from both the US Army Air Force and Bell to help complete the design of the XS-1 and prepare it for test flights.

Douglas D-558

Initially Douglas proposed to build six D-558 transonic research aircraft for the US Navy, each powered by a General Electric TG-180 turbojet engine and equipped with alternative

The first Douglas D-558-2 Skyrocket BuNo.37973 seen at the roll-out ceremony in 1947. The high-speed canopy was so narrow that the flight test pilots had to wear helmets covered with chamois leather to prevent scratching the surface of the perspex. Photo was taken on 10 September 1947 the Skystreak being trucked to Muroc on 10 December. (Douglas 711-15-16)

The second Douglad D-558-II Skyrocket BuNo.37974, in its all-rocket configuration, seen housed under the mother-ship, a Boeing P2B-1S BuNo.84029. On 31 August 1951 this Skyrocket was handed over to the NACA becoming NACA 144 and the Boeing P2B-1S becoming NACA 137. (Douglas SM 11233)

wing and and nose-duct configurations. Maximum speed would be in the region of Mach 0.90. Phase 2 of the programme would involve altering two of the aircraft to be powered by the smaller Westinghouse 24C turbojets with supplementary rocket propulsion units. These modified aircraft would gather aerodynamic data from Mach 0.89 to Mach 1. In phase 3, Douglas would use results acquired earlier to construct a combat version of the D-558. The estimated total cost of the three-phase programme was just under $7 million. Desired altitudes would be in the region of 40,000ft and Douglas designed a jettisonable nose section, including the cockpit, from which the pilot could bale out once the nose section reached slower speeds and lower altitude after its separation from the crippled aircraft.

Following NACA and the Bureau of Aeronautics' favourable review of the Model 558 proposal, Douglas was awarded a Letter of Intent under Contract NOa(s)6850, dated 22 June 1945. The initial D-558-1 project, based on the original straight-wing, turbojet-powered design and intended for transonic research, was to be followed by the D-558-2 with swept-wing and mixed turbo-jet-rocket power for research in the supersonic regime. The design of the D-558-1 was kept as conventional as possible and was completed during April 1947. The 5,000lb thrust J35-A-11 engine — an Allison-built version of the General Electric-designed TG-180 turbojet — was installed. The integral wing tanks had a capacity of 230 US gallons, with provision for two 50-gallon wingtip tanks. High-pressure tyres were fitted and speed brakes were mounted externally on either side of the fuselage just aft of the wing.

Painted a brilliant scarlet and named *Skystreak*, the first D-558-1 (s/n 6564, BuNo.37970) was trucked to Muroc Dry Lake during April 1947 for the first phase of the manufacturers' trials. Early in August the second D-558-1, BuNo.37971, was ready to join the trials programme.

Bell XS-1

Construction of the first Bell XS-1 (46-062) was completed, except for the rocket motor, during December 1945. The terms of the contract with the US Army Air Force dictated that Bell had to test the XS-1 to Mach 0.8 before official acceptance. Both NACA and the USAAF had decided that Bell test pilots should begin by flying the new aircraft in a series of glide tests. These would iron out any problems with the air-launch method and also test the feasibility of operating the XS-1 from conventional airfields, such as Langley, near populated areas. During November 1945 the US Army Air Force selected Pinecastle Field — located in central Florida, near Orlando — as the site for the glide tests. The Agency understood that these initial flight tests would determine the possible use of Langley if operations fitted the safety envelope.

Roll-out of the XS-1 took place on 27 December at Bell's Wheatfield, New York plant. On 19 January 1946 it was flown by a Boeing B-29 *Superfortress* (45-21800) to Pinecastle Field. NACA sent two Langley engineers, Walter C. Williams and Gerald M. Truszynski, to Pinecastle to join Bell test pilot Jack Woolams — who was well known to Langley personnel. Truszynski took charge of the SCR-584 radar tracking equipment, whilst Williams monitored flight test operations and supervised on-the-spot analysis of the resulting data.

The XS-1 glide test programme at Pinecastle lasted approximately three months, from January to March 1946, while Bell readied the second aircraft for powered flight trials. The air-launch method from the B-29 'mother ship' proved very practicable but there were problems landing the XS-1 at Pinecastle — which eliminated NACA's hope that some method of ground launching could be found, so making it viable to operate the XS-1 from Langley. The aircraft showed itself to be aerodynamically sound, with good low-speed handling qualities.

The Bell design team had worked closely with both NACA and the US Army Air Force: indeed, this was the first time that Langley research personnel had become involved in the initial design and construction of a complex research aircraft. The USAAF bore the cost and shared the research expense load, so the collaboration marked a significant departure in NACA procedures. Overall, design issues were amicably resolved although naturally there was much discussion and a few heated exchanges. The wing design was one area of controversy.

A total of ten glide flights by the first XS-1 (46-062) was completed at Pinecastle, and a decision to move to Muroc in California was confirmed in March 1946. The XS-1 was returned to the factory from Florida in order to be heavily modified, with new wings and horizontal tail surfaces. It was painted a bright orange for high visibility. Muroc Army Air Field, known as 'Muroc' after a small family settlement on the edge of Rogers Dry Lake, became the flight-test centre for the US Army Air Force. It covered an area of 300 square miles of desolate waste in the Californian desert north-west of Los Angeles. Originally used as a military bombing and gunnery range, Muroc was an ideally remote location and the concrete-hard lake bed was highly suited for emergency landings, allowing approaches from almost any direction. It was to be the home — and an appropriate environment — for a growing roster of exotic aircraft which have been tested there up to the present day.

The original Langley contingent was called the 'NACA Muroc Flight Test Unit' later the High-Speed Flight Station — HSFS. On 27 January 1950, when Muroc Field's name was officially changed to Edwards AFB, government and NACA personnel alike were quick to adopt the term 'Edwards' in colloquial use. By this time there were 250 NACA employees at Edwards. The base was named formally after Captain Glen W. Edwards of Lincoln, California who, along with four other crew members, were killed on 5 June 1948, in the crash of a Northrop YB-49 flying-wing bomber on the Muroc reservation. In 1959, after the creation of the National Aeronautics and Space Administration — NASA — the High Speed Flight Station became the NASA Flight Research Centre (FRC). On 26 March 1976, NASA renamed it the 'Hugh L. Dryden Flight Research Centre' — DFRC — in honour of the man who succeeded George Lewis as the NACA's director of research. In the 1920s along with L. J. Briggs he had obtained supersonic flow during experiments at the Edgewood, New Jersey, Arsenal with a small free-jet apparatus having a convergent-divergent nozzle.

The NACA team at Muroc was complete by early October 1945 and immediately set up an SCR-984 radar tracking system. The team included telemetry technicians, 'telemetry' being a relatively new innovation which involved on-board instrumentation which could produce signals which were then transmitted from the aircraft to a ground station and a receiver to pick up the signals. The XS-1 had six-channel telemetry to transmit airspeed, control surface position, altitude normal acceleration to the ground station. Two SCR-584 gun-laying radars were added to the tracking equipment.

On 7 October 1946, the second XS-1 (46-063) was delivered by Boeing B-29 *Superfortress* and completed its first glide flight on 11 October, piloted by Bell test pilot Chalmers 'Slick' Goodlin. In preparation for the flight-test programme, US Army Air Force technicans had installed two large liquid-oxygen and liquid-nitrogen tanks. The nitrogen was used to pressurize the XS-1's fuel system, since it burned liquid oxygen and diluted alcohol. A large loading pit was dug so that the XS-1 could be hoisted into the bomb bay of the modified B-29 *Superfortress*.

Bell continued the required contractor flight testing of 46-063 until the middle of 1947. By March 1947, the first XS-1 (46-062) was ready at the Bell factory, and on 5 April it was flown by B-29 to Muroc. On 11 April, with Goodlin at the

controls, it made its first powered flight. A programme of twenty powered flights, as required in the original contract, was completed in May 1947, and with obligations concluded the two Bell XS-1 aircraft were handed over to the US Army Air Forces for continuation of performance flight tests.

During June 1947, at a meeting held at Wright Field, Ohio, the US Army Air Force and the NACA representatives met to discuss the future test programme with the two Bell XS-1 aircraft. A decision was made to allow the USAAF to explore the transonic and supersonic speed envelope using the first aircraft (46-062), and to allow NACA to explore transonic stability and control using the second XS-1 (46-063). The Agency assigned two test pilots: Herbert H. Hoover from the Langley laboratory, and Howard C. 'Tick' Lilly from Lewis, to the project. At a meeting held at NACA Hedquarters in Washington, D.C. on 6 February 1947, attended by US Army Air Forces representatives, a joint agreement for the conduct of all research aircraft projects — XS-1 to XS-4 — was discussed. The Agency would supply its own maintenance and flight crews, and Joseph Vensel — a former NACA Langley test pilot — was designated to supervise NACA flight operations.

On 8 January 1947, during a buffet-boundary investigation, Chalmers Goodlin had reached Mach 0.8 at 35,000 feet — the speed and altitude required by the contract before the USAAF would accept delivery of the XS-1. Three months later Bell began flying the XS-1 (46-062), which had been out of action since its last glide flight at Pinecastle in March 1946

Rare photo depicting the Bell X-1 46063 with its Boeing B-29 'mother-ship' at Muroc and displaying the NACA logo on the fin. The first supersonic flight with the NACA with this aircraft was made on 10 March 1948 piloted by Herbert Hoover. (Via Jay Miller)

It was NACA pilots who flew the Bell X-1B 48-1385 during all the final flight test series. It completed a total of 27 glide and powered flights before retirement, being delivered to the US Air Force Museum at Wright-Patterson AFB, Dayton, Ohio, on 27 January 1959 by which time the NACA had become the NASA. It is depicted at the US Air Force Museum. (US Air Force Museum)

in order to have a rocket motor installed. During May Bell successgully flew the XS-1 (46-063) through a required 8g pullout and final airspeed calibration flight. After a total of twenty-one powered flights — fourteen by 46-063 and seven by 46-062 — the contractors' programme was complete. Since the two XS-1 aircraft now belonged to the USAAF, it was up to their flight engineers and test pilots to 'break the sound barrier' in as few flights as possible.

At a meeting held at Wright Field, Ohio, on 30 June 1947, the air Material Command Flight Test Division of the USAAF discussed test pilots for the XS-1 programme. Colonel Albert Boyd, a highly respected test pilot who directed the division, selected Captain Charles E. 'Chuck' Yeager as project pilot, assisted by Captain Jack L. Ridley as flight test engineer. Lieutenant Robert Hoover was chosen as chase and alternate pilot. 'Chuck' Yeager — a 24-year old fighter ace from Hamlin, West Virginia — was a superlative pilot and an intuitive engineer.

Initially it looked as though the two-pronged flight programme would be delayed on the NACA side. Early in June the XS-1 (46-063) was seriously damaged in a freak on-the-ground accident and had to be ferried by B-29 back to Bell at Buffalo, New York, for repairs. By the time it arrived back at Muroc in July, the activity around 46-062 was so intense that there were usually no technicians left to handle the NACA aircraft. When the NACA XS-1 was finally ready for acceptance test on September 25, it was discovered that neither of the NACA pilots were ready; neither Hoover or Lilly had been checked out on the aircraft!

On 6 August 1947, Yeager completed a familiarization glide flight in the USAAF XS-1, and by the end of the month had completed his first powered flight, reaching M0.85. By early October he was edging past 0.93 towards the 'sonic wall' in the XS-1 which he had named 'Glamorous Gennis' after his wife. On 10 October Yeager reached an indicated Mach 0.94. During the glide recovery frost formed on the inside of the canopy, and despite persistent efforts Yeager could not scrape it off. Chase pilots Bob Hoover and Dick Frost, flying Lockheed P-80 *Shooting Stars*, had to talk him down to a blind landing on the lake bed. Working all night, the NACA technicans analyzed the oscillograph film taken from the XS-1 and discovered that, instead of Mach 0.94, all indications were that the aircraft had actually reached Mach 0.997 at 12,000 metres. This speed worked out at 1,059 kilometres per hour — exceedingly close to the speed of sound.

Sonic Boom

During a pre-flight planning briefing during the morning of 14 October 1947, NACA advised Yeager to fly the XS-1 (46-062) on its ninth powered flight to a maximum speed of Mach 0.9. Data from previous flights was still being analyzed, so that it was felt unwise to go any faster unless the pilot felt it was safe to do so. Technicians had locked the XS-1 into the bomb-bay of the B-29 'mother-ship'. It was then fuelled with 1,177 litres of liquid oxygen and 1,109 litres of diluted ethyl alcohol. At approximately 10.00am Yeager climbed on board the *Superfortress*. It took about twenty minutes to climb to 20,000ft and at 5,000ft Yeager climbed with painful difficulty down the transfer ladder into the tiny XS-1 cockpit. He had two broken ribs, thanks to a fall from a horse. At 20,000ft NACA ground radar cleared the B-29 for launch. Sixty seconds later at 10.26am, Yeager and the XS-1 dropped free. He briefly checked the engine by firing the four chambers of the XLR-11 rocket, closed down two and climbed on the remaining two to 35,000ft, then to 42,000ft. In level flight Yeager switched on the third rocket chamber and watched the Mach meter dial closely as it progressed rapidly to Mach 1. Seconds later, the needle jumped to Mach 1.06 — 700mph. Seeing the Mach number off the scale, Yeager shut down all chambers and jettisoned fuel in a climb. At 45,000ft an unaccelerated stall was made. The descent from 45,000 to 35,000ft was made at Mach 0.7 so that a pressure altitude

survey could be made. On the ground observers had heard the characteristic double bang of a sonic boom. Fourteen minutes after launch, the XS-1 landed back at Muroc; its two chase aircraft and the B-29 followed in turn. The first manned supersonic flight in history had been safely accomplished, and the dreaded 'sound barrier' was a thing of the past.

Walt Williams was soon on the telephone to Langley with the news. Word came in response from NACA headquarters to the effect that the flight accomplishment and future flight tests were to be regarded as TOP SECRET. However, 'Aviation Week' leaked on account of the first supersonic flight in an issue published during December 1947, whilst the US Army Air Force — later US Air force in 1948 — and the Agency did not formally reveal Yeager's accomplishment until 15 June 1948 when General Hoyt Vandenberg, Chief of Staff of the new US Air Force, and Hugh Dryden for NACA confirmed that the XS-1 46-062 'Glamorous Glennis' had repeatedly exceeded the speed of sound, piloted by USAF and NACA test pilots.

Some ten days after the record flight, test pilot Herbert Hoover initiated NACA's flight programme on the XS-1 (46-063) with a familiarization glide flight. On landing, however, Hoover misjudged the approach height and made several hard contacts with the runway, the last of which caused the nose wheel to collapse before the aircraft skidded to a halt on the desert strip with damage to the landing strut. Repairs and bad weather kept the machine grounded for seven weeks.

During the period between October and December 1947, test pilot Howard Lilly made the first two NACA test flights in the Douglas D-558-1 — the US Navy contribution to the programme and the aircraft favoured by John Stack over the Bell XS-1. The D-558-1, BuNo.37970, had arrived at Muroc with a company test team during April. At the end of the summer BuNo.37971, the aircraft planned for extensive NACA service, arrived: this was the aircraft which Howard Lilly flew in November 1947. On 16 December 1947 NACA commenced systematic flight tests of Bell XS-1 46-063 and Herbert Hoover became the first NACA pilot to fly a rocket-powered aircraft; he reached Mach 0.71. By the end of January 1948 Lilly had been checked out and Hoover had accomplished six more powered flights in the XS-1, working up to Mach 0.925. On 10 March Herbert Hoover achieved NACA's first supersonic flight. Three weeks later, Lilly repeated Hoover's supersonic performance.

By the end of 1947 Bell XS-1 had flown to Mach 1.35 at 1,490 kilometres per hour. This success resulted in the US Army Air Force placing an order with Bell for four advanced versions, of which three were eventually completed. These were the X-1A, X-1B and the X-1D. The 'S' was dropped from the designation during 1948 in a revision of aircraft nomenclature following the formation of the US Air Force. The USAAF phase of the programme had not only been a success, but it continued to be one: Chuck Yeager reached the maximum flight speed of the XS-1 on 26 March 1948 by achieving Mach 1.45 — 1,540 kilometres per hour.

By early 1948, the busy NACA unit was responsible for three aircraft at Muroc: the two XS-1s and the second Douglas D-558-1 *Skystreak*. The workload imposed on the Agency's personnel posed a serious problem — especially on the instrumentation staff, since the agency believed in thoroughly instrumentating and calibrating all its research aircraft. Herbert Hoover had become the first NACA staff member and the first civilian to fly faster than sound, subsequently receiving the Air Medal from President Harry Trueman for the feat. The two NACA test pilots were subsequently joined by two others — Robert Champine in late 1948, and a year later by John Griffith.

The National Aeronautic Association selected John Stack of the Langley laboratory to share in the 1947 award of the Robert J. Collier Trophy, the association's annual prize for the greatest achievement in American aviation. In a ceremony at the White House. President Harry Truman presented Stack the award citation, which read:

'To John Stack, Research Scientist, NACA, for pioneering research to determine the physical laws affecting supersonic flight and for his conception of transonic research airplanes; to Lawrence D. Bell, President Bell Aircraft Corporation, for the design and construction of the special research airplane X-1; and to Captain Charles E. Yeager, US Air Force, who, with that airplane, on 14 October 1947, first achieved human flight faster than sound.'

Stack insisted that he should not have been singled out for a share of the Collier award. The NACA contribution to the supersonic flight of the XS-1, he said, had been a team effort. Yet supersonic flight had never been Stack's original interest. His idea for the research aircraft progamme had been restricted to obtaining information in the transonic speed range. The Agency emphasised the co-operative nature of the entire research aircraft. During a public presentation during June 1949 John Stack said:

'The research airplane programme has been a co-operative venture from the start among the Air Force, Navy, the airplane manufacturers, and the NACA. The extent of this co-operation is best illustrated by the facts that the X-1, sponsored by the Air Force, is powered with a navy sponsored rocket engine, and the D-558-1, sponsored by the Navy, is powered with an Air Force-sponsored turbojet engines.'

On 15 April 1947 Eugene F. May piloted the Douglas D-558-1 (BuNo.37970) for the first time but had to abort due to partial power loss. On landing, the *Skystreak*'s left wheel brake disintegrated. Six days later a second attempt ended with similar troubles and the unlucky prototype was grounded for a month. On return to flight status, it was plagued with persistent undercarriage retraction problems. Finally the various snags were resolved, and by mid-July 1947 the aircraft was ready for general handling and high-speed trials. By early August 1947, it had reached a top speed of Mach 0.85. At the same time the second D-558-1, BuNo.37971, was ready to join in the trials, and military pilots shared the flying task. As the flight hours were built up, confidence in the *Skystreak* increased — so much so that the US Navy decided to use the aircraft for an attempt on a new world airspeed record.

Commander Turner Caldwell, USN, the US Navy project pilot, flew 37970 on 20 August 1947 and averaged 640.663 mph during four passes over a low-level 3-kilometre course, breaking the record of 623.738 mph set a month earlier by Colonel Albert Boyd, USAAF, in the specially modified Lockheed P-80R *Shooting Star*. The record gained by Caldwell was held for only five days since on 25 August 1947, Major Marion Carl of the US Marine Corps, flying *Skystreak* 37971, averaged 650.796 mph.

Following these record flights, the D-558-1 was returned to its contractor programme and completed a total of 101 flights prior to being delivered to the Agency on 21 April 1949. However, BuNo.37970 was never flown by the NACA but was used by the agency to provide spares needed to support the trials of the third D-558-1. The second *Skystreak*, the record holder, completed twenty-seven US Navy, US Marine Corps and Douglas flights before being handed over to NACA on 23 October 1947. After being heavily instrumented, this aircraft completed two further flights in November 1947 with NACA's test pilot Howard Lilly at the controls. Winter flooding of the runway and the need for engine maintenance then resulted in the aircraft being grounded for eleven weeks. The intensive flight trials programme began on 31 March 1948, preceded by a single flight in February, but this was marred by tragedy when within five weeks the *Skystreak* crashed and Howard Lilly was killed. This accident, the first in which a NACA pilot lost his life while on duty, occurred on 3 May 1948 when the engine compressor disintegrated shortly after take-off.

The loss of the second *Skystreak* delayed the NACA research programme by almost a year, although the third D-558-1 (BuNo.37972) was delivered to Muroc during November 1947 and was flown on four occasions by Douglas test pilots prior to hand over to NACA. The third D-558-1 remained grounded until April 1949 whilst investigation into the loss of 37971 was being completed. The inquiry exonerated the aircraft and 37972 returned to flight status, making its first flight with the Agency on 22 April 1949. Seven NACA test pilots accomplished a total of seventy-seven additional flights, enabling a great deal of useful data on high-subsonic handling to be recorded. The 82nd and final flight was made on 10 June 1953, completing four-and-a-quarter years of active research test flying.

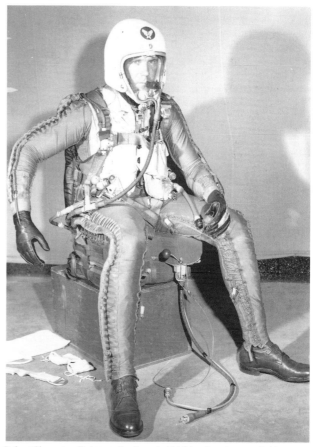

Robert Champine, flight test pilot from the NACA at Langley, seen in his X-series pressure suit. On 7 December 1949 he completed a stability and control investigation at Mach 0.855 in the Douglas D-558-II BuNo.37974 at the NACA High-Speed Flight Station at Edwards AFB. (NACA 573182)

The problems of transonic flight stimulated the development of important new free-flight and ground-based test techniques; the wing-flow, drop-body, and rocket-model methods. The success achieved in all these areas boosted the morale and self-confidence of all NACA employees, at Langley and elsewhere. Planning and monitoring the extensive flight-testing of the Bell XS-1 and the Douglas D-558-1 at Muroc was neither a small or simple task. It entailed long-range supervision from the parent laboratory, which was 2,500 miles away from the engineers doing the work. Concern for proper management led the Agency to create a special research aircraft project panel in 1948, followed in 1949 by the establishment of a larger NACA High-Speed Flight Research Station — HSFRS — at the outpost in California. Langley continued to provide management for this station until 1954, when NACA headquarters decided to make it an autonomous field installation, the NACA High-Speed Flight Station — HSFS. During 1958, this installation became NASA's Flight — later Dryden — Research Centre.

CHAPTER SEVEN
Shapes for the Future

Despite the continuing success of the high-speed flight programme at Muroc, there was still a need for high-speed wind tunnel tests. At Langley, in the 7×10ft tunnel, the NACA technicians built a hump in the test section so that as the air stream accelerated over the hump, models could be tested at Mach 1.2 before the 'choking' phenomenon occurred. A research programme was introduced with the idea of absorbing the shock waves by means of longitudinal openings, or slots, in the test section. The 'slotted-throat' tunnel became a milestone in wind tunnel evolution, permitting a full spectrum of transonic flow studies.

Another high-speed test programme was introduced involving rocket-propelled models or remotely piloted vehicles (RPV) built at Langley but launched from Wallops Island, to the north, on the coast of Virginia. This was now the home of the Pilotless Aircraft Research Division (PARD), established during the autumn of 1945. During the next few years PARD used rocket boosters to make high-speed tests on a variety of models representing new aircraft under development. These included most of the subsonic and supersonic aircraft flown by the US armed services during the decades following World War II. In the 1960s Wallops facilities would support NASA's Mercury, Gemini and Apollo programmes.

In 1946 the Flight Engineers section at Ames ended its de-icing studies (with research into carburettor icing problems being taken over by the Lewis Flight Propulsion Laboratory) and Ames proposed a continuing programme of research in the general field of aircraft operating problems. This proposal

was rejected by the NACA management. During 1947, Harry Goett, in charge of the flight engineering department, suggested to the Agency's headquarters in Washington, D.C. that Ames set up a pilotless-aircraft test facility similar to the RPV operation established by Langley at Wallops Island. Dr Hugh Dryden was opposed to any duplication of efforts in developing rocket-launching and telemetry techniques, which were already in hand at Langley and the PARD at Wallops Island. However, there was no objection to the proposal for launching of recoverable models from a high-flying aircraft. As the icing research work was phased out in 1947, Ames prepared to develop a technique for recovering test models.

By means of air brakes and parachutes, aerodynamically clean instrumented test models dropped at high altitudes from an aircraft could be recovered intact. If dropped from a sufficiently high altitude, they would traverse the transonic range and indeed even reach low supersonic speeds before they had to be braked for landing. The aircraft types selected by the Ames engineers for the drop operation were two Northrop P-61 *Black Widow* nightfighters and one Northrop F-15 *Reporter* photo-reconnaissance aircraft which was fitted with turbo-superchargers. Apparently there was a relative scarcity of surplus P-61 aircraft and one was borrowed from the Smithsonian Museum in Washington, D.C. This was ERF-61C-1-NO 43-8330 which arrived at Ames on 5 February 1951, becoming 'NACA 111'. It served until 8 October 1954 when it was returned to the museum. The first of the type to arrive at Ames was F-15A-1-NO *Reporter* 45-59300, later redesignated RF-61C. It was delivered on 6 February 1948, and according to the aircraft records it was also given the NACA registration 'NACA 111'. It also served until October

Due to the scarcity of surplus Northrop P-61 aircraft this ERF-61C-1-NO 3-8330 NACA 330 had to be borrowed from the Smithsonian Museum. It later became NACA 111. It arrived at Ames on 5 February 1951 serving until 8 October 1954 when it was returned to the Smithsonian. (NACA A-18407)

1954, when it was salvaged by the US Navy. The third Northrop aircraft arrived on 13 July 1951, being an EF-61C-1-NO *Black Widow* 43-8357 which became registered, in turn, 'NACA 130' and 'NACA 146'. It departed on 8 August 1954, being also salvaged by the US Navy. Actually the P-61 was no stranger to Ames as P-61A-5-NO 42-5572 was flight-tested between 20 April and 16 November 1944.

During the drop operation, the supercharged aircraft was flown to altitudes in the region of 42,000 feet. As it was not pressurised, the physical stamina of the flight-test pilots was sorely taxed. But a large amount of useful data was obtained, whilst the development of the recovery technique (a notable accomplishment in itself) was helpful to other agencies later in attempts to recover expended missiles. The greatest contributors to the programme, including the development and use of the drop technique at Ames, were Alun Jones, James Selna, Bonne Look and Loren G. Bright. Among the pilots who flew the unique Northrop aircraft during the drop-test programme were George Cooper, Rudolph Van Dyke and Robert Whemper (all based at Ames), whilst Joseph Walker from the NACA High-Speed Flight Station at Edwards joined in the flight-test programme.

The wind-tunnel personnel at Ames were involved in work on some very advanced aircraft configurations, none of which had yet arrived with the flight research group for testing. It was not until 29 August 1949, that a swept-wing North American F-86A-5-NA (48-291) arrived at Ames for flight tests. Until then, the fastest aircraft in the research stable had been the straight-wing Lockheed P-80A *Shooting Star* which in a dive, as test-pilot Larry Clousing discovered, could reach Mach 0.88. Extensive flying-quality tests were carried out on the P-80A, and these were well written-up by Seth Anderson, Frank Christofferson and Lawrence Clousing. The latter's outstanding work in flight research was recognised in 1947 when the Octave Chanute Award was conferred on him by the Institute of the Aeronautical Sciences. Further flight research at Ames during this period represented a continuation of earlier efforts to confirm in flight the stability and control of aircraft as predicted by wind-tunnel tests. There was also an application of the wing-flow technique to transonic studies both of straight wings, such as used on the Douglas X-3 *Stiletto*.

Larry Clousing was involved in the programme on wing and tail loads. The Ames flight research section was also investigating buffeting — an unpleasant, if not dangerous, phenomenon by high-speed aircraft in the pull-out. the conditions under which it occurred were investigated on a variety of aircraft, and it was whilst testing one of these aircraft — a North American P-51 *Mustang* — that George Cooper, the pilot, thought he could actually see the shock wave on the wing. This phenomenon was observed on the wings of two other aircraft and the conditions under which it would appear — the angle of the aircraft with respect to the sun, etc. — were determined and recorded.

A most important contribution by the Ames flight research group during the period was the development of a 'variable-stability' aircraft. The idea was not new, having been mentioned in German research literature, but it was one of those subjects the practical accomplishment of which requires greater genius than the original conception. The propeller-driven Grumman F6F-3 *Hellcat*, BuNo. 42874, which became 'NACA 158' was used in the first attempt to produce a variable-stability flying simulator. It arrived at Ames on 22 June 1945, being salvaged on 9 September 1960.

The *Hellcat* variable-stability aircraft was flown by five different pilots under conditions which simulated approaches to land, cruising and high speed. A powered servo mechanism in the aircraft which could deflect the ailerons in proportion to the angle of yaw was installed, and by adjustment in flight, effective dihedral angles between $-18°$ to $+28°$ could be obtained. Rudolph Van Dyke was one of Ames pilots involved and he flew the aircraft on several occasions.

Unfortunately Van Dyke was killed on 1 June 1953, in a crash involving Grumman F8F-1 *Bearcat* BuNo. 94819.

The results of this first use of variable stability by the NACA at Ames were promising. The next step was to drive the rudder as well as the ailerons in accordance with yaw angle and also perhaps to introduce roll and yaw rate as inputs in addition to yaw angle.

This Northrop F-15A-1-NO Reporter 45-59300 was delivered to Ames on 6 February 1948. It is depicted carrying an instrumented test vehicle which was dropped from as high as 42,000 ft. It served at Ames until October, 1954, when it was salvaged by the US Navy. (NACA A-14294)

The North American F-86 *Sabre* was the first operational aircraft capable of reaching supersonic speeds in a dive. Ames received F-86A-5-NA 48-291 (later 'NACA 116') on 29 August 1949 and it was soon involved in spinning tests and prolonged dives at very high speed. Rudolph Van Dyke and George Cooper were the pilots. Soon after the programme commenced, the local newspaper began reporting mysterious explosions, the source of which was unknown. The headlines became more prominent — for example, plates had been shaken off the plate rack in the International Kitchen near Niles — but investigation by a sheriff's posse had no luck. Another explosion occurred in the region of the Calaveras reservoir, causing damage to a house. The explosions had been caused by the F-86 diving beyond the speed of sound. As a matter of policy, NACA did not publicize the situation, but some months later the sonic boom was heard elsewhere and became associated with supersonic flight. It soon became commonplace. Ames, it is thought, was first to encounter and recognize the phenomenon.

Flight Test Hazards

Flight test pilots faced many hazards. In the dive test programme with the F-86, George Cooper and Rudolph Van Dyke were probably flying through the transonic range more frequently than any other aviator in the USA. On 9 October 1944 Ames took delivery of a Douglas — XSB2D-1 *Destroyer*, BuNo. 03551 — a carrier type which was also known as the BTD-1. On 24 May 1946, George Cooper and Welko Gasich were airborne in the *Destroyer* over the countryside near Las Gatos. The engine backfired, which started a fire in the induction system. Not being able to return to Moffett Field, George crash-landed the aircraft in a prune orchard. The aircraft was a wreck by the time it came to rest but the crew walked away from it, badly shaken but not seriously hurt.

On another occasion George Cooper was flying a Republic P-47 *Thunderbolt* investigating the effect on the stability and control of the aircraft of reversing the propeller pitch whilst in a dive. His engineering colleagues had assured George that by pushing the button, the propeller would snap back to normal pitch. During one flight Cooper wished to discontinue the flight test, so he pushed the button. Nothing happened, and the propeller continued to have a braking effect; consequently he was faced with making a forced landing with the propeller in reverse pitch. Carefully making a high approach to land, Cooper was some two hundred feet from the runway when suddenly the propeller snapped into normal pitch. Quickly he opened the throttle, overshot the runway and tree tops to go around and make a normal safe landing.

Incidents of the kind mentioned would fill many volumes and were all in the day's work for the NACA flight test pilots, often accompanied by a flight test observer or engineer. Unhappily they did not all turn out well. On 17 May 1948, Ryland D. Carter from Ames was flying North American P-51H-10-NA 44-64691. The wing was equipped with a 'glove' and the trial was in conjunction with flow tests. For reasons unknown the *Mustang* disintegrated in mid-air and its fragments were scattered far and wide over farmland near Newark, California. The pilot was thrown clear but he was unable to activate his parachute and was killed.

This conservative propeller-driven Grumman F6F-3 Hellcat BuNo.42874 NACA 158 was the first attempt to produce a variable-stability flying simulator, being flown by many flight test pilots. It served at Ames between 22 June 1945 and 9 September 1960. (NACA A-22147)

Further Advances and Area Rule

The relaxation of world tensions after the end of World War II did not last long. There was a long cold-war period when relations with both the Soviet Union and China became strained. The Berlin blockade by the Soviets was followed by the Korean War, the first conflict involving the United Nations, from 1950 to 1953. US aircraft production was again on the increase, with new and advanced types of aircraft and missiles. Russia exploded its first H-bomb in the same period, and tensions continued to mount.

There were many technical advances in the field of aviation during this period. During December 1953 — the 50th Anniversary of powered flight — the Douglas Aircraft Company at their El Segundo plant built the new XF4D-1 *Skyray* and North American produced the F-100 *Super Sabre*. This was the first of the 'Century' series of fighters and it could reach approximately sonic speeds at 754mph in level flight. At Edwards AFB, research aircraft were flying even faster. The first aircraft to fly at twice the speed of sound — the Douglas D-558-2 *Skyrocket* — was dropped from its B-29 'mother-ship' and reached Mach 2.0 in level flight on 20 November 1953: the pilot was Scott Crossfield. Less than a month later this record was exceeded when the Bell X-1A went to Mach 2.5 at a speed of 1,612mph.

NACA became interested in the problems of flight beyond the earth's atmosphere and in July 1952 instructed the laboratories to initiate studies of means for the development of a hypersonic, high-flying research aircraft as a joint project between itself and the military services. This became the North American X-15 research aircraft.

Richard T. Whitcomb arrived at Langley during 1943, a graduate in mechanical engineering from Worcester Polytechnic Institute, Massachusetts, and worked initially under John Stack in the 8ft High-Speed tunnel section. The young engineer was keen to work in aerodynamics; the building and

testing of model aircraft, the mania of many boyhoods, still fascinated him. Although the 8ft tunnel did not go transonic until equipped with the slotted throat in 1950 it had reached Mach 0.95 after its repowering in 1945. Whitcomb started to gain a feel for transonic aerodynamics at least five years prior to his commencement of research which led directly to his conception of the area rule.

During this period Whitcomb looked into the main problem facing designers of supersonic aircraft — the large increase in drag, associated with the formation of shock waves, which occurred at transonic speeds. He had access to the data collected from laboratory wind tunnel tests and from the flight of the Bell X-1, and knew that small, lightweight rocket-powered configurations with limited mission cababilities could overcome the transonic drag problem. Operational turbojet aircraft would have to be considerably heavier than the X-1, posing a critical problem. If flying up to and through Mach 1 involved gradual acceleration because of high drag, there would be insufficient fuel left for the aircraft to sustain supersonic flight for long after achieving it.

During 1949 and early 1950, Langley tested models incorporating Whitcomb's sweptback wing and body combination at high subsonic (Mach 0.95) and low subsonic (Mach 1.2) speeds. Results were disappointing, showing very little favourable effect in reducing the drag. There was significantly higher total drag than transonic theory predicted for the drag of the wing and body combined. Whitcomb needed to know more before any fruitful work on the major design problem of supersonic aircraft could begin. The basis of 'area rule' concerned the cross-sectional areas of a wing and body combination. If these areas obeyed the rule by having the proper relationship to each other, the resulting shape should enjoy minimum transonic drag. To verify the rule by experiment, Whitcomb designed models with variously pinched waists and tested them in the 8ft high-speed tunnel. By the end of April 1952 the accumulated data indicated that significant reductions in drag could be obtained by contouring the fuselage. Whitcomb wrote a formal report confirming the theory and the Agency published it as a research memorandum in September 1952, making it available immediately to the American aviation industry but classifying it as SECRET. However at least one aircraft manufacturer — Convair — had heard something about area rule.

Data from the new transonic tunnel had suggested that the new Convair YF-102, a fighter-interceptor required for the defence of the United States, could not fly supersonically as planned since its transonic drag was higher than expected. Convair were naturally worried as the designers had given the aircraft a bullet shaped fuselage, knife-edge delta wings and the most powerful jet engine then in existence, the Pratt and Whitney J-57. An assembly line had been laid down at San Diego in preparation for a massive production programme. It was unfortunate that NACA test results indicated that their best-laid plans were insufficient.

When Convair engineers visited Langley, prior to publication of Whitcomb's report, the latter revealed his surprising discovery of a new rule-of-thumb concerning transonic drag. Over the next few months he worked with Convair to apply the area rule to the YF-102 configuration. In October 1953 NACA reported that a modified aircraft designated YF-102A met the US Air Force specification for a supersonic fighter-interceptor. In late 1953 and early 1954 the aircraft was test flown, but its performance mostly confirmed NACA's pessimistic wind tunnel evaluation. Convair had built two prototypes, followed by a further eight with detail differences. The US Air Force called a halt to the assembly line and advised the company to retool immediately for manufacture of a new YF-102A. With Whitcomb's help, Convair designed a model according to the area rule, and it took less than seven months for a new prototype to be built. In addition to the 'wasp' or 'coke-bottle' waist, the YF-102A (53-1787) was given a lengthened fuselage with sharply pointed nose, a modified canopy, tail fairings and a cambered wing leading edge. It was powered by an uprated J-57 engine. During a flight test five days before Christmas 1954, the YF-102A not only went past the sound barrier but kept going. The area rule had helped to increase the top speed by an estimated twenty-five per cent. The US Air Force contracted Convair for over 1,000 F-102 *Delta Dagger* aircraft, whilst the later F-106 *Delta Dart* flew as a vital part of the US Air Defence Command into the early 1980s. Both types were later used by NACA, later NASA.

Convair had solved their problem and were followed by Chance Vought, who re-designed its F-8U *Crusader* carrier-based interceptor according to the area rule. Grumman and eventually Lockheed followed. In April 1956, the latter's area-rule-based F-104 *Starfighter* was the first jet to exceed Mach 2.0 in level flight. Having received Whitcomb's report, Grumman sent a delegation to Langley to obtain further information. At the time they were under contract to the US Navy for a supersonic carrier-based fighter. In February 1953 Whitcomb visited the Grumman plant at Bethpage, New York, to see the final design layout of the area-rule-based F9F-9 *Tiger* prior to slotted-tunnel and rocket-model tests at transonic speeds. On 16 August 1954 the F9F-9 *Tiger* (later redesignated F11F-1) flew at sonic speed in level flight without the use of an afterburner.

A month later the journal *Aero Digest* published a story on the successful application of the area rule to the Grumman *Tiger* in violation of the journal's written commitment to withhold publishing anything about this work until the veil of secrecy had been officially lifted. It was a year later, in September 1955, when NACA released information on the area rule. Five weeks after the public announcement, the National Aeronautic Association awarded Whitcomb its coveted Collier Trophy for the greatest achievement in aviation in 1955. The citation read:

'Whitcomb's area rule is a powerful, simple, and useful method of reducing greatly the sharp increase in wing drag heretofore associated wtih transonic flight, and which constituted a major factor requiring great reserves of power to attain supersonic speeds.'

As evidence of the value of the area rule, the citation asserted that the concept was currently being used in the design of all transonic and supersonic aircraft in the United States.

Muroc

Joe Vensel, a veteran NACA test pilot, was in charge of flight operations maintenance of the aircraft and also planning and monitoring the daily flight programme at Muroc. Described as crusty but fatherly, he ruled with an iron hand and was somewhat hard of hearing from many flight hours in biplanes with open cockpits. The team of test pilots under him maintained a close link with the NACA flight test engineers, and were deeply involved in all the many flight-test reports produced by the agency.

The High-Speed Flight Test Unit — HSFTU — took over a Beechcraft UC-45 *Expeditor* transport (44-47110) from Langley during May 1947; it became 'NACA 105' in the NACA numerical system. The *Expeditor* was used as a general utility transport for the unit. On 25 August 1950, North American P-51D-25-NT *Mustang* (44-84958) 'NACA 148' was delivered in from Langley and was used for dive tests and pilot proficiency.

On 27 January 1950, the US Air Force held a special dedication ceremony at Muroc for the renaming of the desert facility to 'Edwards Air Force Base'. This ceremony symbolised the increasing emphasis being placed on flight testing, an emphasis which led up to the designation of Edwards during 1951 as the Air Force Flight Tests Centre — AFFTC — with responsibility for flight testing aircraft, operating other test facilities and providing the necessary support and services for

contractors and other government agencies — including NACA. The Korean War in the 1950s stimulated expansion at Edwards, from 795 square kilometres in 1952 to over 1,214 square kilometres by mid-1955, making it the largest flight test centre in the free world.

In nine years after 1950, the Agency at Edwards worked at an intense level and increased its complement accordingly. There were 132 personnel employed by NACA in January 1950, rising to 332 by December 1959. the budget rose from $685,000 in FY 1950 to $3.28 million by FY 1959. On 27 September 1959 Edwards was redesignated the NASA Flight

The North American F-86 Sabre was the first operational aircraft capable of attaining supersonic speeds in a dive. This F-86A-5-NA 48-291 was delivered to Ames on 29 August 1949 becoming NACA 116. It was involved in spinning tests plus prolonged dives at high speeds. It is depicted flying in the bay area near Moffett Field. (NACA A-15024)

Research Centre — FRC — making it coequal with the other NASA centres. The core of the centres — which remains today — was a complex costing $3.8 million, and on 26 June 1954 it moved from its make-do office and hangar space on the south base to the new complex. There were four branches established, later known as divisions which were administration, research, operations and instrumentation.

The Agency's flight-testing programme during the 1950s involved research on the X-series aircraft and research support on various military aircraft programmes. Three second-generation Bell X-1 aircraft were built, although four were requested in the original US Air Force contract (W33-038-AC-20062 dated 2 April 1948). These were the X-1A 48-1384; X-1B 48-1385; and X-1D 48-1386. The X-1C was cancelled prior to construction.

Skyrocket

In August 1945 the US Navy and NACA asked the Douglas Aircraft Company for initial design studies for a more advanced research aircraft capable of supersonic speed. As a result of research into captured German aerodynamic reports by Douglas engineers who had joined the US Naval Technical Mission to Europe in May 1945, the new aircraft — D-558-2 — was designed around swept wings. To supplement the insufficient thrust of available turbojets, a rocket engine with sufficient fuel for two minutes operating at 4,000lb thrust was to be installed.

On 27 January 1947, an order for three D-558-2 *Skyrockets* was finalised under Contract No. NOa(s)6850. On 10 December 1947, the first D-558-2 (s/n 6507, BuNo. 37973) was trucked to Muroc Dry Lake where its J-34-WE-40 turbo jet was fitted. It was delivered to NACA on 31 August 1951, becoming registered 'NACA 143' but was not flown by the Agency, being stored until 1954. At that time it was returned to El Segundo to be modified to the air-launch, all-rocket configuration first devised for the second D-558-2. In this form, powered solely by a 6,000lb thrust XLR-8-RM-6 rocket

When the High-Speed Flight Test Unit was established at Edwards AFB, a number of NACA aircraft were transferred from Langley including this North American P-51D-25-NT Mustang 44-84958 NACA 148. Delivered to Muroc on 25 August 1950 it was used for dive tests and pilot proficiency. It went to Langley during July 1945. (AP Photo Library)

This Republic YF-84A-5-RE Thunderjet 45-59490 later NACA 134 was one of the Langley aircraft transferred to the High-Speed Flight Station at Muroc. It served Langley between 22 August 1949 and 18 November when it was transferred. It was involved in vortex generator research until 1954. (NACA 62957)

engine, the *Skyrocket* was intended for use by NACA for testing the shape of external stores at supersonic speeds. However, this aircraft was only flown once by the Agency, on 17 September 1956, before being officially retired during March 1957.

The second D-558-2, BuNo. 37974, was also initially delivered minus its rocket engine and Douglas test pilot Gene May completed two demonstration flights for the company on 2 and 7 November 1948. The aircraft, still only powered by a J34-WE-40 turbojet, was delivered to NACA on 1 December 1948, and the NACA pilots made twenty-one turbojet-powered flights between 24 May 1949 and 6 January 1950. It was then returned to Douglas to be modified for high-speed, air-launched flights and fitted with an XLR-8-RM-6 rocket engine with additional rocket fuel tanks replacing the turbojet.

A decision to modify *Skyrockets* BuNo. 37974 (and later, BuNo. 37973) to the air-launched configuration was made by NACA and the US Navy on the basis of the flight-test results with BuNo. 37975. This aircraft, the third D-558-2, had initially been fitted with both the J-34-WE-40 turbojet and XLR-8-RM-5 rocket engine and had made its first flight, on turbojet power alone, on 8 January 1949. A little over six weeks later, on 25 February, it was used for the *Skyrocket's* first turbojet — rocket flight. Testing of the mixed-power third D-558-2 continued until November 1949 and, in the

course of its fifteenth flight in the flight trial programme with Douglas, the aircraft exceded Mach 1.0 for the first time on 24 June 1949. There were problems with the conventional take-off system as the heavily-laden aircraft full of high-explosive rocket propellants was very dangerous, Fuel burnt during the take-off and climb to altitude restricted endurance, resulting in a limited top speed of Mach 1.08 — 720mph at 40,000ft — before running out of fuel. The US Air Force's Bell X-1 had demonstrated the practicality of air-launched operations, so during November 1949 the US Navy Bureau of Aeronautics instructed Douglas to modify BuNo. 37975 to air-launch configuration with the original turbojet installation. *Skyrocket* BuNo. 37974 was modified to air-launch mode with rocket engine only. The launch aircraft selected was a US Navy-operated Boeing P2B-1S *Superfortress* BuNo. 84029 (ex-US Air Force B-29-95-BW 45-21787, later becoming 'NACA 137').

By September 1950, the third *Skyrocket*, its modifications limited to the fitting of retractable mounts for the air launch hooks, was ready to make captive flight from the fuselage of the *Superfortress*. On 8 September 1950 with test-pilot William B. Bridgeman, D-558-2 was launched at 34,850ft at a speed of 225mph from the 'mother-ship' piloted by George Jansen, for a flight using the turbojet only. Prior to handover to the Agency, three more flights were made on turbojet power and two using the turbojet-rocket combination, again piloted by Bridgeman. On 15 December 1950, the *Skyrocket* was handed over to the Agency, becoming 'NACA 145'. This third D-558-2 was flown 66 times between 22 December 1950, and 28 August 1956, by NACA, US Air Force and US Marine Corps pilots performing a variety of tests including evaluation of high-speed handling with wing slats fully or partially opened, with slats replaced with chord extension over the outer 32 per cent of the wing panels, and with external stores. At the completion of its NACA test programme the aircraft had flown a total of eighty-seven sorties.

Early in 1950, the second D-558-2 (BuNo. 37974) was extensively modified by Douglas. The original turbojet and rocket engines were removed, the air intakes and exhaust for the turbojet faired over and a 6,000lb thrust XLR-8-RM-6 four-chamber rocket engine installed. The aircraft's endurance had to be increased since it was intended for research at maximum speed, so it was fitted with additional liquid oxygen and diluted ethyl alcohol tanks giving a total capacity of these two chemicals of 345 gallons, and 378 gallons respectively, Upon completion of its modification programme, the *Skyrocket* was ferried from El Segundo to Edwards under the fuselage of the US Navy *Superfortress*. During December 1950, three unsuccessful launch attempts were made. On 26 January 1951, the pilot, Bill Bridgeman, and the D-558-2 almost came to grief on the fourth flight when a communication failure between pilot and the 'mother-ship' captain resulted in an accidental launch. Despite a drop in fuel pressure, Bridgeman climbed the aircraft to 40,000ft and then dived back to base. In the process the *Skyrocket* reached a top speed of Mach 1.28 at 38,890ft the highest Mach number it had achieved.

During six additional flights in the manufacturers' test programme Bridgeman began to explore the *Skyrocket's* maximum performance and successfully reached a number of milestones: Mach 1.79 at 60,000ft on 11 June 1951; Mach 1.85 at 63,000ft on 23 June 1951; Mach 1.88 at 66,000ft on 7 August 1951; and a maximum altitude of 74,494ft on 15 August 1951. By the time BuNo. 37974 was handed over to NACA on the last day of August 1951, it had flown the highest and the fastest. The Agency registered it 'NACA 144' and during a 63-month programme it accomplished a further 75 flights, being piloted by NACA, US Marine Corps and US Air Force pilots. Most were programmed research flights, but two resulted in records.

On 21 August 1953, Lieutenant Colonel Marion Carl, the US Marine Corps officer holding the world's speed record with the second *Skystreak*, reached an altitude of 83,235ft to set a new unofficial world record. Three months later, on 20 November, NACA pilot Scott Crossfield pushed the D-558-2 into a dive from 72,000ft and reached a record speed of Mach 2.005 — approximately 1,291mph at 62,000ft — becoming the first man to travel at twice the speed of sound. Much more useful data came from subsequent flights, but these were less spectacular and culminated on 20 December 1956, with the 105th flight of the second Douglas D-558-2 — the last *Skyrocket* to be flown.

During June 1953, Kermit E. Van Every, chief of the aerodynamic section at the Douglas El Segundo plant, launched a design study for a Mach 9.0 hypersonic research aircraft tentatively designated D-558-3. Powered by a rocket engine, its maximum design speed and altitude were respectively 8,870ft per second — approximately 6,050mph — and 750,000ft. However, lack of funds and the multi-agency agreement to pool efforts on the new North American X-15 programme resulted in cancellation of the study by the Office of US Naval Research.

The achievements of the D-558 test programme in general, and in particular those of the D-558-2, were both highly useful and truly spectacular. The *Skyrocket* programme won for Ed Heinemann, the Douglas engineer-designer, the 1951 Sylvanus Albert Reed award of the American Institute of Aeronautical Sciences, and for test pilot Bill Bridgeman the 1953 Octave Chanute award.

The X- series of research vehicles involved two categories of interesting aircraft. There were those with unique and unusual design shapes, such as the Douglas X-3 *Stiletto*; Northrop X-4 *Bantam*; Bell X-5; the Convair XF-92A and, to a lesser extent, the Douglas D-558-2 *Skyrocket* and the Bell X-2. Then there were the supersonic aerodynamic research vehicles having rocket propulsion and being air-launched from modified 'mother-ships' such as the Boeing B-29 and B-50

Superfortress. These included the advanced Bell X-1 family, the all-rocket D-558-2 and the Bell X-2. The Douglas X-3 was planned as a Mach 2 testbed but failed because of the problem of insufficient thrust developed by the two Westinghouse XJ-34-WE-17 turbojets. The earlier test aircraft were rarely flown beyond Mach 1, most being transonic in performance. It was the rocket-powered supersonic research aircraft which first exceeded Mach 2 and Mach 3.

In the mid-1950s the research aircraft programme continued to expand rapidly, with a look ahead to the future. The High-Speed Flight research involved studies of aerodynamics, stability and control, and handling qualities covering four basic wing configurations: swept-wing, semi-tailless, delta wing, the variable sweep wing, and the low-aspect-ratio thin wing. Each of the problems involved the aircraft industry, which was busily involved in the design of new combat aircraft — different in configuration and speed potential from those already in service but soon due for replacement. The pilots of the US Air Force and the US Navy needed to know and be thought of, thus any understanding of these difficulties would be welcomed as they would have to fly and possibly fight in the new designs.

It will be noted from the aircraft table that, with the exception of the Douglas X-3 *Stiletto*, each of the other four exhibited some degree of pitch-up problem, and only the D-558-2 had generally pleasant flying characteristics. The Agency made modifications to three aircraft to improve their behaviour, but due to cost and complexity considerations, such modifications were not attempted with the Douglas X-3 and Bell X-5.

Rare photo taken at Edwards AFB, depicting the second Douglas D-558-2 Skyrocket BuNo.37974 NACA 144, parked in front of the NACA Boeing B-47A Stratojet 49-1900 NACA 150. The last flight of this Skyrocket, its 105th — took place on 20 December 1956. (Boeing via Peter M. Bowers)

The NACA test pilot complement at Edwards was increased as the intensity of the research test flying programme built up. Robert A. Champline and John Griffith were joined by A. Scott Crossfield, Walter P. Jones, Joseph A. Walker, Stanley P. Burchart, John B. McKay and Neil A. Armstrong. All were qualified aerodynamic engineer-pilots with many flight hours in their log books, and all approached the research aircraft types with great caution. The following is an excerpt from a test flight report from a Bell X-5 flight carried out by Joe Walker on 31 March 1952:

'As the airplane pitches, it yaws to the right and causes the airplane to roll to the right. At this stage aileron reversal occurs; the stick jerks to the right and kicks back and forth from neutral to full right deflection if not restrained. It seems that the airplane goes longitudinally, directionally, and laterally unstable in that order.'

During one flight Joe Walker lost 6,000 metres whilst recovering from a stall; fortunately the stall had occurred at 12,000 metres.

Bell X-5

During late 1948 the programme for the Bell X-5 was initiated following many years of indigenous and foreign research into swept and variable-sweep wings. Following a thorough study of the captured German Messerschmitt P.1101 fighter, with its ground-adjustable variable-sweep capability, the Bell engineering team presented a proposal to the US Air Force calling for the design and construction of no less than twenty-four interceptor aircraft incorporating in-flight variable sweep. Unfortunately US Air Force interest was only minimal, and the evaluation by the Engineering Division of the Air Material Command proved unfavourable. Despite this setback Bell went ahead with plans for a variable-sweep-wing testbed, and on 1 February 1949, the company proposed that a research aircraft based on the design of the German P.1101 be built and flown. On 26 July 1949, the official contract for the construction of two prototype X-5 aircraft under Project No. MX-1095 was signed. The contract number was W-33-038-ac-3928.

On 20 June 1951, the Bell X-5 50-1838 made its first flight, and on 25 October 1955 NACA pilot Neil Armstrong flew the 133rd and last flight before the aircraft was retired to the USAF Museum. The Bell X-5s were the very first high performance aircraft in the world to fly while utilising a variable-geometry wing. (Office of History — USAF Edwards)

The Bell X-5 was designed to investigate the aerodynamic results, in free flight, of varying degrees of sweepback from 20° to 60°, and consequently to determine its applicability to future production aircraft. Two research aircraft were built, these being allocated US Air Force serials 50-1838 and 50-1839. Bell test pilot 'Kip' Ziegler initiated taxi runs on the first aircraft at the Bell Wheatfield plant, and during April 1951 the company made plans for the aircraft and support team to be transported to Edwards. The US Air Force conducted its engineering inspection on 22/23 May, and on 9 June this first X-5 was crated and flown by Fairchild C-119 *Flying Boxcar* to Edwards.

During the morning of 20 June 1951, Bell company test pilot Ziegler flew 50-1838 for the first time and on 21 August — shortly before contractor testing had been completed — Brigadier General Albert Boyd, Commander of the Air Force Flight Test Centre at Edwards, received permission from the authorities at Wright Field to fly the X-5. Bell released the aircraft for this one unusual evaluation flight and on 23 August became the first US Air Force pilot to fly the type. Following the conclusion of the contractors' programme in October 1951, the X-5 was grounded for installation of a NACA instrumentation pack. In December 1951 the US Air Force completed a brief evaluation programme involving six flights. As data was collected, these flights were considered part of the overall NACA effort and so went in the records as joint USAF-NACA sorties. The second Bell X-5 (50-1839) was operated only by Bell and the US Air Force and was lost in a spinning accident on 14 October 1953, killing Major Raymond Popsen, the USAF pilot.

The Bell X-5s were the very first high-performance aircraft in the world to successfully fly whilst utilizing a variable-geometry wing. The vicious spinning tendencies of the X-5 proved one of its most important contributions to the sciences of aerodynamics. The aircraft proved to be a remarkable test bed because it successfully performed research and experimental flights envisaged in the original US Air Force and NACA programme plans. It was one of the few reseach aircraft to achieve this distinction.

X-3 Stiletto

Initially conceived as a research vehicle to obtain experience for the design of future high-speed combat aircraft, the X-3 had one of the longest gestation periods of any Douglas-designed aircraft. As early as 1941, the company — at the request of the US Army Air Forces — had begun to investigate supersonic flight. On 30 December 1943, the company was specifically requested to determine the feasibility of designing an aircraft capable of flying at sustained speeds beyond Mach 1. Despite the Douglas proposal being endorsed by the USAAF on 30 May 1945, and a contract being signed on 20 June 1945 by the Secretary of War, the X-3 did not fly until 20 September 1952. The aircraft was developed under Contract No. W33-038-ac-10413 and Expenditure Order No. R-430-26, later changed to R-403-13.

The X-3 was designed to explore high speed aerodynamic phenomena to speeds of Mach 2 for sustained periods of not less than thirty minutes. It was designed to take off and land under its own power. Additionally, it was to prove the feasibility of using low aspect-ratio wings with high wing loading and the use of titanium in the construction of aircraft. Only one (49-2892) of the two X-3s originally ordered was completed. the second aircraft (49-2893) was cancelled as a result of numerous programme difficulties including limited funding, serious powerplant inadequacies and deficient performance as a result and was eventually cannibalized to furnish spare parts for the first. The aircraft was allocated the Douglas model number 499C, and its USAF project designation was MX-656. Power was to be two Westinghouse XJ-46-WE-I turbojets with afterburner, and the X-3 would be fitted with a jettisonable nose and pressurized cockpit section to facilitate emergency pilot escape at high speed.

With the engines changed to two Westinghouse XJ-34-WE-17 turbojets, 49-2892 was transferred to Edwards. Due to a number of problems involving leakages in the fuel system, flight trials were delayed a further year following completion of the aircraft. On 20 October 1952, Douglas test pilot William E. Bridgeman took the X-3 for an eighteen-minute maiden flight. Few technical problems were encountered during the flight test period but, due to insufficient thrust from the XJ-34-WE-17 turbojets, the aircraft failed to reach its design maximum speed and was found to be only slightly supersonic in a dive, reaching Mach 1.21. As little or no hope of improving the X-3's performance existed, the US Air Force cancelled the programme. After only six flights the aircraft was transferred to NACA, who completed twenty research flights (see table) between 23 August 1953 and 23 May 1956. All flights were made by the NACA test pilot Joe Walker.

It is not generally known that the X-3 contributed to the science of aircraft tyre technology. Because of its high take-off and landing speeds, it had a habit of shedding its tyres with some regularity and the tyre manufacturers went to considerable lengths to solve the problem; the techniques they evolved are heavily used today. A more significant contribution to the advancement of the state of the art was the pioneering by Douglas of fabrication and instruction techniques with titanium. it is true to say that the use of titanium alloys in parts of the X-3 pioneered use of this metal, which was (and remains) difficult to machine and form.

Convair XF-92A

NACA flight-tested the Convair XF-92A (46-682) during 1953 in a programme which followed initial tests by the manufacturer and the US Air Force between 1948 and 1953. Project pilot for the NACA flight test programme was the High speed Flight Research test pilot, A. Scott Crossfield.

The Convair 'Model 7002' research aircraft was originally built as a flying mock-up to test the delta wing configuration as a phase in the development of the XF-92 jet and rocket-powered supersonic fighter with a Mach 1.5 capability. The original contract for the XF-92 was cancelled, but the Model 7002 was later given the designation XF-92A. Full-scale wind-tunnel models were used for this unique test aircraft, and scale models were tested and flown at Wallops Island.

The aircraft's delta wing had a 60 degree sweepback on its leading edge and was fitted with elevons on its straight trailing edge. The large vertical fin and rudder gave ample directional control and stability. Power was by a single Allison J-33-A-29 turbojet equipped with afterburner. One aircraft 46-682 was completed to become the first powered delta wing aircraft to fly. A further two aircraft (46-683/684) were cancelled. Span was 31ft 3ins. Length 42ft 5ins and height 17ft 8ins. The design load weight was approximately 15,000lb.

Scott Crossfield completed the first flight of 46-682 on 9 April 1953, with familiarization on type and static longitudinal stability investigations. The second flight seven days later covered static and dynamic stability and control. Similar flights were carried out until the fourteenth flight on 3 July 1953, which was the first sortie with wing fences fitted. Two flights later, on 22 July, the new fences of a modified design buckled in flight. The next flight was aborted due to an engine malfunction and was followed by a series of tests of low-speed lateral and directional control with fences. For the twenty-fourth flight on 14 October, the latter exercise was completed minus the wing fences. On the same day, after completing a repeat flight, the nose gear collapsed during the landing. This proved to be the XF-92A's last flight, and it was subsequently retired.

Northrop X-4

For many years the Northrop Aircraft Corporation had been involved in building tail-less aircraft, so it was natural that they had been awarded the contract for the XS-4 diminutive semi-tail-less research aircraft. The US Army Air Force contract (No. W33-038-ac-14542 dated 11 June 1946) covered two aircraft. Signing of this contract had been preceded by both Wright Field and NACA being involved in studies which had concluded that a tail-less aircraft might prove significantly less prone than conventional designs to the problems of supersonic aerodynamics. These studies had developed from research in the United Kingdom which had resulted in the development and successful flight testing of the de Havilland DH.108 *Swallow* research aircraft and also from German

research that had led to the development of the extraordinary Messerschmitt Me.163 *Komet* rocket interceptor.

Research scientists believed that many of the undesirable compressibility effects experienced near Mach 1 were caused, in part, by the horizontal tail surfaces of conventionally designed aircraft. The XS-4 was designed to test the semi-tail-less configuration at transonic speeds of approximately Mach 0.88. As the XS-4 flight test programme progressed, it was discovered that its stability criteria in pitch, roll and yaw were directly relevant to the dynamics of forthcoming high-performance fighters. It was also discovered that the lift-to-drag ratio of the XS-4, when its unusual speed-brakes were utilized, could be varied to such an extent that it was possible for the aircraft to be used to simulate the landing characteristics of other machines.

The US Army Air Force, and the US Navy transonic research aircraft programme so nebulously born during 1944 eventually provided funding for several projects that were not necessarily intended to break new ground in terms of performance. Instead, these aircraft were created to serve as research tools — more to prove or disprove a theory than to expand existing performance envelopes.

On 5 April 1946, the US Army Air Forces Air Technical Service — later Air Material Command — awarded a contract to Northrop calling for the construction and preliminary flight testing of two XS-4 research aircraft under project MX-810. These were allocated the US Army Air Force serials 46-676 and 46-677. The XS-4 was one of the smallest aircraft ever built for the USAAF. It was the first US effort to explore the transonic flight characteristics of tail-less aircraft. The wing span measured only 26ft 10ins and the overall length was 23ft 4ins. Powered by two Westinghouse XJ30-WE-7 turbojets, and later two J30-WE-9 engines rated at 6,000lb thrust each, the XS-4 was thought to have sufficient power to explore the transonic speed range but not to crack the sound barrier. Because of the comparatively economical performance of these engines, the XS-4 was expected to be capable of making longer flights than its sister rocket-powered research aircraft such as the Bell XS-1.

The Douglas X-3 research aircraft was built for the US Air Force by Douglas. The X-3 48-2892 was transferred to Edwards AFB on 20 September 1952 and transferred to the NACA whose pilots made 20 research flights between 23 August 1953 and 23 May 1956. It is today preserved in the USAF Museum. (Douglas A110-15-1)

With its close-coupled dimensions and low wing loading, the XS-4 was also expected to be very manoeuvrable — an assumption which was proved to be correct during the flight test programme. Pilots equated the XS-4 with a highly responsive sports car in comparison to the other high-performance test aircraft under evaluation at that time. The XS-4 also found favour because it was equipped with extremely effective brakes on each wing panel. These constituted a powerful control during landing and in the event that a pilot had to 'back-out' of any difficulty at high speed. the aircraft featured the Northrop-originated elevons which combined the functions of elevators and ailerons as used on the company's flying-wing aircraft. All controls were power boosted.

By June 1948, the first X-4 (46-676) was complete and ready for acceptance engineering inspection. The boards' recommendations delayed the delivery from June to November with the aircraft arriving at Muroc on 15 November. On 16 December 1948, piloted by Northrop test pilot Charles Tucker, the XS-4 made its first flight. No serious problems were encountered. In the meantime the second X-4 (46-677) was delivered to Muroc. Unfortunately, winter rains flooded Rogers Dry Lake and further flight testing was delayed until April 1949. The second X-4 made its first flight, with Charles Tucker at the controls, on 7 June 1949, and it had completed six additional flights by 7 July 1949. NACA subsequently requested Air Material Command to order Northrop to install anti-spin parachutes on both aircraft. Both X-4s were unusual in having an all-white paint finish. During November 1950 the Agency initiated its flight test programme, with John Griffith making the first NACA flight on 7 November.

In order to monitor airflow across and around the second X-4, 46-677, and extensive tuft programme was undertaken. Airflow anomalies were easily observed by chase pilots, who also documented the flow patterns on film. The tufts were simply cotton string segments strategically taped in position. US Air Force pilots Chuck Yeager, Pete Everest and Richard Johnson flew the X-4, while John Griffith and A. Scott Crossfield officiated for NACA.

The Northrop X-4 Bantam 46-676 seen in unusual markings, possibly red stripes, applied in order to facilitate tracking and photography requirements. It was fitted with an anti-spin chute. This X-4 46-676 is currently on display at the USAF Academy, Colorado Springs, Colorado, whilst 46-677 is in the USAF Museum it only completing 10 flights before being grounded to provide spares for 46-676. (Office of History — USAF Edwards)

During Crossfield's first flight in the X-4 some interesting phenomena were experienced. He took off on 6 December 1950, accompanied by Pete Everest in a North American F-86 *Sabre* chase aircraft. At Mach 0.88 Crossfield reported the X-4 as 'capricious and delicate to handle.' As he increasd speed, he heard shock wave-induced sounds 'like a freight train running over a steel bridge.' Shortly afterwards, the compressibility effects became unmanageable. Crossfield quickly deployed his dive brakes to reduce speed and then realised that both engines had flamed out. Diving the X-4 to regain flying speed and help restart the engines Crossfield succeeded in relighting only one but it was enough to bring the aircraft safely back to base.

Northrop's diminutive X-4 represented the first US attempt to explore the transonic flight characteristics of tail-less aircraft. Though not designed for supersonic speeds, the X-4 nevertheless proved that conventional tail-less swept-back configurations were not truly suitable for sonic performance. Pitch, roll and yaw instability were quite noticeable at speeds of approximately Mach 0.88 and there was no conventional solution to the problem.

Both X-4 aircraft survived the flight test programme, and there was not a single serious accident during nearly four years of continuous flight testing.

Second Generation Bell X-1's

The second generation of X-1 research aircraft were created to investigate aerodynamic phenomena at speeds greater than Mach 2 and at altitudes above 90,000ft. They were to serve as the next logical step beyond the original three X-1s and would be optimized to explore the more esoteric segments of the high-Mach, high-altitude flight envelope with nearly twice the performance potential of their predecessors. The original flight programme called for the X-1A and X-1B to be used for dynamic stability and air load investigations, and for the X-1D to be used for heat transfer research. The stillborn X-1C was intended as a high-speed armament systems test bed.

On 2 April 1948, the official contract (No.W33-038-ac-20062) for all four X-1 airframes under US Air Force project No.MX-984 was signed. Only three X-1s were built, these being the X-1A (48-1384), the X-1B (48-1385), and the X-1D (48-1386). The fourth aircraft, the X-1C, would have been 48-1387. Included in the flight test programme would be research into aerodynamic heating, pilot-activated reaction control systems, supersonic armament problems and instigation of a series of miscellaneous high-speed, high-altitude aerodynamic anomalies related generally to stability and control.

The Bell X-1D was the first of the second-generation aircraft to roll out from the Buffalo, New York factory, and it made its debut at Edwards suspended from the bomb bay shackles of Boeing EB-50A *Superfortress*, 47-006A during July 1951. With Bell test pilot Jean Ziegler at the controls it was launched over Rogers Dry Lake on 24 July 1951, taking nine minutes to glide back to earth. Upon landing the nose gear failed and the X-1D slid somewhat ungracefully to a halt. Repairs took several weeks and it was mid-August before a second flight could be scheduled.

On 22 August 1951, with the X-1D attached to its B-50 'mother-ship' everything initially went as planned. But as the pair climbed through 7,000ft, Lieutenant Colonel Frank Everest, the X-1D pilot, noted upon entering the cockpit that the nitrogen source pressure indicator was giving a very low reading. After discussion with the Bell engineers, the decision was made to abort the mission and jettison the propellants in the X-1D. Shortly after initiation of the jettison process, an explosion rocked the aft end of the aircraft. This was immediately followed by flames which were visible from the chase aircraft, which was close underneath the B-50. Everest quickly vacated the X-1D cockpit and moments later Jack Ridley, an engineer aboard the B-50, pulled the handle which released the shackles holding the aircraft in place. Less than a minute later, the highly advanced research aircraft was a toasted pile of wreckage in the desert, some two miles west of the south end of Rogers Dry Lake.

The Bell X-1A (48-1384) arrived at Edwards on 7 January 1953, shackled to EB-50A 47-006A, and just over four weeks later on 14 February 1953, Jean Ziegler successfully completed the first glide flight. This was followed by a second some six days later, and by the first powered flight the following day — 21 February 1953. After contractor flights in April the aircraft was returned to the Bell factory for modifications. Following a return to Edwards on 16 October 1953, the X-1A resumed flight operations with a powered flight on November 21, in the hands of Chuck Yeager. A further flight was achieved eleven days later. On 8 December, Yeager took the X-1A out to Mach 1.9 at 60,000 ft while carefully exploring the aircraft's stability and control envelope. A further high-speed flight was scheduled for 12 December.

On the morning of the scheduled flight, the mated aircraft were fuelled and checked over. The take-off went smoothly and launch of the X-1A and ignition of its XLR11 rocket engine proved problem-free, as did the climb to pitch-over altitude. After attaining 70,000 ft Yeager levelled the aircraft and accelerated. In a matter of seconds, Mach 2 had been reached and passed, with the Machmeter continuing to move. As the X-1A passed through Mach 2.4, Yeager noticed with some concern that the aircraft suddenly had begun a gentle roll to the left. Corrective action in the form of right aileron and mild rudder damaging followed, but this only resulted in an exaggerated roll to the right. Moments later, Yeager and the aircraft were completely out of control. Though the throttle had been cut, violent tumbling followed the initial roll series and continued for no less than 36,000 ft. During the rapid descent, Yeager was thrown around in the cockpit and knocked into a state of semi-consciousness. Once the X-1A had entered denser atmosphere around 35,000 ft is stabilised in a subsonic inverted spin. Yeager recovered his senses 6,000 ft later and within seconds he had analyzed his predicament and initiated standard inverted-spin recovery procedures. The X-1A rolled upright and shortly afterwards was heading back towards Edwards some sixty miles distant.

Post-flight analysis showed that the X-1A had experienced a high-speed phenomenon known as 'roll-coupling' at 1,612 mph at an altitude of 74,200 ft. The possibility of this occurring in aircraft flying at high speeds had been predicted by a number of aerodynamicists, but Yeager's flight was the first occasion on which it had been encountered. Following this flight the US Air Force declared that no further high-speed trials above Mach 2 would be undertaken and that the X-1A would now be used to explore flight at very high altitude. As a result of this, NACA were requested to postpone the forthcoming accession so that the USAF could complete its proposed high-altitude programme. Major Arthur Murray was assigned to fly the altitude missions, and no less than fourteen flights proved necessary to accommodate programme requirements, with only four of these being successful. The highlight of the four was a record-setting flight on 26 August 1954, during which a maximum altitude of 90,440 ft was achieved. This record was retained for some two years, until Captain Iven Kincheloe in the X-2 reached 125,907 ft on 7 September 1956.

During September 1954, the US Air Force handed over the X-1A to NACA, who had the aircraft flown back to Bell for installation of an ejector seat. This was a natural reaction to the acknowledged stability and control failings of the aircraft at high Mach numbers. Several additional modifications were carried out. The X-1A was returned to Edwards during mid-1955 and was soon involved in a series of exploratory missions. The first NACA-sponsored flight, which resulted in a speed of Mach 1.45 at an altitude of 45,000 ft, took place on 20 July, with NACA pilot Joseph Walker at the controls. The second NACA flight followed on 8 August but unfortunately resulted in the loss of the aircraft and a narrow escape for Walker. An explosion had wrecked the X-1A during the climb to altitude and attempts to save the aircraft proved futile. The X-1A was eventually jettisoned in the Edwards AFB bombing range. A post-accident investigation revealed the cause of the explosion to be the incompatibility of Ulmer leather and liquid oxygen.

The third and final member of the second-generation X-1 family, the X-1B (48-1385), arrived at Edwards on 20 June 1954. By this time the X-1A had already demonstrated the type's maximum speed and altitude capabilities, so the US Air Force programme directors decided to use the X-1B primarily for pilot familiarization flights. Following these it

was to be turned over to NACA. The first X-1B glide flight, originally scheduled as a powered flight, was completed on 24 September 1954, with Lieutenant Colonel Jack Ridley in command. This was followed by flights on 6 and 8 October, and by the first familiarization flight on 13 October. Five additional 'famil' flights took place during the following six weeks, these ending with two flights piloted by Lieutenant Colonel Frank Everest who was programmed to fly the Bell X-2.

On 3 December 1954, the US Air Force turned over the X-1B to NACA. Shortly afterwards it was loaded on board Boeing B-29-96-BO *Superfortress* 45-21800 and flown to the Langley facility for the installation of test instrumentation. The original carrier (an EB-50A-5-BO, 46-006) had been lost with the third X-1. The X-1B stayed at Langley for almost eight months and on 1 August 1955 it was returned to Edwards. Renewed flight testing of the aircraft, with an initial sortie to verify the X-1B's airworthiness following the modifications, took place on 14 August. John McKay, a NACA test pilot was in the cockpit, and he was to remain the assigned pilot for the following twelve missions. Eventually he would pass the reins to a then little-known NACA test pilot named Neil Armstrong. The last four X-1B missions were flown by the latter and he had the honour of making the last landing ever in a second-generation Bell X-1.

Most of the NACA X-1B flights up to this point had been conducted for the purpose of aerodynamic heating research. Thermal sensors and associated recorders had been installed at Langley, and the several flights completed between August 1956, and July 1957 had been primarily to accumulate data in this part of the flight envelope. The last three X-1B missions were flown with extended wingtips and a rudimentary hydrogen peroxide-fuelled reaction control system. The latter was never actually used in exo-atmospheric flight, but the technology proved of great value in designing a similar system for the forthcoming North American X-15.

The Northrop X-4 research vehicle was one of the smallest aircraft ever built. The underlying idea was that the diminutive airframe would be most sensitive to the slightest aerodynamic reaction at transonic speeds. Depicted is the second X-4 46-677 with flight calibration markings on the fuselage.
(Office of History — USAF Edwards)

Following the completion of the X-1B's seventeenth flight by NACA, a decision was made to ground the aircraft temporarily in order to install a small set of ventral fins to improve directional stability at high speeds and altitude and to equip it with a new XLR11 rocket engine. Unfortunately, during an inspection conducted whilst the aircraft was grounded, fatigue cracks were discovered in the X-1B's liquid oxygen tank. An attempt at welding the cracks failed, and during June 1958 a decision was made to cancel the remainder of the flight test programme. Following this decision, the reaction control system was removed and installed in a US Air Force Lockheed NF-104A *Starfighter*. On 27 January 1958 the X-1B was turned over to the USAF Museum located at Wright-Patterson AFB, Ohio, for permanent preservation.

Bell X-1E

The Bell X-1E was born out of a requirement to explore potential performance improvements made possible by the use of extremely thin airfoil-section wings and the incorporation of a significantly more efficient engine turbopump. With these changes the aircraft was expected to achieve speeds of Mach 2.5. Only one X-1E was built and it was a modification of the second first-generation X-1 46-063. A follow-on contract for this aircraft was signed during April 1952, the contract number being W33-038-ac-9183. The X-1E was produced as much out of desperation as legitimate need. During early 1951 the second X-1 was still flying for NACA. The US Air Force, expecting the new X-1D to arrive at any time, retired the first X-1 to the Smithsonian Institution. the Agency was also expecting to receive a new aircraft, the third X-1, with its new turbopump and increased fuel capacity. the arrival of this aircraft would allow NACA to retire the second X-1.

The Bell X-1D 48-1386 was jettisoned to destruction following an inflight explosion and fire on 22 August 1951 over Edwards AFB, having completed only one glide flight prior to its loss. The bomb bay dimensions of the B-29 'mother-ship' played a critical part in the design of this second-generation research aircraft. (Bell Aerospace Textron)

All these plans fell by the wayside when the X-1D and the third X-1 were destroyed in accidents before their respective flight test programmes could be completed. To compound NACA's problems, it was discovered during 1951 that the high-pressure nitrogen spheres in the second X-1 were prone to explode after 700 to 800 cycles. One fill-up and emptying was considered a cycle. In the hope of correcting the problem, three nitrogen spheres were removed from the first X-1 residing in Washington, D.C. for testing but two of these burst after a few hundred cycles.

There were now no flyable X-1 aircraft. A hurried decision to assure the Agency's continued participation in high-speed research produced a programme to modify the second X-1 into what was effectively a new aircraft. The second X-1 had been retired from the NACA high-speed flight stable following the fifty-fourth and last NACA mission on 23 October 1954. During March 1954 the revamped X-1 was officially designated X-1E, and by mid-1955, most of the modification work — which included the new cockpit and canopy configuration — required to accommodate the NACA-specified ejection seat and the new wing had been completed at the NACA facility at Edwards. Several months were involved in ground-checking the aircraft, and by late November it had been cleared for flight test.

Following an abortive first launch attempt on 3 December 1955, the first X-1E glide flight was completed successfully on 15 December, with NACA test pilot Joseph Walker in the cockpit; he was to remain the pilot for the X-1E, flying a total of twenty missions. With Walker at the controls, the X-1E explored its flight envelope in a series of test flights. On 7 June 1956 the aircraft reached a speed of Mach 1.55. This was the first X-1E flight over 1,000mph and the aircraft's first supersonic flight since modification. Additional sorties culminated in the first Mach 2 flight on 31 August 1956, and a maximum speed flight of Mach 2.24 — approximately 1,480mph — on 8 October 1957. Following the installation of twin ventral fins during December 1957, the X-1E was again declared ready for flight. On 14 May 1958, it successfully completed its eighteenth mission.

A minor landing accident following a flight on 10 June 1958 gave the NACA engineers a chance to incorporate a performance-improving engine modification, allowing engine combustion-chamber pressures to be increased from 250psi to 300psi. This — coupled with an experimental and significantly more powerful propellant known as 'Hidyne' or 'U-deta' — 60 per cent unsymmestrical dimethylhydrazine and 40 per cent diethylene triamine — was expected to give the X-1E a near-Mach 3 speed potential. On 17 September, Walker made his last X-1 flight, handing over the X-1E programme to NACA test pilot John McKay two days later. McKay successfully completed the remaining four X-1E flights after which the aircraft was grounded, this time for replacement of the pilot emergency ejection system. During the grounding, X-ray inspections of the fuel and oxidizer tanks were undertaken. When the negatives were returned from the laboratory they revealed a serious crack in the fuel tank. This, coupled with the imminent arrival of the new North American X-15, resulted in a final decision to terminate X-1E flight-test work.

Summary of X-Series Research Aircraft

Aircraft	Configuration	Propulsion	Sponsor	Project No.	Year	First Flight	No.	Lost	Fatalities	Launch system
Bell XS-1 (X-1)	Straight wing	Rocket engine	USAAF	MX-653	1945	1946	3	1	0	Air drop
Bell X-1 (Advanced)	Straight wing	Rocket engine	USAF	MX-984	1948	1953	3	2	0	Air drop
Bell XS-2 (X-2)	Swept wing	Rocket engine	USAAF	MX-743	1945	1955	2	2	2	Air drop
Bell XS-3 (X-3)	Straight wing (low aspect ratio)	Two turbojets	USAAF	MX-656	1945	1952	1	0	0	Ground
Northrop XS-4 (X-4)	Swept semi tail-less	Two turbojets	USAAF	MX-810	1946	1948	2	0	0	Ground
Bell X-5	Variable wing sweeping	One turbojet	USAF	MX-1095	1949	1951	2	1	1	Ground
Convair XF-92A	Delta wing	One turbojet	USAF	MX-813	1946	1948	1	0	0	Ground
Douglas D-558-1	Straight wing	One turbojet	US Navy	Nil	1945	1947	3	1	1	Ground
Douglas D-558-2	Swept wing	Jet + rocket	US Navy	Nil	1945	1948	3	0	0	Both

Second X-1 subsequently became the X-1E. X-1A, X-1B, X-1D only — X-1C cancelled. Pilot plus one crewman lost with *Bell XS-2*.
One *Bell X-3* cancelled when almost complete.
Douglas D-558-II — first with jet, then first and second modified to all rocket. Designed for mixed jet and rocket propulsion.
Douglas D-558-II — All-rocket first and second air launched from modified Boeing B-29. Third aircraft could operate from ground or be air launched.

The huge dimension of the Boeing NB-52B 52-0008 with 185 ft wing-span, are apparent in this photo showing the 50 ft long North American X-15 fitted under the starboard wing. This Stratofortress is in service with the NASA today at the Ames-Dryden Flight Research Facility at Edwards AFB.
(Office of History — USAF Edwards)

Northrop M2-F3 NASA 803 Lifting-body seen ready for launch from the Boeing NB-52B 52-008. Flown 1966–67. (Office of History — C-453-70 — USAF Edwards)

The Bell XV-15A NASA 702 Tiltrotor is leased to the manufacturer for transportation studies. It has been demonstrated at the White House in Washington DC. It is depicted in flight over New York landmarks and marked NASA-Army-Navy and is finished in a new livery. (Bell Helicopter Textron 027402)

This unique Ryan VZ-3 Model 72 Vertiplane VSTOL research aircraft 56-6941 was developed under a joint US Navy/Army programme. It was powered by two 1,000 shp T-53-L-1 engines. Originally it had a tailwheel undercarriage, modified in 1958 with a nose-wheel, ventral fin and other modifications. Following a crash in February 1959 it was rebuilt for NASA as 'NASA 235' with a lengthened fuselage and other changes. (AP Photo Library)

Sikorsky Model S.69 high-speed research helicopter, powered by a PT6T-3 turbo twin pac, driving three-bladed coaxial rotors. A joint US armed services and NASA project, the XH-59A 71-1472 was used by the US Army for flight testing the advancing blade concept rotor system. Depicted is 73-21942 the second YH-59A-SI which incorporated several changes. (NASA 79-H-212)

This Bell UH-1B NASA 415 ex-63-8695 on floats provides a search and rescue capability for the aeronautical research programme conducted at the Wallops Flight Facility. It is also used as an application research platform and is current on the NASA inventory. (NASA WI 82-231-2)

The Northrop X-4 46-676 seen shortly after its roll-out at the Hawthorne facility late 1948. Markings were limited to the national star and bar, serial number, and USAF on the wings. An overall white finish was adopted. Seven NACA pilots flew the X-4, with Scott Crossfield completing 29 flights. (Northrop 68-04691)

Fitted with a modified raised cockpit with a flat V windshield, and the height of the fin increased by 14 inches, Douglas D-558-2 Skyrocket BuNo.37973 is seen shortly after convential take-off from Muroc. It was later converted for air-launch and all-rocket engine configuration. It later became NACA 143 and today is with the Planes of Fame Museum, Chino, California. (Douglas DAC 12439)

The X-1E was the first aircraft in the world to fly supersonically with a four per cent thickness-chord-ratio wing, and the first to prove that high-Mach capability and adequate stability were possible using such thin airfoil sections. In designing the four per cent wing Stanley Aviation pioneered a number of construction techniques, not the least of which was an extremely thin, multi-spar main load-bearing aerofoil structure. NACA and Reaction Motors Inc. also pioneered an improved power-plant propellant system with the development of their low-pressure turbopump for the XLR11 variant utilised as the engine for the Bell X-1E. When the X-1 was modified to X-1E configuration, the new and totally revised stepped windscreen and hard canopy were in fact modelled after those of the aesthetically pleasing Douglas D-558-2 *Skyrocket*. The ejector seat fitted was a surplus from the second Northrop X-4, 46-677.

On 14 October 1947 the Bell Aircraft Corporation-built X-1 experimental transonic research aircraft flew into history. With Charles E. 'Chuck' Yeager at the controls, it became the first manned aircraft ever to penetrate the mysterious compressibility 'barrier' and travel at a speed literally faster than that of sound. Somewhat less auspiciously, the X-1 heralded the arrival of the X-plane family — a highly varied collection of research and test aircraft which soon would prove to be perhaps the most valuable research tools in all aerospace history.

NACA Test Pilot Flight Record

Early Research Aircraft

Pilot's name	X-1 46-063	X-1 46-064	X-1A 48-1384	X-1B 48-1385	X-1E 46-063	X-3 49-2893	X-4 46-677	X-5 50-1838	D-558-1 37971	D-558-1 37972	D-558-2 37973	D-558-2 37974	XF-92A 46-682
Armstrong, Neil A.				4				1					
Butchart, Stanley P.							4	13			12		
Cannon, Joseph		1											
Champine, Robert	13									9		12	
Cooper, George							1						
Crossfield, A. Scott	10						29	10			13	53	25
Griffith, John	9						7				15	9	
Hoover, Herbert	14												
Jones, Walter, P.							14	8		5			
Lilly, Howard	6								19				
McKay, John B.				13	5		1	6		8	1	12	
Reeder, John P.								1					
Walker, Joseph A.	2		1		21	20	2	78		14		2	

Note: First flight of the Bell X-1 46-064 took place on 20 July 1951. Glide flight and nose wheel collapsed on landing. On 9 November during preparations for captive flight with Boeing B-50A 46-006, destroyed in postflight explosion. Bell X-1 lost plus B-50 and Cannon injured.
George Cooper made one NACA pilot check during 1953. Cooper was based at Ames. Flew X-4.
John P. Reeder made one NACA pilot check in the X-5. Reeder was based at Langley. 16 July 1952.

NACA Research Aircraft

Aircraft	Configuration	Speed	Flight	Research problem	Behaviour problem
Douglas D-558-2 BuNo. 37975	Swept-wing	Mach 1.0	66	Swept-wing pitch-up during manoeuvring	
Northrop X-4 AF 46-677	Semi-tail-less	Mach 0.9	82	Pitching oscillation of increasing severity approaching Mach 0.95	Poorly damped 'hunting' motion about all three axes: 'Washboard road' motion
Convair XF-92A AF 46-682	Delta-wing	Mach 0.9+	25	Delta pitch-up during manoeuvring	Sluggish and underpowered
Bell X-5 AF 50-1838	Variable-sweep	Mach 0.9+	133	Unacceptable stall-spin behaviour: swept wing pitch-up during manoeuvring	Dangerous stall approach and spin tendencies
Douglas X-3 AF 49-2892	Low aspect-ratio thin-wing	Mach 0.95	20	Coupled motion instability during abrupt rolling manoeuvres	Sluggish and very underpowered

Modifications: *Douglas D-558-2* BuNo. 37975. Various wing slat and wing fence combinations, leading edge extensions.
Northrop X-4 AF 46-677. Increasing the thickness of the trailing edge of the wing and elevon.
Convair XF-92A AF 46-682. Various combinations of wing fences.

CHAPTER EIGHT
Advanced Research Aircraft

On 24 June 1952, a NACA committee resolved that the agency should expand its research aircraft programme in order to explore flight characteristics of atmospheric and exo-atmospheric designs capable of speeds of up to Mach 4 and Mach 10 at extending high altitudes. Earlier that year the NACA research engineers at Langley, Wallops Island, Ames and Edwards had studied papers which embodied a proposal for a large supersonic aircraft which would launch, at Mach 3, a small manned second-stage vehicle to accelerate to hypersonic speeds. Another proposal for a hypersonic test vehicle was to equip the Bell X-2 research aircraft with reaction controls and add two jettisonable solid rocket motors as boosters. With such booster rockets, it was claimed, an air-launched X-2 could be flown at a speed of about Mach 4.5 to orbital altitude. This latter proposal was accepted as the more practical. In October 1953, NACA announced that the time was ripe for looking into the feasibility of procuring a manned hypersonic research aircraft.

At a meeting held in Washington, D.C. on 4/5 February 1954, a NACA panel declared that the X-2 was too small to use for hypersonic research. What was needed, it said, was a completely new and larger vehicle built specifically for hypersonic research extending into the upper atmosphere and into space itself. NACA headquarters told its laboratories to explore the requirements of a possible hypersonic research aircraft, and to form a special group of researchers to investigate different systems.

Historical line-up of NACA research aircraft taken at the Flight Research Centre, Edwards AFB, depicting four models involved in high-speed research. The six aircraft include the Bell X-1 6063, Northrop X-4 Bantam 6676 and 6677, Douglas D-558-1 BuNo.37970 and 37971 plus a Douglas D-558-2 Skyrocket.
(Office of History — USAF Edwards)

The Ames group concerned itself solely with sub-orbital long-range flight and ended up favouring a military-type air-breathing (rather than rocket-powered) aircraft in the Mach 4/5 range. The High-Speed Flight Station at Edwards suggested a larger, higher-powered conventional configuration generally similar to the Bell X-1 or Douglas D-558-1 research aircraft with which it was familiar. The staff at the Lewis Flight Propulsion Laboratory at Cleveland, Ohio, recommended against a new manned research aircraft, arguing that hypersonic research could and should be done by expanding the Wallops Island rocket-model technique. It reminded NACA that previous aircraft programmes had been unduly burdened by anticipated military applications.

The conclusions of Langley's hypersonic aircraft study group of 1954 differed substantially from those voiced at Ames, Edwards and Lewis. The original intention of the Langley group was to determine the feasibility of a hypersonic aircraft capable of a two- to three-minute excursion out of the atmosphere into space. The idea was to create a brief period of weightlessness in order to explore its effects on space flight. Chairman of the Langley group was John V. Becker, chief of the Compressibility Research Division and principal designer of the pilot 11-inch hypersonic tunnel. The group's members included flight test pilot James B. Whitten.

By the end of April 1954 the Becker group had completed a tentative design of the winged aircraft it had in mind, as well as an outline of proposed experiments. In the absence of the rapid development of a major new engine, propulsion to hypersonic speed was to be provided by a combination of three or four smaller rocket motors. Launch of the aircraft would be by the proven air-drop method developed at the High-Speed Flight Station for the X-1 and refined during the flight test programmes of the subsequent research aircraft.

At this stage John Becker and Peter F. Korycinski from the Compressibility Research Division ran head-on into a major technical problem. At Mach 7, the critical speed on return from orbit, re-entry at low angles of attack appeared impossible because of disastrous heating loads. Renewed tests in the 11-inch hypersonic tunnel provided the engineers with at least one clue to the nature of this problem. If the proposed hypersonic vehicles' angle of attack and associated drag were increased, deceleration would begin at a higher altitude. Slowing down into the thinner, low-density, atmosphere would make the heat transfer problems much less severe. Throughout 1954 the heating problems of high-lift, high-drag re-entry required considerable attention from researchers at Langley. Becker's group came up with a vehicle concept which was really little more than an object of about the right general proportions and the right propulsive characteristics to achieve hypersonic performance. No one knew whether any structure could be devised which could survive the anticipated air temperature — estimated at 4000°F — affecting a winged vehicle during re-entry. To make matters even worse, there were still the problems of stability and control of such a vehicle.

Langley had been forewarned by the NACA High-Speed Flight Station of the difficult problems of hypersonic stability. The X-1A had been pushed far beyond its normal transonic speed range to about Mach 2.5 in December 1953. In September 1956, US Air Force test pilot Captain Milburn G. Apt was killed in the Bell X-2 over the Mohave desert. The cause of this tragedy was similar to that of Chuck Yeager's near-disaster in the X-1A. The stability problems facing the Becker group were even more severe than those of the Bell X-1A since the new vehicle was designed for Mach 7. By late 1954 the laboratory had an engineering answer — location of the horizontal tail somewhere other than far above or well below plane of the wing. In the final design of the X-15, North American would place the horizontal tail just slightly below the wing plane. One of several generic X-15 configuration tests in the wind tunnel at Langley had a conventional cruciform-style horizontal and vertical tail surface — others utilised a less orthodox X-type configuration.

Finally, at the end of June 1954, the Becker Group reached a stage where it felt it could produce a convincing case for the feasibility of a Mach 7 research aircraft. The NACA headquarters agreed that it was time for the US armed forces to receive a unified presentation. Representatives of the US Navy and the Scientific Advisory Board of the US Air Force assembled at NACA in Washington, D.C., on 9 July 1954, for what became the first of many presentations on the possible new research vehicle. Three months later, on 5 October, the NACA Aerodynamics Committee met in executive session at the High-Speed Flight Station hoping to come to a final decision about the desirability of a manned hypersonic research aircraft.

During December 1954, NACA, the US Air Force and the US Navy signed a 'memorandum of understanding' setting up a new 'research airplane committee' to assume responsibility for technical direction of the 'X-15' project. On 30 December, the Air Material Command invited aircraft manufacturers to a bidders' briefing to be held at Wright-Patterson AFB on 18 January 1955. At this briefing the Agency and the armed forces informed potential contractors of the design and operational requirements of the hypersonic aircraft.

North American Aviation was awarded a contract in September 1955 to develop the X-15, and another in June 1956 to build three prototypes. The Reaction Motors Division of Thiokol Chemical Corporation received a contract for development and production of the rocket engines. The original X-15 (56-6670) would make its first flight, a powerless glide, on 8 June 1959, with Scott Crossfield in the cockpit, eleven months after dissolution of NACA and the establishment of NASA. The flight test programme for the X-15 would begin in March 1960, and one of the first NASA pilots

to fly the aircraft was one Neil Armstrong — who, within the decade, would be the first man to walk on the moon.

At the bidders' briefing, representatives from the Bell Aircraft Corporation., Boeing Airplane Co., Chance Vought Aircraft Inc., Consolidated Vultee Aircraft Corp. (Convair), the Douglas Aircraft Company, McDonnell Aircraft Corp., North American Aviation Inc., and Republic Aviation Corp., met the NACA and US Air Force personnel to discuss the competition and the basic design requirements for Project 1226 — to be designated X-15. Another briefing took place with the four prospective propulsion system manufacturers — Aerojet, General Electric, North American and Reaction Motors — on 4 February.

Only the first three of nine North American YF-107A aircraft ordered were completed. On 11 June 1957 this JF-107A 55-5118 went to the NACA at Edwards AFB becoming NACA 207. After only four flights it was grounded, being replaced by 55-5120. (Office of History — USAF Edwards)

The airframe manufacturers were requested to design their aircraft around any one of the following rocket engines: the Bell XLR-81, the Aerojet XLR-73, the Reaction Motors Inc. XLR-10 or XLR-30, or the North American NA 5400. Engine choice would be up to the airframe manufacturer, and if the NACA service panel chose another, this would not effect the final design evaluation. Deadline for entries was 9 May 1955. Between February and May, five of the intitial nine airframe contenders dropped out of the competition. Those remaining were Bell, Douglas, North American, and Republic. During this period, representatives from these companies met NACA personnel on numerous occasions at the different laboratories and reviewed technical information on various aspects of the forthcoming X-15 design. The Agency provided the contractors with further information gained as a result of wind-tunnel tests in the Ames 10 × 14in supersonic tunnel and the Langley Mach 4 blowdown jet tunnel. By 9 May 1955 all the Project 1226 bidding contractors had submitted their proposals. Bell submitted their Model D-171; Douglas their Model 684, and Republic their Model AP-76. Finally North American submitted their design, marked 'Secret' and numbered ESO 7487. Between 26 and 28 July 1955, the US Air Force, the US Navy and the NACA evaluation teams met to co-ordinate their separate results. The Agency concluded that the North American proposal met their requirements but if the US Air Force found it to be unsuitable, a Douglas design would be acceptable. The US Air Force agreed completely with NACA's position whilst the US Navy decided not to be put in the situation of casting the dissenting vote and finally accepted the decision of the US Air Force and NACA. As a result the three agencies designated North American as the prime contractor in the Project 1226 competition.

Unfortunately on 23 August 1955 the North American office in Dayton, Ohio, verbally informed the US Air Force that the company wished to withdraw its Project 1226 bid because the expenditure of engineering manhours on other projects would not allow the company to meet the desired deadline. Seven days later North American sent a letter to the US Air Force requesting that the company be allowed to withdraw from Project 1226. The second-place bidder was the Douglas Aircraft Company, but the feeling was that rather than award the contract to Douglas, the competition should be reopened. On 23 September, R. H. Rice, Vice-President and Chief Engineer of North American, wrote explaining the reason for the company withdrawing from the competition but submitted proposals for two alternative courses of action, one of which was to extend the schedule over an additional eight-month period. On 6 December 1955, North American Aviation's Los Angeles, California division received a letter contract from the US Air Force (AF 33(600)-31693), calling for the design, construction and development of three X-15 aircraft.

The unique McDonnell XP-85 Goblin 48-3886 was tested at Ames in March 1946 and again in January 1948 being installed in the 40 × 80 tunnel. It was a single-seat parasite fighter designed to be carried by a Convair B-36 bomber. Only two were built. (NACA A-12703)

Flight Test

NACA had a long history of involvement with the armed services and other US government agencies in research into new aircraft development projects. As early as the 1920s it was not uncommon for the NACA to participate in development projects, or to participate in flight tests of military aircraft, and or on occasion to test-fly for a contractor. For example, Bill McAvoy from Langley flew some of the hazardous spin and dive tests on the Grumman XF3F for both the manufacturer and the US Navy. During World War II, the Agency was heavily involved in many flight investigations related to improving the combat potential of US military aircraft such as the Republic P-47 *Thunderbolt* fighter, the Curtiss *Helldiver* dive-bomber and the North American P-51 *Mustang*. These aircraft and many others, were extensively studied in the NACA wind tunnels — a support which became tradition.

This co-operation role of support continued after World War II, encouraged by several factors including an official US Air Force policy of 'concurrency' testing whereby a large number of production-model aircraft were tested at the NACA laboratories and out-stations scattered around the

United States. This accelerated the testing process, and reduced the chances of encountering problems which could arise should an experimental design be committed to production on the basis of merely a few tests on the prototype(s). With the advent of a new generation of transonic and supersonic fighters and bomber aircraft, there was a pressing need to deliver pre-production or early production models to NACA for evaluation and research into possible defects. As will be seen from the aircraft listing at the end of this volume, NACA either returned the aircraft or retained them in Agency service, often flying them on a variety of research opportunity tasks for many years. The majority of service aircraft appeared at the High-Speed Flight Station at Edwards, as this was the base for the US Air Force Flight Test Centre.

As early as 1950, the Agency at Edwards had participated in a US Air Force development programme in which engineers provided assistance with the XF-91 experimental interceptor. This was a radical departure from conventional design, being a single-seat supersonic interceptor powered by a 5,200lb static-thrust J-47-GE-3 and a 6,000lb st XLR11-RM-9 rocket engine. It had an inversely-tapered wing planform with broader chord at the wingtips than at the roots. Two aircraft were ordered, one being tested with a V-type tail assembly. In 1952 the Agency made a major contribution to saving a military aircraft which was in serious trouble — the Northrop F-89 *Scorpion* tandem two-seat all-weather jet fighter, which was required for a high-priority air defence programme. In the early months of 1952, six F-89s lost their wings in flight. With more than a thousand on order the US Air Force faced a serious crisis. The Agency at Edwards loaned a team to Northrop, who installed strain gauges on an experimental F-89 and then studies the data acquired from test flights. A serious weakness was discovered in the wing structure, which was re-designed to strengthen it. The *Scorpion* subsequently went on to a long and useful service career. The NACA assistance on this programme enhanced the reputation of the agency among the armed services and the flight testing community within industry.

The F-89 *Scorpion* problem was followed by a major investigation involving another US Air Force aircraft, the Boeing B-47 *Stratojet* bomber. However, the B-47 was not in any difficulty — NACA had requested the loan of one aircraft to study aerolastic wing flexing. The B-47, a six-jet strategic medium bomber, was a shoulder-wing monoplane with a very thin swept wing. An aircraft with such a wing could display peculiar aerodynamic and a structural load behaviour as a result of interaction between wing and tail deflections and transonic airflow changes. Two of the NACA laboratories were very interested in the B-47: Langley wished to study the effect of aeroelasticity upon structural loads, while Ames wanted to study the relationship between aeroelasticity and dynamic stability. Because of runway length restrictions the operation of the B-47 from either centre was marginal so a Boeing B-47A-BW (49-1900) was delivered to Edwards during May 1953, where it flew until 1957. Flight testing revealed some serious design deficiences, including buffeting problems which restricted the bomber to speeds no greater than Mach 0.8. NACA requested that the US Air Force loan them a Boeing B-52 bomber as soon as possible, so that research could be extended through Mach 0.9+ and up to 15,000 metres in altitude. The Agency never did receive the B-52, but secured permission to install instrumentation in a B-52 *Stratofortress* being flown by Boeing.

During 1957 the High-Speed Flight Station continued large jet-aircraft studies using a Boeing KC-135 tanker loaned by the US Air Force. However, flight tests were suspended after a near-disastrous mid-air collision between the tanker and a jet trainer from the US Air Force Test Pilots School at Edwards. The KC-135, piloted by NACA's Stan Burchart, made a forced landing on Rogers Lake but the trainer — whose civilian pilot instructor apparently never saw the tanker — crashed, killing the student. The US Air Force delivered a second Boeing KC-135A *Stratotanker* (55-3124) on a ninety-day loan, and this aircraft completed a number of useful flights before being returned to the US Air Force in 1958. The NACA research involved high-altitude cruise performance and landing approaches, including instrument approaches, during which the effect of the wing spoilers on the glide path during the approach were recorded.

The Century Series

The introduction of the 'Century' series of fighters and interceptor aircraft into US Air Force service involved support from the NACA High-Speed Flight Station and its major flight testing facilities. Initially the North American F-100 *Super Sabre*, the Convair F-102 *Delta Dagger* and the Lockheed F-104 *Starfighter* went to NACA in support of early armed services development with the types. The McDonnell F-101 *Voodoo* and Republic F-105 *Thunderchief* also appeared briefly, in order for the NACA flight test team to familiarize themselves with the characteristics of the aircraft. Mention must be made of the Republic F-103 which was a projected Mach 3 interceptor to have been powered by a 2,000lb static thrust YJ67-W-3 turbojet and a 37,400lb XRJ55-W-1 ramjet. The initial design had a flush cockpit and a periscope for the pilot. Armament would have consisted of six GAR-Falcon missiles in retractable fuselage side bays and 36 2.75-inch rockets, with a later provision for two GAR-1 and two GAR-3 missiles, the latter with nuclear warheads. It never left the design stage, but nevertheless was an intriguing project. The North American F-107 programme was an abortive attempt to develop a Mach 2 fighter-bomber, replaced by the F-105 which received a large production order. NACA acquired two of the three F-107 aircraft built, the first and third, which proved most useful in a study of design features in support of the forthcoming North American XB-70 and X-15 programmes. The F-107 was built as an all-weather interceptor/fighter-bomber version of the F-100, with a 23,500lb st YJ75-P-9 turbojet fed by an air intake above and behind the cockpit. There was provision for four 20mm cannon in the nose and up to 1,000lb of underwing stores. The wing had 45° sweepback. Nine were ordered but only the first three YF-107As were completed, flying during the autumn of 1956.

The first of the 'Century' series of fighters and the world's first operational fighter capable of supersonic performance was the F-100. The first YF-100A (52-5754), powered by a XJ57-P-7 engine, was first flown by George Welch at Edwards AFB on 25 May 1953. George flew the first production F-100A (52-5756) on 29 October 1953 on the same day as the YF-100A set a world speed record of 755.149mph.

The McDonnell F-101 *Voodoo* was designed to serve with the US Strategic Air Command as a long-range escort and pentration fighter, but was subsequently developed for both tactical and air defence roles. At the time of its introduction into US Air Force service it was the heaviest single-seat fighter ever accepted.

Convair's F-102 *Delta Dagger* was the first delta-wing aircraft in the US Air Force inventory. The design, commenced in 1950, was closely related to the experimental XF-92A, which Convair had built in 1948 to provide data for the proposed F-92 Mach 1.5 fighter designed in consulation with Dr Alexander Lippisch, whose research on delta wing designs in Germany was well known. The first YF-102 (52-7994) flew at Edwards AFB on 24 October 1953 and the second on 11 January 1954, both powered by the J57-P-11 turbojet. These aircraft were deficient in performance and Convair embarked on a major investigation and redesign programme. The latter occupied only 117 days, and on 20 December 1954 the first YF-102A made its maiden flight. The most prominent feature was the longer fuselage with a 'coke-bottle' waist — the first application of area rule developed by Richard Whitcomb of the NACA. In the redesign, the leading edge of the wing was cambered, the cockpit canopy was redesigned and a J57-P-23 turbojet fitted.

The Lockheed F-104 *Starfighter* was the first operational interceptor capable of sustained speeds above Mach 2, and the first aircraft to hold both the world speed and altitude records for aircraft. Design began in November 1952 and the US Air Force ordered two prototypes in March 1953. The design concept was radical, with a long, needle-nose fuselage tightly tailored around a single large turbojet. It had small thin wings with no sweepback and a T-tail. With a span of only 21ft 11in the *Starfighter* was one of the smallest aircraft ever produced for the US Air Force. Tony LeVier, Lockheed's test pilot flew the first XF-104 (53-7786) on 7 February 1954. This and the second prototype were powered by the 10,000lb static thrust afterburning Wright XJ65-W-6 turbojet. With this engine an XF-104 reached Mach 1.79 on 25 March 1955. Mach 2 was first reached in a YF-104A piloted by Joe Ozier on 27 April 1955. The high-altitude capabilities of the *Starfighter* were further enhanced in 1963 when three F-104A aircraft were modified to NF-104A configuration. This involved the addition of a 6,000lb st Rocketdyne AR-2 rocket engine fitted in the tail above the jet pipe, a larger fin and rudder, extended wing tips and reaction jet controls at the nose, tail and wing tips. These NF-104As were capable of reaching altitudes up to 130,000ft and were used by NASA and the US Air Force Aerospace Research Pilots School based at Edwards AFB. The NASA aircraft were later involved in astronaut training.

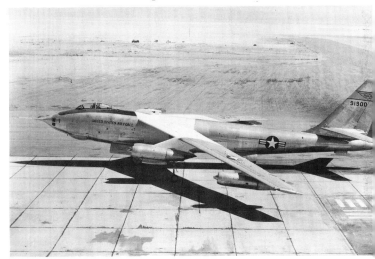

Boeing B-47A 49-1900 was delivered to the Flight Research Centre at Edwards in May 1953 where it flew until 1957. It was actually operated by the NACA between 11 July 1952 and 28 February 1958. The NACA at both Ames and Langley were interested in studies involving aeroelastic wing flexing, and it was runway lengths that dictated it be based at Edwards. It later became NACA 150. (NACA A-19582)

First contracts for the Republic F-105 *Thunderchief* were placed during 1954 and it was the first supersonic tactical fighter-bomber developed from scratch. The wing was moderately highly swept and incorporated low-speed ailerons and high-speed spoilers for lateral control together with a droop-snoot leading edge. An unusual feature of the design was the arrangement of the speed brakes as four segments of the rear jet-pipe fairing. With a gross weight of approximately 45,000lb, the aircraft required a powerful engine. The first two aircraft were propelled by the J57-P-25, later F-105s being powered by the J75-P-3. The first *Thunderchief* (designated YF-105A, 54-0098) first flew on 22 October 1955, piloted by Russell M. Roth, and exceeded Mach 1.

As a result of the Air Force's procurement policy stressing 'concurrency' testing in the early 1950s, the NACA received prototypes of new service aircraft for its own evaluations in support of the US Air Force development effort. It was assumed that if a design appeared to warrant production, a number of prototype aircraft should be built and tested extensively, and any changes incorporated on newly-emerging production aircraft. This avoided the time delays which would be expected if only a handful of prototypes were built. It was an expensive exercise, but concurrency testing meant aircraft could be assigned to weapon, engine, system, and flight aerodynamic testing. During 1954 the first two of the 'Century' series aircraft arrived at the NACA High-Speed Station Flight; the North American F-100A *Super Sabre* (52-5778) and the Convair YF-102 *Delta Dagger* (53-1785). Whilst NACA was interested in using the YF-102 to extend the data on delta wing performance already derived from the XF-92A, they eagerly awaited the arrival of the area-rule version, the F-102A. During 1956 a Convair F-102A (54-1374) arrived at Edwards for use by the NACA team.

During 1954 a series of mysterious crashes involving the North American F-100A fighter claimed the lives of several US Air Force pilots as well as that of George Welch, North American's chief test pilot. His F-100A had suddenly yawed more than 15° and broken up whilst making a rolling pullout from a dive at supersonic speed. Pete Everest, the US Air Force project pilot on the F-100, had recommended that the aircraft be modified to overcome supersonic directional stability. Unfortunately he was over-ruled following a series of evaluation flights by pilots from US Tactical Air Command. With a large production order pending, the US Air Force had no choice but to ground the aircraft until investigators could discover what had happened and so modify the design. NACA became involved and indeed were already concerned with evaluating the general stability and control of the F-100. From October to December 1954, the NACA flight test pilot Scott Crossfield flew F-100A 52-5778 on a series of flights to define the 'coupling boundaries' of the aircraft. In December 1954 NACA added a larger vertical fin to the F-100A which gave 10 per cent more surface area. Eventually North American installed an even larger fin having 27 per cent greater area as well as wing tip extensions, and the F-100 series went on to a long and distinguished life. 52-5778 remained at Edwards and was used for a variety of NACA research projects. When the first F-100C (53-1712) arrived in 1956 the Agency evaluated the behaviour of a pitching motion damper. As expected, the damper further increased the aircraft's resistance to coupling. The second F-100C (53-1717) arrived with the NACA in 1957 and was used for general research support, including chase flights and pilot proficiency.

The Convair YF-102 did experience some inertial tendencies, the first encountered on a delta aircraft, but not as serious as those found on the F-100. The Agency used the YF-102 (53-1785) primarily to extend the data acquired on the basically similar XF-92A. Research engineers did a complete drag survey of the aircraft, especially under various conditions of lift, this data greatly assisting research engineers interested in correlating results taken from flight testing with results from wind tunnel tests of the configuration. NACA flight-tested the F-102A (54-1374) delivered to Edwards during 1956 for comparisons between two generally similar configurations, the F-102A having area rule, and the other — YF-102 — lacking it. Before NASA returned the F-102A in 1959, flight test pilots Jack McKay and Neil Armstrong flew a series of landing approaches under various lift-to-drag ratios and power conditions, in preparation for the ill-fated Dyna Soar programme.

Since 1954 the NACA High-Speed Station Flight had requested a Lockheed F-104 *Starfighter* for evaluation. It was during the late summer of 1956 when they received a pre-production YF-104A (55-2961), which later became registered 'NASA 818'. The aircraft had some similarity to the Douglas X-3 in configuration. It was an alluring Mach 2 design with a high T-tail, long fuselage, narrow wing-span, and troublesome J79 jet engine, posing numerous challenges. There were roll coupling problems and the tail hinted ominously at pitch-up, though Lockheed had designed a stick shaker or 'stick kicker' into the controls to prevent an unwary pilot from getting into

pitch-up difficulties. These were the areas in which NACA had a vital interest. The company and service flight-test programme for the F-104 had suffered many setbacks, mainly because of turbojet problems and equipment failure. It took twice as long as expected, with a number of accidents and incidents, including fatalities. At one stage Lockheed had lost all its instrumented test aircraft and the NACA YF-104A was the only one left. The US Air Force asked for its return, but the Agency proposed that they run the test programme for Lockheed using the YF-104A, with a team of NACA flight-test pilots. The roll coupling study commenced during May 1957, with NACA test engineer Thomas Finch working closely with the Lockheed test team and the company aerodynamicists on studies of the expected rolling characteristics of the F-104 to predict what might happen in flight.

The Agency's flight test engineer-pilot Joe Walker flew the trials, and over the next nine months the YF-104A completed more than sixty roll investigations which showed the aircraft to be generally acceptable. Flight test results and analogue studies indicated that transonic and supersonic rolls near zero 'g' 'entry' conditions could lead to autorotation, a tendency for the aircraft to continue rolling despite the pilot applying corrective aileron, with accompanying pitching and yawing motions. It was recommended that if this occurred, the pilot use the stabiliser to damp out any tendency of the F-104 to couple. The NACA further recommended that Lockheed limit the aileron travel at transonic and supersonic speeds, only permitting full aileron authority with the aircraft in the low-speed landing gear-and-flaps-down configuration. Lockheed built mechanical limits into the F-104, added a yaw damper and put cautionary comments in the pilots notes.

NACA — later NASA — flew the YF-104A on a variety of research tasks. Equipped with reaction controls, it flew as a training aircraft for X-15 pilots. Before being finally retired in 1975, it performed a series of aerodynamic investigations including boundary-layer noise trials for Ames. The Agency at Edwards flew three other F-104s, two single-seat aircraft and one two-seater which came from Ames. F-104A 56-749 was lost in 1962, the pilot Milton Thompson ejecting safely. An in-flight asymmetric flap deployment caused uncontrollable rolling, fortunately at height, and Thompson stayed on board through four rolls of increasing rapidity and coupling tendencies before ejecting. Air Traffic Control at Edwards reported smoke in the direction of the bombing range. The pilot landed safely in the desert, rolled up his parachute and thumbed a lift in a passing pick-up truck for the ride back to NACA Base Operations!

North American YF-107A 55-5118, the first built, was received during 1957 and became registered 'NACA 207'. Unfortunately, it was reported as 'completely unreliable' and completed only four flights before the Agency grounded it. The third aircraft built, YF-107A 55-5120, completed forty flights during 1958–59 before being destroyed in take-off accident. The pilot, Scott Crossfield, escaped with no injury. On the basis of F-107 flight testing, North American refined the design of the side stick planned for the X-15. the side-stick programme was one of NACA's major accomplishments with the F-107. Proposed inlet and fin studies were cancelled after the retirement of the first aircraft, but the F-107 certainly helped X-15 development proceed more smoothly.

Flying the US Air Force machines was often as hazardous for the NACA flight test pilots as flying the regular X-series research aircraft. On the first flight of the NACA F-100A (52-5778) Scott Crossfield had to make a 'deadstick' landing following an engine fire warning. This was something North American test pilots doubted could be done since the early *Super Sabre* lacked flaps and landed very fast. Crossfield followed up the flawless approach and landed by coasting the aircraft off the lakebed, up the NACA ramp and then through the front door of the NACA hangar, frantically attempting to stop the F-100 which by this time had used up its emergency brake power. Crossfield missed the hangared X-fleet, but crunched the nose of the *Super Sabre* through the hangar wall. It is reported that Chuck Yeager proclaimed that while the sonic wall had been his, the hangar wall belonged to Crossfield!

For various reasons, service testing at the agency's Edwards centre was greatly reduced after 1958. In 1959 the High-Speed Flight Station had been renamed the NASA Flight Research Centre and was now heavily committed to the new X-15 programme, so the engineering staff lacked the manpower, resources and time to become involved in other projects. Acquisition and development of military aircraft declined, so there was less for NASA to do in service-related flight testing. When the pace of acquisition stepped up again in 1960, the Flight Research Centre once more actively supported military aircraft projects with flight testing.

On the first day of October 1958, NACA employees Doll Matay and John Hedgepath put up a ladder and took down the NACA winged shield emblem from over the entrance door over the station headquarters building. The National Aeronautics and Space Administration had arrived in the Californian desert, bringing with it a new era of space projects, soaring budgets and a lot of publicity. For the past five years, advanced planners had devoted an increasing amount of time to study of possibility of hypersonic aircraft flying at Mach 5+ and winged spacecraft. There was a heavy commitment to the new X-15. On the last day of NACA, the station complement at Edwards was 292; the new National Aeronautics and Space Agency (NASA) employed 8,000 civil servants, of which 3,368 worked at Langley. The peak at Edwards was reached in 1965, with 669 personnel employed at the desert centre.

Walter Williams and the personnel at the High-Speed Station Flight were the first to recognise that their activities had broadened considerably since the early days of the Muroc Flight Test Unit. There was a new and major role to play with the pending X-15 programme, which — together with the Project Mercury programme — represented a two-pronged approach to studying the problems posed by manned spaceflight. Williams had always sought laboratory status for the

station, making it an equal with other NACA laboratories, despite the workload and size being far smaller than either Langley or Ames. Independence from Langley had been achieved during 1954. With the introduction of the NASA, the traditional laboratories — Langley, Ames and Lewis — were redesignated research centres. Walter Williams finally got his wish for on 27 September 1959, NASA headquarters redesignated the High-Speed Flight Station the 'NASA Flight Research Centre' — FRC. That name held until it was renamed the Dryden Flight Research Centre — DFRC — in 1976 in honour of Dr Hugh L. Dryden. However by the time the Edwards station became a centre, Walter C. Williams had gone, joining Project Mercury as its operations director. He was greatly missed, having had a great influence on the local community and encouraging station personnel to take part in civic affairs.

Paul F. Bikle replaced Williams, and in fact was the latter's choice for the job. Bikle was well known in military flight test circles having been on the staff at Wright Field, and was serving as technical director of the US Air Force Flight Test Centre at Edwards when he was invited to join the NASA on 15 September 1959. He remained as an excellent administrator for the next twelve years. Bikle was equally at home with the research engineers, the flight test pilots, the crew chiefs and the mechanics on the flight line.

Bikle accepted the challenge presented by the X-15 programme and requested eighty new positions for the growing X-15 team. The budget for the centre increased from $3.28 million in 1959 to $20.85 million in 1963, and to a high of $32.97 million in 1968. The facilities expanded with a new communications building in 1963, a runway noise measurement system in 1964 and a high-temperature loads calibration laboratory in 1966, which would prove very useful during the Lockheed YF-12 programme. A special high-speed flight test corridor had been built for the X-15.

The unique Convair XP-92A Delta 46-682 is seen in flight during NACA tests conducted at the Flight Research Centre at Edwards AFB. It was projected for a top speed of Mach 1.5 and both Convair and the NACA obtained valuable data from this aircraft. (Office of History — USAF Edwards)

North American X-15

The X-15 was designed and built to explore the identifiable problems of space and atmospheric flight at very high speeds and altitudes. the speed to be achieved tentatively was set at Mach 6.6 or greater. Similarly, the altitudes to be aimed at were 250,000ft or greater. Its origins can be traced back to 1952, when a NACA committee resolved that the agency should expand its research aircraft programme in order to explore flight characteristics of atmospheric and exo-atmospheric designs capable of Mach 4 to Mach 10 and an altitude range of twelve to fifty miles.

As mentioned earlier, the laboratories at Ames, Langley, Wallops Island, Lewis and Edwards had formed groups to tackle the problems of hypersonic aerodynamics. The Langley group chose to deal with the problem in more depth than the other laboratories; after the research objectives were determined, the team turned their attention to the questions of propulsion and launch techniques. Lewis had favoured an unmanned rocket that could be launched from Wallops Island, and only Langley and Edwards submitted proposed configurations, the former being documented in greater detail and being more useful for planning purposes. On 23 December 1954, NACA and representatives from the US Air Force and the US Navy singed a Memorandum of Understanding, with NACA having technical control of the project. The US Air Force and the US Navy would fund the design and construction phases. Upon completion of contractor testing, the aircraft would be turned over to NACA — later NASA — which would conduct the flight testing and report the results. The memo concluded that 'accomplishment of the project is a matter of national urgency.'

The three parties created a Research Airplane Committee, an inter-agency body of senior executives to supervise the project, with Dryden representing the Agency. The design was left up to the laboratories, Langley, the High-Speed Flight Station and the contractor. This committee existed until 26 October 1957, when the Office of Advanced

Research & Technology — OART — closed it down. Its last significant action had been on 18 February 1964, when it approved the Langley-developed Hypersonic Ramjet Experiment — HRE — for the North American X-15-2. By mid-December 1954, the specification panel had stipulated that the new research X-15 should be capable of attaining an altitude of 76,000 metres and an airspeed of 2,000 metres per second — Mach 6.

Preparations for the X-15 programme were extensive and involved building the three research aircraft, modifying two Boeing B-52 *Stratofortress* bombers 52-003 and 52-008 to air-launch the X-15 and developing a powerful fully reusable rocket engine for the aircraft. A special full-pressure flight suit was required for the X-15 pilots, as was a special motion simulator connected to analogue computing equipment. All X-15 pilots spent between eight and ten hours in the simulator, equivalent to a ten to twelve-minute flight. A special aerodynamic test range running from Utah to Edwards, across the Nevada and Californian deserts, had to be formulated and equipped. All went smoothly, with the exception of the development of the powerfful 57,000lb thrust rocket engine. The Thiokol XLR-99 encountered so many delays and problems that North American were forced to substitute two of the older XLR-11 engines, as used in the earlier Bell X-1 series, until the larger rocket engine was ready towards the end of 1960.

One of the most important features of the X-15 programme, involving the High-Speed Flight Station, was the co-ordination of the tracking range required, known as 'High-Range' or 'High Altitude Continuous Tracking Radar Range'. NACA and the US Air Force co-operated in the planning, with the High-Speed Station Flight instrumentation team under Gerald Truszynski involved in the layout and location. It had been determined as early as November 1955 that the range should be at least 640km long and served by three radar tracking stations able to provide precise data on aircraft position, re-entry prediction, geometric altitude and ground speed. Other vital requirements included an air-launch site to be located over an emergency dirt landing area, intermediate dirt landing sites, intermediate launch drop sites, available airfields that could assist with radar cover, and a reasonable straight flight course.

This Convair F-102A Delta Dagger 56-1358 was delivered to Ames on 23 December 1957 for fire control tests plus autopilot manoeuvres. It was used until 21 March 1960. Another F-102A 56-1304 was also used by Ames arriving on 10 April 1957 for research and development tests. (NACA A-24027)

It was concluded by Truszynski and his team that the best flight course lay on a line from Wendover, Utah, to Edwards, with tracking stations located at Ely and Beatty in Nevada and at Edwards. The range would take the X-15 over some of the most beautiful, rugged and desolate terrain in the Western Hemisphere. It would fly high over Death Valley before a steep descent down over the Searles basin to a final landing at Rogers Lake. Construction commenced during 1956 on the High Range, and by July 1959 it was complete. It measured some 780km long, with a flight corridor of 80km. The three tracking stations had radar and telemetry tracking with oscillograph recording, magnetic tape data collection and radar console monitoring services. Each maintained a local plot of the X-15, Edwards having a master plot, with interstation communication facilities via radio and telephone landline link. On every flight, 87 channels sampled ten times per second relayed information from the X-15 to the ground stations. The cost was high. The Edwards master unit cost $4,244,000; the remaining two together cost about the same. In all the US Air Force spent $3.3 million on construction of the High Range — which subsequently proved beneficial to later NASA high-speed, high-altitude research aircraft programmes.

During the years prior to the arrival of the X-15, the High-Speed Flight Station supported the design and development stages of the programme; it was involved in reaction control studies on the Bell X-1B (48-1385) and later the Lockheed YF-104A (55-2961) together with the sidestick evaluation on the North American F-107 aircraft. It was an interesting and exceedingly busy time, with the NACA team reviewing development progress on the X-15 aircraft, attending meetings with the contractor, participating in inspection of the mock-up and putting forward both criticism and suggestions. When the first North American X-15 (56-6670) arrived by road on 17 October 1958, the station's technical staff was more than ready to be involved.

The static test programme at Edwards took some five months to complete. During this period North American converted the B-52A 52-003 and B-52B 52-008 to be capable of launching the X-15: they became designated NB-52A and NB-52B respectively. *Stratofortress* 52-003 had been moved to Palmdale on 4 February 1958, and 52-008 on 6 January 1959. It must be mentioned that other types were considered as possible carrier aircraft for the X-15. These included the Boeing B-50 and B-52, and the Convair B-58 and B-36. The B-52 became the primary carrier aircraft because of its performance, extraordinary lifting power, reliability and its relative ease of major modification. In addition to the two Boeing B-52 'mother ships', a stable of chase aircraft were made available. These were to prove most useful in the launch and landing areas whilst performing various visual checks to ensure flight safety.

Following five months of ground testing, the first X-15 (56-6670) — with North American test pilot Scott Crossfield at the controls — completed its first captive flight under the wing of its B-52 on 10 March 1959. Several additional captive flights followed, culminating in the first glide flight on 8 June. Just prior to landing the aircraft began a series of increasingly wild pitching motions. Engineers subsequently modified its boosted control system to improve low-speed control authority and made changes to linearize hinge moments. On 17 September the X-15 completed its first powered flight, when Crossfield flew the second aircraft (56-6671) to a maximum speed of Mach 2.11 and a maximum altitude of 52,341ft. Crossfield had resigned from NACA to join North American and the X-15 programme.

This Lockheed TV-1 BuNo.33868 has a dual personality being also designated P-80C 48-379. It later became registered NACA 206. It was delivered to Ames on 12 October 1953 and used on Bullpup missile simulator trials. It was retained until February 1960. (NACA A-21015)

The third flight, again with the second X-15, was also successful but the fourth flight was not. During the latter, on 5 November, an inflight explosion and fire in the engine compartment necessitated engine shutdown and an emergency landing on Rosamund Dry Lake, one of the several designated emergency landing sites. Propellants were jettisoned during a steep descent, but this process was not fully carried out and the landing weight was higher than normal. When the nosewheel made contact with the lakebed, the fuselage buckled just aft of the cockpit and bent severely enough to cause the bottom to scrape the ground. In order for repairs to be undertaken the X-15 was returned to Los Angeles, remaining there from November 1959, to February 1960.

Whilst the second X-15 was being repaired, the first aircraft became the primary flight test vehicle, following modifications and the installation of an XLR11 rocket engine. On 23 January 1960, with Scott Crossfield as the pilot, 56-6670 made its first powered flight and in the process set a new speed record for the programme along with a new altitude record of 1,669 mph and 66,844 ft respectively. The exciting flight test programme now began to accelerate. Joe Walker, the primary NASA X-15 test pilot, flew the aircraft on its first agency-sponsored sortie on 25 March, and Captain Robert White became the first US Air Force pilot to fly the X-15 on 13 April. By this time a total of six NASA and US Air Force test pilots had been assigned to the X-15 programme. These included Joe Walker, Neil Armstrong and John McKay for NASA and Captain Robert White and Major Robert Rushworth for the US Air Force. One 'odd man out' was Commander Forrest Peterson of the US Navy. Later the NASA team would be joined by Milton Thompson and William Dana, whilst three US Air Force pilots would include Captain Joe Engle, Captain William Knight and Major Michael Adams.

By August 1960 each flight had expanded the envelope of the X-15 in speed and altitude to the point where new records were being set during virtually every flight. On 4 August Joe Walker flew the first X-15 to 2,196 mph, so breaking the absolute speed record of 2,094 mph set some four years earlier by the Bell X-2. This was followed by several altitude flights, including Robert White's on 12 August 1960; an epic which took the aircraft to 136,500 ft, thereby breaking the four-year-old X-2 record of 126,200 ft. On 7 February 1961, Robert White reached 2,275 mph in the first X-15.

Biomedical Aspects

The remaining flights with the X-15 powered by the XLR11 rocket engine served to familiarize new pilots with the aircraft's unique flight characteristics. In addition, important biomedical baseline information was gathered which had a direct bearing on the special MC-2 full-pressure suit developed by the David Clarke Company and its immediate successor, the A/P22S-2. The standard MC-2 was only a partial pressure suit which did not provide pressurisation for the pilot's feet. The biomedical aspects of the X-15 programme proved of great importance and would be useful for the future space programme in which the Agency was to become involved. Pilot biological parameters were closely monitored via an electrocardiogram device and an oxygen flow sensor, whilst a differential sensor monitored both suit and cockpit, plus suit and helmet pressure. This data was telemetred to the ground whilst the X-15 was in flight. Additionally, significant research was conducted on weightlessness since the X-15 was possibly the first aircraft to expose the pilot to this environment for any significant period of time. The last series of flights with the XLR11 allowed installation and low-speed flight testing of the new Northrop-developed 'hot nose' sometimes referred to as the 'Q-Ball' nose. It provided angle of attack, sideslip and dynamic pressure information to the pilot via cockpit instrumentation.

Some twenty-five flights were eventually completed by NASA, the US Air Force and North American before the first flight-rated XLR99 rocket engine arrived at Edwards for installation. This engine — the most powerful, most complex, and safest man-rated throttleable rocket propulsion system in the world — had undergone an exhaustive test programme. Despite being long in gestation, the XLR99 was soon to prove exceptionally reliable and extraordinarily safe. On 28 March 1960, Thiokil delivered the first XLR99 to Edwards for installation in the third X-15 (56-6672) after which the aircraft was moved to the static test area for an intensive ground-test programme. On 8 June, however, a malfunctioning relief valve and pressurizing gas regulator caused a catastrophic explosion which totally destroyed the engine and severely damaged the X-15. On 17 June, with 56-6672 back at North American for repair, the second XLR99 engine arrived at Edwards for installation in X-15 56-6671. After extensive ground testing, the aircraft was prepared for its first flight, this being accomplished without any major difficulties on 15 November 1960 with Scott Crossfield at the controls.

Crossfield was to fly the X-15 on only two more occasions. These last two flights — on 22 November and 6 December 1960 — were the last of the obligatory contractor demonstration flights, and they served effectively to terminate North American's participation in the X-15 flight test programme. At this time the first X-15 was returned to North American for installation of the XLR99 engine. It departed from Edwards on 8 February 1961, returning to join the second X-15 on 10 June. With the completion of the contractors' participation it was now the task of NASA, the US Air Force and the US Navy to embark on a systematic programme that was designed to fly the X-15 to the extremes of its performance envelope.

This programme involved four broad objectives: verification of predicted hypersonic aerodynamic behaviour and hypersonic heating rates, study of the X-15's structural characteristics in an environment of high heating and high flight loads, investigation of hypersonic stability and control problems during atmospheric exit and re-entry, and investigation of piloting tasks and pilot performance. By late 1961 these four areas had been generally examined, although detailed flight research was to continue with the first and third X-15s. Physiological researchers discovered that the heart rates of X-15 pilots varied between 145 and 180 beats per minute during a flight, compared with a normal 70 to 80 beats per minute for test flights in other aircraft. Aeromedical experts eventually concluded that prelaunch anticipatory stress, rather than actual postlaunch physical stress, influenced the heart beat. They used these rates as baselines for predicting the physiological behaviour of future NASA pilot-astronauts.

The first agency flight took place on 7 March 1961, involving the second X-15 (56-6671) powered by a XLR99 and with Robert White at the controls. It accelerated to 2,905 mph whilst climbing to 75,000 ft. On 30 March, Joe Walker achieved a new altitude record in this aircraft, 169,600 ft. He was wearing for the first time the new A/P22S-2 full pressure suit. Mother nature then had an affect on the X-15 programme since flights had to be suspended during most of February and March due to rain and snow covering the dry lakes.

By the autumn of 1961, the X-15s had achieved speeds in excess of 3,500 mph and altitudes in excess of 215,000 ft. During October this speed record was improved upon with a flight to 3,920 mph, and in November with one to 4,093 mph. These were the first flights to reach the aircraft's design Mach number. On 29 October 1961, the third X-15 (56-6672) which had been almost completely rebuilt was returned to NASA. It had been fitted with a Minneapolis-Honeywell MH-96 self-adaptive flight control system which was designed to make the X-15 easier to control with either the aerodynamic or jet control system, or a combination of both. Indeed, it was possible for the MH-96 to control the X-15 without assistance from the pilot. The system had been flight-tested aboard a Lockheed F-94C, a McDonnell F-101A, a Cessna 310 and in the X-15 simulator at North American.

Whilst preliminary flight test activity was being undertaken with the third X-15 (56-6672), on-going exploratory flights continued using the first aircraft 56-6670. Design altitude was acheived in this aircraft by Joe Walker on 30 April 1961, when 246,700 ft was reached. The flight test programme for the third aircraft was progressing smoothly; it had completed its first flight with Neil Armstrong at the controls on 20 December. The MH-96 was proving to be of benefit to X-15 pilots during high altitude flights, and underscoring this was a flight with Robert White on 21 June — again to 246,700 ft. On 17 July, the Walker and White record was broken with White in the third X-15 who flew to 314,750 ft, this being the first sortie after which astronaut's wings were awarded to an X-15 pilot. On the following day the prestigious Collier Trophy was presented by President Kennedy to Crossfield, Peterson, Walker and White.

On 3 August 1961 the first X-15 was returned to North American for modifications required by follow-on flight requirements. Changes included the installation of a nose-mounted camera for optical degradation tests and fitting of a new inertial guidance system which was originally intended for the Boeing X-20. Wing-tip pods were installed for transport of miscellaneous experiments.

The flight-test programme had been marred by few serious accidents and even fewer injuries, and any damage sustained to aircraft had been repaired. On the 74th flight in the programme, on 9 November 1962 with the second aircraft piloted by John McKay, a post-launch malfunction occurred at 53,950 ft which prevented fuel jettison; an over-gross weight emergency landing was made at Mud Lake, Nevada. The aircraft was badly damaged when the gear failed, and

John McKay was seriously injured. He managed to jettison the canopy, but could not escape, becoming distressed by ammonia fumes. A rescue helicopter blew the fumes clear and McKay was cut free from the X-15, which had turned turtle on landing. The aircraft was set aside for rebuilding, giving birth to the high-performance X-15A-2.

One further accident would, unfortunately, mar forever the X-15's otherwise outstanding safety record. This occurred with the third aircraft on 15 November 1967. US Air Force pilot Major Michael Adams was making his seventh flight, which was his third at the controls of the MH-96-equipped aircraft. Following a normal launch, Adams achieved the flight speed and altitude objectives of Mach 5.2 and 266,000 ft. Included in the flight plan was wing-rocking in order to facilitate optical tracking of the aircraft's exhaust plume. Ground personnel monitoring the wing-rocking exercise noted that the X-15 was greatly exceeding the required bank angles. Less than 30 seconds later the aircraft had veered some 90 degrees to its ballistic flight path and had begun entry into a spin — still while travelling at speed in excess of Mach 5. The spin continued from 230,000 ft down to 125,000 ft, at which point a conventional dive was entered. High-frequency pitch oscillations were encountered which led to command saturation of the MH-96 system computer and a resulting increase in the severity of the pitch control problem. Structural loads by this time had increased to plus and minus 15 g — far beyond the X-15's plus 7.33 g and minus 3 g design limits. Seconds later the aircraft disintegrated and Major Michael Adams did not survive.

The second X-15 at the time of the accident (56-6671) had accumulated a total free-flight time of 4 hrs 40 mins 32.2 secs. Following the decision to rebuild the aircraft, it was transported by truck to North American at Los Angeles where major modifications took place. The aircraft was redesignated X-15A-2 and made its first flight at Edwards AFB on 25 June 1964. The programme was to continue.

Century Series Aircraft
Operated by NACA/NASA

Manufacturer	Model	Serial	No.	Mach	Period	Notes
North American	F-100A	52-5778		1.3	1954/60	Inertial coupling studies
	F-100C	53-1712		1.4	1956/57	
	F-100C	53-1717		1.4	1957/61	
McDonnell	F-101A	53-2434	NACA 219	1.7	1956	Transferred to Langley 22 August 1956
Convair	YF-102	53-1785		0.98	1954/58	
	F-102A	54-1374		1.2	1956/59	
Lockheed	YF-104A	55-2961	NASA 818	2.2	1956/75	Completed 1,439 research missions in a 19-year period
	F-104A	56-734		2.2	1957/61	
	F-104A	56-749		2.2	1959/62	Lost in accident — pilot Milton Thompson ejected safely. Inflight asymmetric flap deployment caused uncontrollable rolling
	F-104B	57-1303	NASA 819	2.2	1959/78	Transferred from NASA Ames
Republic	F-105B	54-102		2.0	1959	Pilot familiarization
North American	YF-107A	55-5118	NACA 207	2.0	1957/58	
	YF-107A	55-5120		2.0	1958/59	Take-off accident. Aircraft damaged beyond repair

CHAPTER NINE
Transition to NASA

The research conducted at Ames during the period between 1950 and 1953 had a strong trend toward the more fundamental. There were great efforts to develop the theory required for deeper understanding of transonic, supersonic, and hypersonic flows. The term 'hypersonic' referred generally to a speed regime of Mach 5 and over, where linear theories — dependent on small disturbances, two-dimensionality, and constant gas properties — broke down. Flight-test research was quickly becoming increasingly scientific and sophisticated.

One of the most significant developments of this period was the discovery by Richard Whitcomb of transonic area rule, for reducing the drag of aircraft. The resulting shape had a constriction like a contemporary Coca-Cola bottle and the terms 'Coke-bottle' or 'Marilyn Monroe' in connection with the fuselage shape were often heard. At the time of its discovery the transonic area rule had a very limited theoretical basis, a condition which personnel on the Ames staff took steps to correct. One of the early contributors to the theory was R. T. Jones, who published a NACA report and introduced the concept of 'supersonic area rule'. This rule made it possible to minimize the drag of an aircraft for any chosen Mach number. Jones's report was a substantial contribution to the theory and application of the general area rule.

The Helldiver NACA 147 was fitted with an extra raised rear cockpit with a manually controlled periscope, both clearly visible in this photo. The Helldiver arrived at Ames on 18 December 1948 remaining until salvaged by the US Navy in June 1955. (NACA A-17119)

A large part of the task given to the flight research branch fell into the field of guidance and control. With the advent of automatic control for aircraft and missiles, the dynamic behaviour of aircraft was being examined in a much more sophisticated and scientific light than ever before. The flight-test pilot, with an enormous ability to compromise and adapt, had always been able to compensate for the imprecision of existing design knowledge regarding aircraft dynamics. But the new electronic and mechanical servo systems which were to guide aircraft did not have the abilities of the human pilot. The problem would require the merging of two very diverse disciplines — flight aerodynamics and electronics and servo mechanisms. In other words, there was a requirement for a combination of human and electronic guidance. Fighters with computing gunsights and interceptors with automatic or semi-automatic missile fire-control systems were on the horizon. Flight research had taken a quantum stride in complexity and sophistication, and there were exotic new trends.

The Grumman F6F-3 *Hellcat*, (BuNo. 42874 and now 'NACA 158') variable stability aircraft had been modified so that a wider range of lateral-directional stability variables might be investigated. In a programme the handling qualities in this area were evaluated by twelve different pilots including two from the US Air Force, four from the US Navy, five from NACA and one from Cornell Aeronautical Laboratory — which had commenced similar work with variable-stability aircraft about the same time as Ames. Pilot opinion in quantitative form was obtained.

The US Navy had developed a radio-operated control system by means of which the flight of a test aircraft could be remotely controlled from another aircraft or, for take-off, from a ground control station. Such a system was installed in a Curtiss SB2C-5 *Helldiver* dive-bomber whilst a Grumman F6F-5 *Hellcat* was equipped to serve as airborne controller. The pair of aircraft were turned over to Ames for research and they became the laboratory's entry into a new and important field — automatic or radio control. The US Navy had apparently lost interest in this particular remote-control system, so the SB2C-5 (BuNo.83135) drone became available to Ames while becoming 'NACA 147' in the fleet. The F6F-5 *Hellcat* (BuNo.79669) became 'NACA 208'.

The Agency installed additional equipment in the *Helldiver* drone to allow it to simulate a radar-controlled interceptor. The radar was represented by an optical system — a manually-controlled periscope pointed at a manoeuvring target — represented by another aircraft — and the SB2C-5 was then flown directly towards the target. Photographs of the target taken simultaneously through the periscope and through a camera gun rigidly mounted on the aircraft were later developed and comparisons made to determine tracking accuracy. Using a Reeves analogue computer, ground studies included simulation and calibrations for accuracy. Data for later involvement in automatic control systems, including autopilots, were obtained from this interesting exercise.

Flight-test research, though now more sophisticated, was nonetheless hazardous, as mentioned earlier. On 1 June 1953, whilst putting a Grumman F8F-1 BuNo.94819 *Bearcat* through some step-control manoeuvres, Rudolph Van Dyke crashed into San Francisco Bay and was killed. The cause of the crash was never determined, but it was conjectured that since the crash had been preceded by a long steep glide in which no evidence of an attempt at recovery had been apparent, Van Dyke had perhaps failed to connect his oxygen mask and had lost consciousness from a lack of oxygen. Rudy had served as a research test pilot at Ames since 1947 and had made important contributions to many of the laboratories' flight research trials. His passing was a serious loss to Ames and caused much sorrow.

Aerodynamic heating was of prime importance to the designers of high-speed aircraft in the 1950s. The heating to be experienced by future hypersonic aircraft would not only be relevant to skin friction and thus the overall drag of aircraft but ultimately would melt the leading edges of the vehicle unless some form of thermal protection was provided. Without this protection, the surfaces — particularly the leading edges — could reach structurally intolerable temperatures at Mach numbers as low as 2 or 3. Ames scientists were involved in tests and investigations which resulted in technical papers covering such subjects as skin friction in the laminar boundary layer of a flat plate at Mach 2.4 and a comparison with existing theory. Test data from several wind tunnels and from flight trials confirmed the validity of the Reynolds analogy and, more particularly, the modified analogy proposed by Rubesin. Morris Rubesin was at Ames and produced a technical paper providing a modified Reynolds analogy which allowed for compressibility effects in the boundary layer. He was also involved in one of the early studies of transpiration cooling.

Technical Fields

The rapid growth in aeronautical knowledge between 1954 and 1957 was characterized by expanding horizons. There was considerable concern over the many impressive technical developments in Soviet Russia and it was recognised in high places that a race was on that could have the gravest consequences. The NACA chairman, James H. Doolittle, was quick to point this out to Congress in January 1957, and implied that the levelling-off of NACA's appropriations over the past few years was scarcely a rational response to the challenge from the USSR.

Fighters and bombers capable of Mach 2 were now a reality, whilst a new research aircraft — the North American X-15 — was being designed to fly at Mach 5 or more. Despite the armed services devoting an ever-increasing portion of their budget to guided missiles, the aeronautical scientists were gradually turning towards space. The Bell X-2 research aircraft had reached an altitude of over 126,000ft in 1956, and the new X-15 when completed was expected to fly more than twice as high. There was a race developing between the United States and the Soviets to launch the world's first satellite, but not until late 1957 was it revealed that the winner was the USSR. The Russian *Sputnik I* was launched successfully on 4 October 1957, and *Sputnik II*, with a live dog on board, was launched a month later on 3 November.

During this period the Agency found itself at a severe disadvantage in the keen competition for technical manpower. The national demand for engineers and scientists had mushroomed beyond all expectations and the supply, particularly of personnel with above-average qualifications, was exceedingly short. The US was in the midst of scientific explosion, and there was widespread use by the military services of research and development contracts. Both the aircraft companies and other research and development organizations were busily expanding their technical staffs, and new research and development groups were attempting to become established. There was a large manpower market, offering salaries well beyond those which the Civil Service Commission would allow NACA to pay. The activities of the Agency had been restricted by manpower quotas imposed by Congress but now, because of the low salary scale, the quotas — low as they were — could not be filled. The agency was also losing senior staff members, some key men, owing to the highly attractive salaries offered by industry. Despite the Agency's hard fight to improve its salary position, it never achieved equality with industry. The organization continued to operate under a severe manpower handicap.

Two Curtiss SB2C-5 Helldivers were used at Ames including this aircraft BuNo.83292 delivered on 22 December 1948, just four days after BuNo.83135 which became NACA 147. The Helldiver in the photo remained for only 23 days, departing on 14 January 1949, yet it received the NACA winged emblem.
(NACA A-15943)

Between 1954 and 1957, a number of key personnel were lost to Ames either by transfer to industry or retirement. William McAvoy, the NACA senior test pilot, retired on the last day of July 1954, having survived exciting incidents and near mishaps during his thirty-five years of flight test work at Langley and Ames. He was replaced as Chief of the Flight Operations Branch by George Cooper.

One of the important research and development programmes undertaken at Ames between 1954 and 1957 concerned boundary-layer control — BLC — as tested on a North American F-86 *Sabre*. This programme covered an investigation of three different kinds of boundary-layer control — boundary-layer removal by suction at the leading edge of the wing and the flaps, and the use of a jet just ahead of the flap to energize the boundary layer. All forms of BLC were intended to improve the landing characteristics of the aircraft and all were intended to exploit the high mass flow of the turbojet engine, which can be exploited in high-drag high-power regimes such as during the approach for BLC purposes.

This programme jointly involved the 40 × 80 ft tunnel and the Flight Research sections. In the wind tunnel, details of the flow system were worked out through the use of a model incorporating a jet engine and the wings of the F-86. The design features developed were then applied to an actual *Sabre* aircraft and their performance was checked by flight testing. In these, the landing approach with and without BLC was evaluated by a number of US Air Force, US Navy, North American and NACA pilots. On 2 October 1956 a North American F-100A *Super Sabre* (53-1585A), which later became 'NACA 200', arrived at Ames for BLC tests. It stayed until 15 February 1960.

North American F-86A-1-NA 47-609 later NACA 135, being handled gracefully and carefully into position in the huge 40 × 80 ft tunnel at Ames. Both wing surfaces are heavily tufted for flight test purposes. It arrived at Ames on 10 April 1950. (NACA A-15957)

This production Republic F-84F-5-RE 51-1364 Thunderstreak was delivered to Ames on 31 October 1953, later becoming NACA 155. The records indicate it crashed on 7 March 1957. Earlier the first production F-84F 51-1346 was delivered to Ames on 1 March 1954 being out again ten days later. (NACA A-18856)

During the mid-1950s a number of interesting investigations were made of complete configurations representing flyable aircraft. One such investigation, run in the 12 ft tunnel, was involved with the aerodynamic problems involved in the design of a large turboprop-powered, swept-wing aircraft of the bomber or transport type. Turbojet engines dominated the aeronautical scene but there was still a great deal of interest in the use of turboprop engines for long-range high-speed transports. Stability and control problems were envisaged, and tests were carried out on a powered model of what the Ames engineers considered a practical — perhaps an optimum — configuration for such an aircraft. The research engineers were later reassured about the validity of their work when the Russians introduced a turboprop bomber, the Tupolev Tu-20 *Bear*. This closely resembled the configuration tested in the 12 ft tunnel.

Great interest in aircraft having vertical take-off and landing (VTOL) and short take-off and landing (STOL) capabilities began to develop during this period and a number of investigations of VTOL or STOL devices were made in the 40 × 80 ft tunnel. These devices included a full-scale helicopter rotor, a wing having a propeller mounted in a circular hole cut through the chord plane and the McDonnell Model XV-1 *Convertiplane* which was a combination of helicopter and aircraft. The aircraft (53-4016) arrived at Ames during 1952 for tests in the 40 × 80 ft tunnel.

Flight Research

A new trend was introduced in flight research, aiming toward the use of simulators for the study of aircraft guidance and control problems. There were both ground and airborne simulators, and work on the latter had begun several years earlier with the development of variable-stability aircraft, the first being the Grumman F6F- *Hellcat* 'NACA 158'. This useful aircraft served until 9 September 1960 and was supplemented by a North American F-86E *Sabre* (50-606A) on 30 June 1955, which became 'NACA 157', and was converted for variable-stability work. Ames' involvement in this field was responsible for the use of negative dihedral on the new Lockheed doube-sonic F-104 *Starfighter*. The concept of negative dihedral (usually referred to as 'anhedral') was quite foreign to the theory of aircraft designers. Apparently even Kelly Johnson, Lockheed's chief designer, was astonished to hear that an F-104 simulation at Ames had shown anhedral to be distinctly advantageous. Tony LeVier, chief flight test pilot at Lockheed, visited the laboratory and flew the variable-stability aircraft and so confirmed the facts discovered at Ames. The result was that anhedral became one of the distinctive features of the outstanding F-104 fighter. On 10 March 1958 Ames took delivery of a Lockheed F-104B *Starfighter* (57-1303A) which amongst other tasks was engaged in variable-stability trials, whilst a North American F-100C *Super Sabre* (53-1709A) was delivered to Ames on 9 April 1956 for variable-stability trials. This aircraft was probably the first to be numbered in the new NASA system, becoming 'NASA 703'. The 700 series was introduced at Ames and remains in use today.

The NACA, and later NASA, was one of the first agencies to appreciate the transport capabilities of the ubiquitous Douglas C-47 *Skytrain*, built in large numbers during World War II and serving the Allies on all battle fronts. A large number served with the US Navy in the R4D- designated series. Under a US Department of Defence Directive dated 6 July 1962, the R4D- designations allocated to US Navy variants of the C-47 were standardized with those of the many US Air Force variants. On 14 December 1948 a Douglas R4D-6 (BuNo. 99827) arrived at Ames and was registered 'NASA 701'. It was used on general utility duties as well as being involved in research and development trials into gust phenomena. Three days later on 17 December — coincidentally the anniversary of the first flight of the DC-3 in 1935 — a second C-47 (ex-US Navy but unidentifiable) arrived. The C-47

'NASA 701' served at Ames until 9 September 1965. Indeed, the Douglas transport served all the laboratories and out-stations within the two agencies.

The North American F-86 *Sabre* became one of the many flight test workhorses operated by Ames over the years, leading up to the transition to NASA. Another flight simulator consisted of an F-86 in which a variable-control system had been installed. This device allowed the flight-test pilot to vary the dynamic characteristics of the aircraft's longitudinal control system over a wide range and enabled studies to be conducted on factors such as control feel response, sensitivity, breakout force, and time constant.

Reference has already been made to the tracking research made by flight engineers at Ames using several aircraft equipped with cameras and fixed gunsights. The next phase of this interesting programme involved tracking trials using an F-86 aircraft equipped with a gyro-mounted electronically-controlled computing gunsight which would automatically provide the proper 'lead' on the target as dictated by its course, speed and distance. This aircraft an F-86E (48-291) was delivered to Ames on 29 August 1949 and became 'NASA 116'.

A target and target-drone aircraft had been in use for a period in tracking tests, but this proved problematic and expensive, and there was always the added risk of a collision between the two aircraft. Ames research engineers had become so adept in their automatic control-systems trials that they felt it would be possible to simulate the manoeuvring target aircraft with sophisticated equipment carried in the tracking aircraft. A veteran Ames flight test pilot, Fred Drinkwater, was involved in these trials, together with engineers Brian Doolin and Allan Smith. The target aircraft appeared as a moving spot of light on a glass screen in front of the pilot's windscreen. It was a rather complex system and the spot of light could not give forewarning of turns — an actual target aircraft could be seen to bank its wings when turning.

This target simulator was later adapted for simulators for the guidance of the radio-controlled *Bullpup* missile developed by the US Navy. This air-launched missile was visually guided by the pilot towards a ground target by means of a radio link.

The first production Lockheed F-94C-1-LO Starfire 50-956 was delivered to Ames on 29 July 1954 becoming NACA 156. It was fitted with a thrust reverser, being initially checked out in the 40 × 80 ft tunnel then used for in-flight thrust reverser trials. According to the Ames records this Starfire crashed on 18 November 1958. (NACA A-20694)

The second North American YF-93A 48-318 NACA 151 seen parked at Ames during flight test trials. It was delivered on 6 May 1951 and along with NACA 139 remained at Ames until 1956. The NACA designed flush intakes are evident in this photo. (NACA A-16545)

This immaculate late model North American F-86D-60-NA 53-787 Sabre Dog, later NACA 216, was delivered to Ames on 17 March 1955. It was used on a variety of flight test presentations remaining at Ames until 1 February 1960. (NACA A-212922)

This type of tracking system could be modelled by a modification of the Ames tracking simulator. The results were highly successful and the resulting simulator proved more than useful in training US Navy pilots in *Bullpup* missile operations. Fred Drinkwater was again involved, along with research engineers Joe Douvillier and John Foster.

One of the ground-based simulator trials at Ames began with a brief investigation of the longitudinal stability and control characteristics to be expected of the new X-15 aircraft as it travelled through the upper reaches of the atmosphere.

Robert C. Innis was a highly qualified engineering test pilot who joined the NACA at Ames in 1954. Along with Hervey Quigley he was involved in boundary-layer control and STOL test flying. Bob Innis is seen climbing on board the first production North American F-100C-1-NA 53-1709 in NACA markings, which was involved in variable-stability trials. It later became NASA 703. (NACA A-21844)

An analogue computer was programmed to represent the dynamics of the aircraft. Another ground-based simulator was used to find out exactly what it was in the stability and control characteristics of an aircraft that caused the pilot to select a certain approach speed during a carrier landing. Here again pilot opinions was a dominant element of the trial, and the opinion of numerous pilots were elicited. Again Fred Drinkwater was involved, together with research engineer Maurice White. The aircraft and carrier deck were represented in simulated form on a oscilloscope screen, and involved four of the new jet fighters of the US Navy. Results confirmed the value of the ground-based simulators as a research tool, and were confirmed by an extensive flying programme in which carrier-approach handling characteristics of no less than 41 different types of fighter aircraft were evaluated together with extensive opinion data from pilots.

Bob Innis and Hervey Quigley investigated the lift, drag and stalling characteristics of this Stroukoff YC-134A NASA 222 at Ames, although it is felt it is a C-123B heavily modified as the serial is given as 54-556. The single YC-134A 56-1627 was delivered to Ames on 30 April 1959 for use as spares. (William T. Larkins)

George Cooper, described as a quiet and competent individual and an excellent engineering test pilot, was much liked and highly respected by his colleagues. It was felt that because of his unassuming character George had never received the credit he deserved, and it was pointed out that back in 1948–49 while the heroic and much lionised pilots of

the Bell X-1 were occasionally breaking the sound barrier at Muroc, he and Rudy Van Dyke at Ames were doing it twice a day in an F-86 without any publicity whatsoever. During 1954 George's fine work received official recognition when he won the Octave Chanute and the Arthur S. Fleming Awards.

Investigations into the dynamic response of aircraft were also carried out during this period. A joint study undertaken by Ames and the High-Speed Flight Station at Edwards involved a single Boeing B-47 *Stratojet* bomber (49-1900) which had been delivered to Langley on 11 July 1952, but transferred to Ames on 17 March 1953. This aircraft was thoroughly instrumented to measure the dynamic response of various parts of the aircraft structure to excitation in the form of pulsed control inputs. The measured response motions were converted to frequency response and transfer functions, which were compared with analytically predicted values. This study gave an insight into the dynamic problems of the highly flexible bomber and transport types being built.

Harvey J. Allen and Alfred J. Eggers had pursued the blunt-nose principle which they recommended for the warheads of ballistic missiles. The idea of a man-carrying ballistic missile had appeared and human physical endurance placed limits on the oscillations and tumbling of a re-entry body. Great interest was being manifest in the usefulness of rocket-launched man-carrying gliders for long-range hypersonic flight in the outer reaches of the atmosphere. NACA had undertaken discussions with the US Air Force on the desirability of constructing an experimental boost-glider vehicle as the next step beyond the X-15 in the research aircraft programme. The US Air Force were naturally interested in considering the military potentials of such an aircraft, and as 1957 came to a close it was about to initiate *Project DynaSoar* which called for the development of a world-girdling, man-carrying, boost-guided vehicle.

The importance of Harvey Allen's blunt-nose principle was now being recognized and his other major contributions to the development of hypervelocity vehicles were also widely appreciated. During 1955 he received the Sylvanus Albert Reed Award of the Institute of the Aeronautical Sciences, and in 1957 he was awarded the NACA Distinguished Service Medal. Allen was also honoured by being invited to present the prestigious Wright Brothers lecture in Washington, D.C., on 17 December 1957 — the 54th anniversary of the first flight by the Wright Brothers. The subject of the paper on this occasion was 'Hypersonic Flight and the Re-Entry Problem'.

One NASA historian reflected at this stage on the rapid increase in the speed of travel by man. In the forty-two years of aeronautical history between 1903 and 1945, man had achieved a speed of approximately 650 mph. In the twelve years between 1945 and 1957, his pace had tripled and at this rate could be expected to grow five-fold or more in the five years 1957 to 1962. As we write this paragraph, the forecast for the next twenty years (1990–2010) is that man will travel at more than 4,000 mph. Designers of new space aircraft are thinking of speeds up to Mach 25.0.

On the Fringe of Space

During 1957 it is estimated that some 50 per cent of the Ames laboratory projects were in some degree related to space flight. The pace of space research was increasing, however, and it acquired additional impetus from the launching or *Sputniks I* and *II*. Both the USA and USSR had announced plans for launching satellites during 1955. However, the appearance of *Sputnik I* on 4 October and of *Sputnik II* on 3 November 1957, came as a tremendous shock to most of the civilised world. The door opened on new vistas for man's exploration, a new and inviting frontier lay open to world scientists and research engineers, and the human spirit was fired by the realization that man had suddenly acquired the power and ability to escape from planet Earth.

Pete Knight seen in front of the second X-15 56-6671 on 10 March 1967 after flying at Mach 6.7 (4,520 mph) at 102,100 ft which became the unofficial world absolute speed record. Capt William J. 'Pete' Knight USAF completed 16 flights in the X-15 programme, flying all three aircraft.
(Office of History — USAF Edwards)

The modified North American X-15-2 56-6671 was fitted with external propellant tanks, one on either side. When empty they were jettisoned and retrieved by parachute, for future use. It first flew in modified form on 25 June 1964, piloted by Major Robert Rushworth, USAF, at Edwards AFB. It is preserved in the USAF Museum. (AP Photo Library)

This Cessna 172K Skyhawk c/n 172-58729 was purchased by NASA on 8 May 1972 as N7029G later NASA 507. It is currently in storage at Langley. Flight testing took place at the Wallops Flight Facility. It is depicted with the tail unit and rear fuselage heavily tufted. (NASA W180-111-2)

A huge 22,000lb CH-47 Chinook helicopter, earlier damaged by fire at the factory, made an excellent candidate for this crashworthiness test at Langley. The huge helicopter was dropped 50ft to the concrete below during 1975. The experiment was a combined NASA-Army programme with the Chinook heavily instrumentated. (NASA 75-HC-198)

Excellent air-to-air photo of a rebuilt Martin B-57 by General Dynamics as WB-57F, registered NASA 926 ex-63-13503. It is one of three operated by the NASA by the Johnson Space Centre. The wing span was increased to 122ft 5in and the service ceiling was quoted as 68,000ft with an absolute maximum altitude of 73,800ft. (NASA 574-34507)

The current NASA aircraft inventory includes four Beech King Air 200 executive aircraft. The Jet Propulsion Laboratory operates NASA 7, whilst Wallops has NASA 8 which is depicted. Marshall has NASA 9 c/n BB-1091, and Ames has NASA 701 c/n BB-1164, which was originally on lease in 1983, later purchased. (NASA WI 81-390-9)

This Grumman F-14A Tomcat BuNo.157991 arrived with the NASA at Dryden
during 1979 on a joint US Navy-NASA programme. This was preceded by
preliminary testing by NASA personnel at the Grumman, Bethpage, New York,
facility. (NASA ECN 11600)

The McDonnell Douglas F-15 Eagle 71-0287 — DEEC — Digital Electronic
Engine Control — aircraft seen in flight over the desert near Edwards AFB. It
was the eighth F-15 built and had been involved in spinning tests, and along with
71-0281 acquired on indefinite loan for a variety of research tasks. Today
71-0287 is on the NASA inventory as NASA 835. (NASA ECN 17710)

The capacity of Russian rockets was far beyond any known vehicles in the free world, and the Soviets were well in the lead for some years to come. Their launch and control systems for large rockets were also very advanced. The challenge was on. The American people required no further convincing that a major space effort, costly though it would be, should be undertaken as soon as possible. There was, however, the question of the composition of the space effort and who would administer it. One of the main contenders for running it was the US Department of Defense — DoD. It was evident that each of the three US armed services appeared willing to take on the task individually, so each was competing with the other two services in addition to interested civil agencies.

The US Navy was already involved in the *Vanguard* satellite project, whilst the US Army had perhaps given more thought to the design of satellites than either of the other two services. It was the US Air Force which was busy building rocket motors most likely to be used in the space programme, and was already in command of the huge Atlantic Missile Range. On the civilian side, the National Academy of Sciences (their IGY Committee organized the *Vanguard* project) had a serious interest in any national space programme. Along with the American Rocket Society it favoured a new civilian-orientated national space research agency. President Eisenhower had indicated a desire to avoid the militarization of space, and was afraid of adverse world reaction to an American satellite launched by the military using an ICBM — Intercontinental Ballistic Missile — rocket.

In January 1958, Dr Hugh Dryden, speaking for NACA, proposed that the US space programme be jointly undertaken by the DoD, NACA, the National Academy of Sciences and the National Science Foundation together with the universities, research institutions and industry of the nation. There was a predominant feeling that the majority of the US space programmes would be scientifically oriented, but best controlled by an aviation agency. The US Department of Defense still appeared to be in the running. Early in 1958 they formed the Advanced Research Projects Agency — ARPA — to supervise the military and hopefully the US space research programme. Whilst the personnel were civilians the agencies used for the tasks were the three armed services. ARPA received the endorsement of the President, and moved ahead with great rapidity.

Naturally the thoughts of the Ames staff were markedly altered by the foreknowledge of a new responsibility for Research into space. However the inertia of the laboratory's ongoing research programme was such that work continued largely unchanged during 1958, but the year was not one of high-productivity research. There were several reasons for this vacuum. The top personnel were busy planning for future space operations, whilst on the shop floor personnel were aware of a hovering shadow which indicated changes in the NACA organisation. A psychological disturbance throughout the agency in general led to resignation of a number of high-ranking research personnel.

X-15 landing-approach studies continued during 1958, with Ames flight test pilots George Cooper and Fred Drinkwater involved. The studies had indicated that the landing-approach manoeuvre could more easily be accomplished if a simple, and reliable method for quickly controlling the thrust of jet engines was available. The thrust response of jet engines was notably sluggish and when the pilot required a little more or less thrust to change the approach angle at a relatively high thrust and then to modulate it by means of a fast-acting thrust reverser, the problem would be solved. The use of a thrust reverser on an aircraft after touch-down was acceptable as feasible, but the research engineers were not happy at the thought of reversing the thrust of the engine whilst in the air. Nevertheless, Ames engineers under Alan A. Faye tested the concept, building a thrust reverser mounted in a Lockheed F-94C-1-LO *Scorpion* (50-956) which became 'NACA 156'. It

was delivered to Ames on 29 July 1954. The aircraft was first checked out in the 40×80ft tunnel after which the principle was demonstrated in flight. This research attracted great attention in aviation circles since it represented the first in-flight use of a jet thrust reverser. In addition to Alan Faye, George Cooper and Seth Anderson were involved. Regrettably, the Ames aircraft records indicate that the F-94 crashed on 18 November 1958.

By early 1958, the Agency became involved in the planning of a space-research programme. Its interests in space were reflected in the subject matter of the papers presented at the NACA Conference on High Speed Aerodynamics held at Ames in March. A major item was the launching into orbit of a manned vehicle and there were a number of interesting possibilities as to what kind of vehicle it should be. A simple blunt-nose capsule like a ICBM warhead, having no lift capability was one candidate. It should be a relatively simple construction of nonsymmetrical shape and elementary control surfaces would be capable of producing small amounts of lift for achieving a degree of re-entry flight-path control. Finally, it could be a winged glider providing considerable control over its re-entry path and landing site. In general, it appeared that Langley research personnel favoured a non-lifting vehicle and the Ames research personnel a lifting vehicle of some kind. Each had certain advantages.

The Bell XV-3 VSTOL aircraft 54-148 arrived initially at Ames for 40×80ft wind tunnel tests on 4 August 1957, remaining until October 28. It returned for flight test trials on 12 August 1959 and is depicted in the hover mode at Ames. It departed back to the manufacturer on 9 June 1965. (NACA A-27737)

The virtues of the non-lifting vehicle seemed to outweight those of the lifting vehicle. The dominant influence was perhaps competition. The USSR had convincingly demonstrated their satellite-launching capabilities as well as the impressive power of their booster rockets. They would not be too long before they put a man in space. The US Air Force had initiated a man-in-space project and was moving ahead with all possible despatch. It was clear to NACA that the watchword was speed. The vehicle that would allow the man-in-space task to be accomplished most quickly was the one to be selected at this time. Later, other types of vehicles could be listed, but the answer now seemed to be simple — a non-lifting vehicle.

At this time the Ames laboratory became very interested in orbits and trajectories. Since the lifting vehicles were expected to be controllable in landing as well as re-entry, their slow-speed characteristics were of great importance. Representative of the slow-speed tasks run on such vehicles were those conducted in the 12ft tunnel. There were many more research programmes involving space. However on the first day of October 1958, the National Advisory Committee for Aeronautics — an agency which had served the United States for forty-three years — ceased to exist.

CHAPTER TEN
Ames and V/STOL Research

In late 1965, Smith De France, who had built, moulded, and directed Ames with a firm hand since the virgin days of 1940, retired. His sucessor was Harvey Allen, who admitted to being more a researcher than a centre director and had also been at Ames since 1940. He owned a vintage Duesenberg, was a gourmet cook, a lover and player of classical music, a Perry Mason fan and an amateur archaeologist. However his reign was short-lived. Unwilling to tolerate increasingly complex management demands or to confront the unavoidable reduction in which everyone knew were approaching, he retired early in 1969 and for a time afterwards the future of Ames was bleak; there was even talk and strong rumours of the centre closing. Fortunately this was an idea which never got beyond speculation.

In the search for a new director Jack Parsons stepped in briefly as acting director, having been associate director since 1952. However, ill health forced his retirement almost immediately. He died barely five months later, leaving a large gap in the administrative hierarchy. Harvey Allen returned briefly as acting director. In February 1969, the NASA announced a new director. Dr Hans Mark was professor of nuclear engineering at the University of California at Berkeley, and head of the research reactor at Livermore. He brought to the centre an outsider's view of Ames's strengths and weakenesses, and an energetic management approach which created an effective variety of research options for the centre. He remained in command until 1977. The new director changed the working atmosphere almost immediately,

On 2 April 1965 the first production Convair Cv.990 Coronado c/n 001 arrived at Ames to be fitted out as a flying laboratory designed to facilitate astronomical studies high above the haze of the atmosphere. This task took the Convair named "Galileo" NASA 711 on international global flights. However on 12 April 1973 it was lost in a mid-air collision with a US Navy Lockheed P-3 with loss of 16 personnel. (NASA A-36864-3)

often working a 12-hour day and expecting his staff to do the same.

On 2 April 1962, the US Navy at Moffett Field transferred 75.6 acres to NASA and this — combined with a further 39.4 acres covering several purchases of adjacent privately-owned land — brought the total owned by the Agency to 115 acres. Ames in 1969 was spread over some 366 acres, operated no less than eighteen wind tunnels, two sets of ballistic ranges and ten flight simulators. There were eleven arc-jet facilities in operation and eight laboratories. Major buildings numbered fifty-five, soon to be increased by the completion of the Flight Simulator for Advanced Aircraft — FSAA — which had been built for testing the characteristics of jumbo jets and supersonic transports. Costing $2.6 million and six stories high, the huge new simulator topped the range of research tools available at Ames in the flight simulation field. During January 1969, just prior to Hans Mark being appointed director, Ames announced the construction of a Space Sciences Research Laboratory. The Californian centre also boasted a growing fleet of unique research aircraft, including a twin-engined Model 23 Learjet 'NASA 701' which had arrived on 17 September 1965 and a huge Convair CV990 *Coronado* 'NASA 711' which arrived on 2 April 1965 and had been named 'Galileo'. This was a flying laboratory, designed to facilitate astronomical studies high above the obscuring, hazy atmosphere at ground level. It had thirteen ports cut into the upper fuselage so that telescopes and other instruments could be pointed upward at the starry panorama during a five-nation eclipse observation programme conducted from Ames during 1965. Another flying laboratory was a Convair 340 'NASA 707' resident since 21 May 1963, whilst a de Havilland (Canada) C-8A *Buffalo* 'NASA 715 was soon to be converted into an augmented-wing jet STOL aircraft.

A psychological blow to Ames was the transfer to the Flight Research Centre at Edwards AFB during 1959 of the flight research aircraft. However, over the next decade Ames slowly regained its variety of research aircraft, especially those involved in V/STOL research. The San Francisco Bay area was ideal for flight research with these new types, with the 40 × 80 ft wind tunnel and the flight simulators providing a variety of facilities for obtaining, checking and expanding V/STOL data. Over the next few years, reseach aircraft would become more and more essential to the activities at Ames. In February 1965, an agreement between NASA and the US Army recognised the agencies' 'mutual interests in aviation technology with a collaboration sought to achieve tangible economies and promote efficiency with respect to continuing research and development of aeronautical vehicles.' In late 1960, the US Army Aeronautical Research laboratory was established at Ames, with the US Army providing staff and operating certain facilities, whilst NASA supplied technical and personnel support. the US Army refurbished and modernised one of the two old 7 × 10 ft wind tunnels and assigned forty-five people to assist in research. Joint research projects were soon under way in that tunnel. As can be seen by the chart of V/STOL projects, mostly sponsored by the US Army, an extensive flight test programme involving converti-planes and early V/STOL types was under way. Some came to Ames on a joint NASA and US Army Aviation programme involving either wind-tunnel testing, flight testing, or both. In the field of astronautics, sounding rockets launched from White Sands Missile Range in New Mexico carried payloads from Ames. Test flights designed both to check the rocket's control system and to map the magnetic field of the flightpath were successful. In co-operation with the US Air Force, the X-24 lifting body had undergone full-scale wind-tunnel testing at Ames. First trials would take place at Edwards during 1969–70.

In 1968, two aeronautical research projects were providing for the future of Ames. One was the developing Convair CV990 *Coronado* programme of flying aeronautical laboratories. Operated as a national facility for scientists, the first was about to be equipped for a wide variety of research tasks including infra-red photography, meteorological testing and forest fire spotting. The other was a new V/STOL research aircraft, the Ryan VZ-3 *Vertiplane* or *Vertifan* which underwent flight tests before delivery to Ames on 24 August 1959. It became 'NASA 235', later 'NASA 705'. This unique aircraft was equipped with control-rotating propellers submerged in the wings and driven by the jet exhaust. These provided lift for vertical take-off, hovering and vertical landing.

The NASA centres at Ames, Langley and the short-lived Electronics Research Centre in Cambridge, Massachusetts, became involved in a five-year programme to improve light-aircraft technology and devise a workable collision-avoidance system. The three NASA centres each had $500,000 to spend on different aspects of the programme. The role of Ames was to flight-test six of more general-aviation aircraft, as well as to perform a series of wind-tunnel tests on its Learjet. By June 1966 the total NASA personnel employed at Ames reached a peak of 2,310.

A major resource was its variety of motion simulators. Approved in 1965, the Flight Simulator for Advanced Aircraft was not put into operations until 1970. This and other similar simulators would help to solve the problem of the future of Ames, a centre involved in a little of everything. Faced with heavy and continuous budget cuts, the NASA headquarters were naturally wondering if Ames was essential. The Vertical Motion Simulator planned during the early 1970s was submitted as part of the projected facilities building programme for 1974. It was finally approved in 1975, as a result of steady efforts on the part of Ames to convince headquarters that it was essential for both VTOL and helicopter research. Hans Mark argued, quite correctly, that Ames was equipped with the best flight simulators in the

world. The 40 × 80 ft wind tunnel, which had been in continual use since its dedication in June 1944, continued to be one of Ames's most important facilities. As low-speed VTOL and V/STOL work gained in importance, the huge tunnel was operated round the clock to accommodate the many demands on it. It was refurbished during 1973 and a few years later an extension was added to create a second test section, an 80 × 120 ft offshoot attached to the original tunnel. Increasing the versatility of the tunnel, the addition greatly strengthened the unique facilities argument.

The Bell X-14A 56-4022 VTOL research aircraft was originally powered by two Bristol Siddeley Viper turbojets. It had Beechcraft Bonanza wings and undercarriage, and T-34 Mentor tail surfaces. It went to Ames on 2 October 1959 serving until 29 May 1981 when our good friend Ron Gerdes was involved in a crash landing with no injuries. (NASA A-29976)

Now preserved in the US Army Aviation Museum at Fort Rucker, Alabama, the Bell X-14B NASA 704, with the fuselage heavily tufted is seen flying in the vicinity of Moffett Field during 1971. Pilot is Ron Gerdes an experienced NASA engineer flight test pilot. (NASA A71-6414)

Research aircraft and their associated projects played a prominent role in the life of the Ames research institution. Between 1969 and 1976 the centre acquired several aircraft. During March 1970 NASA ordered a Lockheed Model L 300-50A C-141A *Starlifter* transport to be equipped as a flying astronomical laboratory and fitted with a 91.5 cm telescope. This aircraft came direct from the Lockheed-Marietta, Georgia production line and was delivered to Ames on 3 February 1972, to be registered 'NASA 714' and known as the 'Gerard P. Kuiper Airborne Observatory'. It became fully operational during 1974. In the same year, in co-operation with the Canadian government, plans were commenced for a jet-powered augmented-wing STOL research aircraft. A de Havilland (Canada) C-8A *Buffalo* 'NASA 716' had been delivered to Ames on 10 June 1967, remaining until 22 September 1981 when it was transferred to the Canadian government in Ottawa. This was the first of a new line of experimental short-haul research aircraft. That same year, in co-operation with the US Army, a tilt-rotor research vehicle was planned; a further example of major V/STOL work involving NASA and an outside agency. During February 1973 the US Air Force signed an agreement with NASA for development of STOL aircraft in conjunction with the facilities at Ames.

The Convair CV990 'Galileo' ('NASA 711', the prototype *Coronado*) became one of the major programmes at Ames, proving invaluable in a variety of co-operative research ventures with other agencies. A medium-altitude research aircraft, 'Galileo' had undertaken a number of weather and earth resources surveys. Tragically, on 2 April 1973 it collided over Moffett Field with a US Navy Lockheed P-3 *Orion* whilst both were making approaches to land. The loss of the entire crew of sixteen was a heavy blow for Ames, but plans were immediately initiated to acquire another CV990 which was delivered on 10 December 1973. It was registered 'NASA 712' and aptly named 'Galileo II'. During 1974 it resumed the task of its predecessor. Another major addition to Ames was the Quiet Short-Haul Research Aircraft — QSRA — for which design plans were completed in 1974. On 2 August 1974 a de Havilland (Canada) C-8A *Buffalo* 'NASA 715' returned from loan — since 1967 — as N326D with the National Science Foundation at Boulder in Colorado. It went to the Boeing factory at Seattle in Washington State for modification to QSRA configuration, returning to Ames on 3 August 1978. This graceful aircraft proved an exciting research tool, its remarkably short touch-downs approaching avian competence.

A full programme of research projects was undertaken by the new Convair CV990, ranging from studying wildlife migration patterns and ice-floe movements to making archaeological surveys of Mayan ruins and studying monsoon behaviour in the Indian Ocean area. On 16 May 1977, the Convair flying laboratory under the joint management of NASA and the European Space Agency (ESA) began a ten-day simulation of a Spacelab mission. A mission specialist and two payload specialists each from NASA and ESA participated. Spacelab, a major element in the Space Shuttle System, was located in the cargo bay of the Orbiter with other facilities and equipment similar to those in laboratories on the ground. Objectives of the simulation included evaluation of management of payload and mission operations to develop low-cost concepts for Spacelab, studies of interactions between Spacelab personnel and principal investigators on the ground, and the development of minimum training requirements. Another prime concern was to involve ESA and NASA Spacelab managers in the same roles they would have during an actual Spacelab flight. 'Galileo II' was to make six-hour flights on each day of the simulation, and the payload and mission specialists would remain confined throughout the ten-day period to work on the experiment payload and sleep in

VSTOL research increased in tempo at Ames during the early 1960s prompted by the requirements of the US armed services. This Ryan VZ-3 Vertiplane 56-6941 was developed under a joint US Navy/US Army programme. It arrived at Ames on 24 August 1959, being retired to the Smithsonian Museum on 20 June 1966. (NASA A-26050)

adjacent living quarters.

'Galileo II' surveyed archaeological sites in Guatemala on 25 October 1977 in an effort to learn more about the Mayan civilisation which flourished three centuries ago. Three different types of radar were used to penetrate the dense tropical foliage to different depths, allowing identification of features not readily disinguishable by other means. Signs of roads, stone walls, agricultural terraces and other man-made structures were sought. The aircraft also carried a scanning infra-red sensor to detect differences in vegetation, seeking clues to the extent and type of farming done by the Mayans. The flight was a co-operative effort among researchers at the NASA at Ames, the Jet Propulsion Laboratory at Pasadena and the University of Texas at San Antonio.

On 2 May 1979 'Galileo II' participated in a summer-long international study of the summer monsoon, which annually brings torrential rains to the Asian subcontinent. Operation MONEX — monsoon experiment — explored the origin of the monsoon winds in order to improve short-range prediction and understanding of the monsoon's role in global weather patterns. The Convair operated from bases in Saudi Arabia and elsewhere in the region, in co-ordination with several other aircraft, ships and a variety of ground-based facilities. The operation was part of a large-scale atmospheric research programme being conducted by the World Meteorological Organisation of the United Nations.

Research

Work on aircraft and flight problems was not neglected, but it was noteworthy that work on spacecraft and space flight predominated. Notable features of research during this period included the acceleration of the development and use of flight simulators and the work building up on V/STOL aircraft. A tragedy occurred on 19 April 1961 when flight test pilot Donovan R. Heinle was killed in the crash of a McDonnell F-101 *Voodoo* at Edwards. Reportedly the aircraft got into a flat spin from which Don was unable to effect a recovery. He was well liked at Ames and had made many contributions to the centres' flight research; his passing was a sad loss.

Manpower problems at Ames became acute. The centre had not been permitted to augment its staff to handle the new research responsibilities thrust upon it, and some of the new fields of research required training and experience of a kind which the old NACA staff did not possess. Disciplines involved in life-science research were obviously outside the scope of NACA's experience and training of one kind or another would be required. Through oral persuasion and the manpower squeeze, the NASA headquarters was forcing Ames to move in the direction of research contracting. The Life Science Directorate clearly and with good reason would use contracting to accompish many of its research objectives. The Space Sciences Division, which, because it was new, suffered most keenly from the manpower shortage, would also be forced to follow the contract route, at least for routine services such as computing and for supporting research and development work.

Many of the aircraft configuration studies undertaken during the 1960s were concerned with the design of the North American B-70 Mach 3 bomber and with preliminary designs for a commercial supersonic transport. The B-70 project commenced formally on 14 October 1954 when the US Air Force issued its GOR 38 — General Operational Requirement — for an intercontinental bombardment weapon-system piloted bomber. The aircraft was to be all-chemical-fuelled and was expected to be in the active Strategic Air Command inventory during the period 1965–75, with a target introduction for the first thirty aircraft by 1963. Issue of the GOR came after some two years of US Air Force, NACA, NASA and industry studies of the feasibility of such a weapon system, which was initially for an aircraft with a Mach 0.9 cruise and supersonic dash capability for a 1,000 mile penetration of enemy airspace. However, by the time the prototypes were ready the aircraft was designed to maintain a supersonic cruise at all times, with a top speed of Mach 3.

Both Boeing and North American competed for the contract, the latter being selected during December 1957. The initial design studies had been launched at Los Angeles as NA-239 on 12 September 1955, but constant changes in the US Air Force directives took the project through NA-259, NA-264, NA-267, NA-274 and finally NA-278, which covered three prototypes ordered on 4 October 1961. The weapon system designation for the bomber which became the B-70 was WS110A. A reconnaissance version was also considered (as WS110L) but this was dropped in July 1956. Some five years later the strategic reconnaissance version was resurrected as the RS-70 at which time the US Air Force was seeking to retain the programme in the face of mounting opposition to the concept of a supersonic unarmed bomber force. The third prototype was deleted from the programme on 3 May 1964 and the construction of two XB-70A aircraft was allowed to proceed only as a research programme.

The SST had become of interest as a national development project and it was felt to be the responsibility of the NASA to take the lead in determining the most promising general configuration from which a successful SST might be developed. Langley sponsored what were known as the supersonic commercial air transport — SCAT — configurations. One design had a wing with controllable sweep, another had a canard configuration with a fixed delta wing. These two designs were studied intensively by NASA in its wind tunnels and also by major aircraft companies. Ames concentrated its efforts on the delta-wing and canard configurations. Both the Boeing and Lockheed aircraft companies were later awarded government contracts for SST designs based essentially on the NASA Langley and Ames configurations.

As mentioned earlier, NASA headquarters ruled in late 1959 that all flight research involving flight testing would be conducted exclusively at the NASA Flight Research Centre at Edwards AFB. One exception to this rule was the flight research on V/STOL aircraft. The transfer of the variable-stability aircraft was a painful blow to Ames since the centre had pioneered the development and application of these highly useful devices. During 1958–9 a North American F-100 *Super Sabre* had been converted into one of the most elaborate and fully automated variable-stability aircraft ever built.

V/STOL Studies

It was acknowledged that there was a deep chasm between aircraft which derived lift and control forces from their motion through the air — aircraft — and aircraft which derived these forces largely from the direct application of engine power to special lifting surfaces — helicopters. Efforts to bridge the gap were now meeting with success as a result of the growing sophistication of aircraft and power-plant designers. However the resulting craft were capable only of short take-off landing — STOL. The general class of V/STOL aircraft was of growing interest, especially to the armed forces, and both the US Army and the US Air Force had let contracts with industry for the development of prototype V/STOL configurations. Both Ames and Langley greatly increased their interest and efforts in V/STOL work. Ames concerned itself with the low-speed aspects of V/STOL research and in particular with the control and handling qualities of such craft. This worthwhile task was undertaken by Seth Anderson, whose work was published as 'An Examination of Handling Qualities Criteria for V/STOL Aircraft'.

On 6 March 1959 a Stroukoff YC-134A twin-engined transport (54-556) was delivered to Ames, to be registered 'NASA 222' and was followed on 30 June 1961 by Lockheed NC-130B *Hercules* 58-712. These two aircraft were investigated extensively by Hervey Quigley and engineer test-pilot Robert Innis, who had joined the Ames team in 1954. Both transports were capable of STOL operation by virtue of boundary-layer-control installations. On the YC-134A the suction form of BLC was applied to the flaps and also to the ailerons, which moreover were dropped on approach to give additional lift. The Lockheed C-130, on which BLC and high lift were obtained by blowing air over the flaps and drooped ailerons, was also capable of STOL operation. The air for the BLC was provided by two jet engines mounted in outboard wing pods. At slow speed the YC-134A was observed to have a bad stall characterised by an uncontrollable roll and large sideslip angles. The NC-130B was closely observed and the lateral and directional handling characteristics were found under certain conditions to be so poor as to render landing the *Hercules* at a speed of less than 65 knots very difficult.

With wings heavily tufted and 'barn door' type flap system very evident, the Ryan VZ-3 56-6941 NASA 235 is seen in flight in the vicinity of Ames. Following a crash in 1959 it was rebuilt for the NASA, later becoming NASA 705. It served until 24 February 1959. (NASA A-29657-8)

Great interest had developed in more radical types such as the Bell X-14A VTOL test-bed prototype vehicle built for the US Air Force and flown extensively by engineer test-pilot Fred Drinkwater in the Ames flight research programmes. Interest in what was eventually to lead to the X-14 was first expressed by the US Navy during 1949. However, Bell had been conducting VTOL design studies since June 1944. Development work between 1949 and 1956 culminated in the signing of the X-14 contract on 24 May 1956, design work being initiated during July 1955. The contract number, dated 14 May, was AF33(600)30837.

North American X-15 research aircraft about to touch down at Edwards AFB on its dual nose wheel and tail skids. The marks under the fuselage are due to frozen condensation. The first X-15 flight, without power, was on 8 June 1959 and the last on 24 October 1968. (Office of History — USAF Edwards)

The Bell X-14 was created to explore the impression of a pilot flying a VTOL aircraft from a normal crew station whilst using standard aircraft flight references. Of equal importance was the research into general VTOL technology, involving both aircraft and engine. The X-14 was the first VTOL aircraft to fly using a jet thrust diverter system for vertical lift. Only one was ordered by the US Air Force and was allocated serial number 56-4022. During April 1953 Bell undertook the design of a testbed aircraft known as the Model 65 Air Test Vehicle using existing aircraft components on completion was registered N1105V. The X-14 was planned to accomplish an extension of the earlier ATV programme, again incorporating existing aircraft components where possible. The wings, landing gear and ailerons were from an early Beechcraft 35 *Bonanza*, and the empennage section including vertical and horizontal tail surfaces was from a Beechcraft T-34A *Mentor*.

No less than 199 flights were completed by the three X-15 research aircraft. The first 56-6670 is in the National Air & Space Museum, the second, 56-6671 is seen after delivery to the USAF Museum at Wright-Patterson AFB. The third 56-6672 was destroyed on 15 November 1967 whilst flying at Mach 5 and 125,000 ft. Major Michael Adams, USAF, was fatally injured. (USAF Museum)

Two Armstrong-Siddeley *Viper* turbojet engines were selected for propulsion. On 2 October 1959 the Bell X-14 (56-4022) was delivered to Ames, and becoming initially 'NASA 234' and later 'NASA 704'. The Viper 8 engines were removed, and replaced by two General Electric J85-5 turbojets rated at 2,680 lbs thrust each. In turn, two J85-GE-19 turbojets rated at 3,015 lbs thrust each were subsequently fitted. These latter engines were acquired from the defunct Lockheed XV-4 *Humming Bird* programme. Following this engine change, the Bell X-14 was redesignated X-14B.

The X-14 concept verified that the vectored-thrust jet V/STOL aircraft was a viable design. The technology later proved of great interest to Sir Stanley Hooker, who utilized data from the NASA trials during the design of the Hawker P.1127, predecessor of the *Harrier* VTOL fighter. Flight tests, employing the X-14's variable stability and control systems, resulted in major contributions to the understanding of V/STOL handling anomalies. The X-14 also proved useful as a testbed for unique V/STOL concepts, such as the NASA direct side-force manoeuvring system. Over 25 test pilots from around the world previewed V/STOL handling qualities of proposed designs in the X-14 at Ames prior to embarking on first flights in actual prototypes. The fact that the X-14 flew continuously and safely for over a quarter of a century is particularly noteworthy. Few accidents and no major injuries marred its distinguished career.

Flight testing continued until 29 May 1981, when a hard landing accident occurred following loss of roll control while in a ramp hover. The problem was traced to saturation of the VSCAS autopilot roll servos. The landing collapsed the main gear and ruptured the fuel tank: the ensuing fire seriously damaged the tail section. The engineer test pilot, Ron Gerdes, escaped injury, possibly due to the fact that the aircraft was always flight-tested at either an altitude of 2,500 ft or higher (so allowing for a manual bail-out) or no higher than 12 to 15 feet! Following the accident the X-14B was not flown again and was placed in storage at Ames: in May 1989 it was unceremoniously shipped to the US Army Aviation museum located at Fort Rucker, Alabama. It is worth recording that Neil Armstrong once flew the X-14B, simulating a lunar landing vehicle.

More radical designs of V/STOL aircraft envisaged the use of lifting propellers built into wings and fuselage. Such applications became more practical as a result of the development by General Electric of gas generators and turbine-driven fans especially designed for the purpose. As a joint project, the US Army, NASA at Ames and General Electric produced conceptual designs and preliminary evaluations of a number of different lifting-fan V/STOL configurations. Models of these were evaluated aerodynamically in the 40 × 80 ft tunnel.

Convertiplanes

Initially designated XL-25 in the liaison aircraft category, then XH-35 in the helicopter class and finally XV-1 as the first type in the new vertical lift category of aircraft designations, the 'Convertiplane' used the unloaded-rotor principle and was the result of an experimental programme undertaken jointly by McDonnell, the US Army, and the US Air Forces Air Research & Development Command. At Ames the XV-1 was tested in the 40 × 80 ft wind tunnel. Two aircraft were built 53-4016 and 53-4017, both now preserved in museums in the USA.

The Bell XV-3 V/STOL aircraft was unique in that it completed no less than four wind tunnel tests at Ames and was also involved in a flight-test programme which lasted from 12 August 1959 to 9 June 1965. During May 1951 the US Army and the US Air Force declared their official commitment to a new tilt-rotor programme by awarding a Phase 1 study contract to Bell Helicopter. Bell worked on a mock-up between July 1951 and May 1952, completing it in time for the first official armed forces inspection in June 1952. Under the terms of the original contract, the new Bell Model 200, was to be a utility-category vehicle. It was designed to carry a crew of two and two litter patients, or a total of four crew members. Additionally the design was optimized for observation and reconnaissance duties and rescue missions. Of greater importance, the Model 200 was to provide design and test data for the development of larger, high-performance tilt-rotor aircraft.

Though Bell used their own designation of Model 200 for the new tilt-rotor for record-keeping and part-number purposes, the armed forces soon came up with an official designation. At first the aircraft was referred to as the XH-33 in the helicopter designation listing which was in no way indicative of its tilt-rotor capability. Several months later, a decision was made to redesignate the new tilt-rotor test bed XV-3 under a new designation system conveniently covering many new designs in this field.

During October 1953 an official contract (AF33(600)-20777) for two full-scale XV-3 'titling-thrust-vector-convertiplanes' was awarded to Bell. Construction of the first XV-3 took from January 1952 to December 1954 to complete. Finally on 10 February 1955, XV-3 (54-147) was officially rolled out of the Bell production facility for the first time. Airframe and static tests lasted from March to August 1955, whilst in April 1955 the second XV-3 (54-148) was completed and rolled out to join its stablemate. With Bell pilot Floyd Carlson at the controls, the aircraft made the first hovering flight of the programme on 11 August 1955, completing the first major flight test objective successfully.

On 18 August 1955, shortly after the first hover flight, a hard landing occurred which caused minor damage to the airframe and rotors. The problem was quickly traced to wing-pylon-rotor instability. Fixes were applied during the rebuild which followed, and some four weeks later the XV-3 was once again ready for testing. A decision was made to initiate a required 200-hour tiedown test series to explore the myriad problems.

It was 11 July 1956 when the first XV-3's first inflight pylon change took place with a transition to five degrees, the pilot noting an immediate ten knot increase in forward speed and no major degradation in controllability. For the first time the XV-3 had demonstrated that it could, indeed, do what it was designed to do. Three Bell test pilots were now conversant with the new vehicle; Floyd Carlson, Dick Stansbury and Elton Smith. The XV-3 was equipped with an unconventional but functional ejector seat powered by bungee cords and released downwards.

By October 1956, partial conversions to 20 degrees had been completed and maximum speeds of 80 knots had been achieved. Unfortunately, 25 October 1956 was a black day in the history of the XV-3 programme. Following recurrence of dynamic instability problems, the first XV-3 (54-147) crashed just a few miles from the Bell factory at Saginaw, Texas. The pilot, Dick Stansbury, survived the accident, though not without sustaining permanently disabling injuries. The loss of the aircraft and pilot was a major blow to the programme.

Several weeks after the accident, a decision was made by the tilt-rotor design team to run a full-scale wind tunnel test using the second XV-3 (54-148) in the Ames 40 × 80 ft tunnel. On 4 December 1956, a meeting was held at Ames to discuss this proposed test. Agreement to go ahead with the development and flight test of a two-bladed rotor system, replacing the three-blade system, was reached. On 4 August 1957 the XV-3 was partially dismantled and loaded aboard a US Air Force Douglas C-124 *Globemaster* for shipment to NACA at Ames. Three weeks later it was ready for mounting in the 40 × 80 ft tunnel. The test programme lasted for six weeks. The end result of these tests was discovery of a pylon-rotor-wing torsional-mode instability at high rpm and conversion angles above 30 degrees. Most of the necessary modifications were made following the return of the aircraft to the Bell factory on 28 October 1957. Ground runs were resumed the day after Christmas. On 21 January 1958, Floyd Carlson, made the first successful hover, this being followed by a quick and cautious flight evaluation test programme.

On 17 February 1958, 54-148 accomplished its first tentative transition, moving the pylons out to five degrees and back. Two days later this was expanded to 15 degrees and 80 knots, and by 7 March to 29 degrees and 115 knots. Confidence was returning to the tilt-rotor team and the envelope was expanded more rapidly. On 21 March 55 degrees was reached, but on 6 May instability suddenly raised its unpleasant head once more. During May 1958 a decision was made by Bell, the NACA and the US Army to go back for a full-scale wind tunnel programme. Unfortunately at this time the 40 × 80 ft tunnel was heavily programmed, so Bell decided to complete an analogue computer study prior to turning the aircraft over to Ames. On 16 September 1958 the XV-3 was partially dismantled and flown to Ames in a US Air Force Lockheed C-130 *Hercules*.

North American X-15 56-6670 seen parked by its 'mother-ship' Boeing NB-52 Stratofortress, showing clearly the 'crutch' type pylon fitted to the starboard wing of the B-52. The striped nose test boom on the X-15 is equipped with both pitch and yaw vanes. (NASA 60-X-14)

Two weeks later the XV-3 was mounted inside the wind tunnel and trials commenced on 7 October, but unlike the preceding tests these were not conducted by remote control but had a pilot in the cockpit throughout the programme. This was a practice always discouraged by NACA and it was finally discontinued in the mid-1970s due to the strong possibility of accidents. Bill Quinlon made most of the XV-3 tunnel runs and the veteran engine test pilot Fred Drinkwater 'flew' the aircraft during at least one wind-tunnel test. The trials were completed on 24 October and the XV-3 was removed from the wind tunnel three days later; it was returned to Bell on 29 October in a Lockheed C-130 transport.

On 16 November 1961 three unique Hiller YROE-1 one-man foldable helicopters, built under licence in the United Kingdom by the Saro company at Eastleigh, Hampshire, were delivered to Ames. They were from a small batch built for the US Marine Corps and had the mysterious small BuNo.'s 4020, 4021 and 4024. Depicted is an early XROE-1 in US Marine Corps markings being test flown at the factory in the UK. (British Hovercraft Corporation H373-B)

The Lockheed XV-4A Humming Bird of the US Army 62-4500 arrived at Ames on 15 July 1964 for tests in the 40 × 80ft tunnel. These were completed by September 16. Depicted is XV-4A 62-4503. At this time the US Army were also evaluating the Ryan XV-5A and the Hawker Siddeley P.1127 Kestrel. (Lockheed RE 7975-2)

The XV-3 successfully converted to 30 degrees on 17 December 1958 with no problems, and on the following day Bill Quinlan accomplished the first full conversion to 90 degrees. At long last the aircraft had done what it was supposed to do. By early April 1959 the programme was progressing smoothly, with rolling take-offs and landings, and on 14 October the XV-3 achieved an altitude of 12,300ft, reaching 11,200ft on its second flight of the day. Phase 1 testing had been completed on 24 April 1959 and the XV-3 was flown by Lockheed C-130 to Edwards AFB for initiation of Phase II testing. This was completed on 9 July, when the aircraft was flown to Ames in a *Hercules*. The total hours on the airframe were just under 125.

At Ames a new XV-3 flight-test programme was begun and lasted for just over two years, ending in July 1962 with airframe hours of 325 having been reached. Following completion of its NASA flight test programme, the aircraft was returned once more to the 40 × 80ft wind tunnel. In May 1965 Bell requested return of the XV-3 and a Douglas C-124 *Globemaster* flew the aircraft back to Texas on 9 June 1965. In May 1966 the XV-3 was back at Ames in the now familiar wind tunnel — for the fourth time. It was unfortunately damaged when a mounting failed, severely damaging the rotors and airframe, and it was determined by both Bell and NASA that a rebuild programme would be prohibitively expensive.

Following a year or so in storage at Wright-Patterson AFB, the XV-3 was placed in storage at Davis-Monthan AFB in Arizona. It could lay claim to an impressive and unquestionably successful flight test history. Among these claims were 270 flights, a total flight time of over 325 hours, 110 full conversions and the tilt-rotor indoctrination of several pilots, including at least three from the US Army, two from the US Air Force, two from NASA, and four from Bell. Power for the XV-3 was provided by a single 450hp Pratt & Whitney R985-AN-1 radial engine. An advanced development of the XV-3, known as the XV-3A and powered by a 600bhp Lycoming XT53-(LTC1B) turbine engine, was projected during the early days of the programme. Essentially this would have been converted from one of the two XV-3 prototypes, modified and updated. This aircraft differed from its piston-engined stable-mate somewhat significantly in the performance envelope. Development of the XV-3A never got beyond the mock-up stage.

The XV-15 Tilting Testbed

It was not by chance that the US Army, the US Air Force and the Federal Aviation Administration started co-operative work with Ames on V/STOL aircraft and helicopters. A new perspective on the centre's role within the NASA framework was slowly emerging, taking advantage of changing national priorities and consciously guided by director Hans Mark. In late 1968, the Senate Committee on Aeronautical and Space Sciences had announced that a new commitment was to be made in federally supported aeronautical research. This would clearly affect Ames, since its research effort within the NASA had always tended toward aeronautics rather than towards space. The 1970 annual report by the director mentioned the strong possibility that Ames would play a leading role in future short-haul aeronautical research, including V/STOL aircraft. Two years later Hans Mark announced that the national budget for research and development in aeronautics had increased from a little over $100 million in 1969 to $161 million.

The decision to transfer all helicopter research from Langley to Ames has remained controversial in some quarters. Langley had been involved in helicopter research since the 1930s, and it was natural that some personnel at that centre were understandably upset at losing that element of its research. After the decision was announced in 1976 it was pointed out that the transfer was logical in terms of Ames's short-haul centre designation, and in view of the relationship which Ames had established with the US Army Air Mobility Research and Development Laboratory. The helicopter research provided tangible proof of the significance of the alliances Ames had built and the ability of the centre's managment to use them for Ames's benefit. Co-operation with the US Army for joint research in VTOL studies dated back to 1965, and was greatly expanded by 1969. The US Army personnel at Ames had doubled, and in 1970 the Army consolidated its flight research and development at the US Army Air Mobility Research & Development Laboratory, with headquarters at Ames. Co-operative work in the large VTOL and V/STOL studies programme led to additional exploratory and advanced work, with the tilt-rotor research aircraft and later the Rotor System Research aircraft — RSRA — coming out of the first joint successes. The US Army's endorsement of Ames would help the centre gain the leading role in helicopter research within NASA. Back in

1973 a similar agreement was reached with the US Air Force, both agencies co-operating on the development of both the US Air Force Advanced Medium STOL Transport prototype and the NASA Quiet Propulsive Lift Flight research programme. Later in 1973 a US Air Force office had been established at Ames to administer joint programmes, with personnel permanently assigned. By 1976 co-operation with the Federal Aviation Adminsitration had expanded to include joint research using the augmenter-wing de Havilland (Canada) C-8A *Buffalo* and establishment of the Air Safety Reporting Systems Office, which conducted continuing analysis of air safety factors.

Neil A. Armstrong, the first man on the moon, completed seven X-15 flights, making the first on 30 November 1960, achieving Mach 1.75 1,155 mph and 48,480 ft. His last flight was on 26 July 1962, in 56-6670 achieving Mach 5.7 3,989 mph and 98,900 ft. Fitted to the X-15 nose is the aircraft's attitude sensor which assists during critical exist and re-entry periods. (NASA 60-X-35)

The US Army Ryan VZ-11/XV-5A 62-4505 NASA 705 seen parked at Ames where it arrived in March 1964 remaining until 30 January 1974. It was involved in both 40 × 80 ft tunnel tests plus a flight test programme. It was damaged during October 1966 and reconstructed as XV-5B. It is today preserved in the US Army Aviation Museum, Fort Rucker, Alabama. (NASA A-41358)

The demise of the pioneering Bell XV-3 tilt-rotor aircraft in 1966 left the company without a full-scale 'proprotor' flying testbed. The interest in tilting proprotor designs remained exceptionally strong, however, and a large number of related advanced technology projects were thoroughly explored during this period of apparent inactivity. In 1965 the US Army established a programme for a composite-capability aircraft which would combine, in one aircraft, the good hover characteristics of the helicopter and the efficient high-speed cruise abilities of the conventional fixed wing aircraft. The result of this programme with Bell was the birth of the Model 266, a twin-tilt-rotor design that was large enough to be considered as a possible replacement for the Boeing Vertol CH-47B *Chinook* helicopter, and the de Havilland (Canada) C-47A *Caribou* aircraft. The basic Model 266 mission requirement continued for a time as the US Air Force Light Intra-Theatre Transport — LIT. Funding constraints brought on by the war in Vietnam, coupled with a questionable need, eventually led to the project's demise in early 1969.

During 1972, it was determined by the US Army and NASA that a full-scale development in the form of a testbed vehicle was indeed feasible, practical and appropriate. The two agencies provided funding for the design, fabrication and testing of two full-scale tilting proprotor vehicles. A design competition took place between the various companies involved in tilt-rotor research, the field being narrowed down to Boeing Vertol and Bell. A three-month study contract was quickly awarded to each company. The Boeing effort was the rather conservative Model 212 Tilt Rotor Research Aircraft, whilst Bell's offering was a significantly improved Model 300 with a gross weight of 12,400lbs and a redesigned 'H' tail offering improved directional stability. NASA requirements involved additional changes, not the least of these being the replacement of the 1,150shp Pratt & Whitney engine with 1,550shp Lycoming engines and incorporation of dual-wheel Canadair CL-84 type landing gear in place of single-wheel units. A changeover from Douglas ESCAPAC 1-E ejection seats to Rockwell-Columbus LW-3B ejection seats was also made, together with the addition of crashworthy fuel tanks,

boosted rudder/elevator controls and a third hydraulic system for redundancy. The Model 300 became an almost totally new aircraft and was redesignated Model 301.

On 17 April 1973 the NASA team announced their decision for Bell to proceed with their proposal, and on 31 July a contract was officially awarded to Bell Helicopter Textron. On 26 September 1973, the US Army and the NASA decided to bless the Model 301 with an official designation, the XV-15 — a title suggesting both its experimental and VTOL nature. The engineering design was completed by March 1975, and final assembly of the first aircraft was completed by May 1976 with an official roll-out on 22 October 1976. Ground runs commenced on 21 January 1977, with first conversion to 75 degrees on 22 March and first full conversion on 31 March 1977. By the first day of May, the XV-15 had completed over 31 successful ground runs with a free hover accomplished on 3 May. A decision had been made by NASA to limit the flight test time of the first XV-15 to three hours, and with this complete on 31 May it was restricted to ground-run testing. On 23 March 1978 the XV-15 was airlifted by Lockheed C-5A *Galaxy* to Ames for wind-tunnel testing which began on 18 May and was complete on 23 June.

Further exploration of the XV-15 envelope was scheduled to be undertaken by the second aircraft. Limited funding became available from the US Army. NASA and Bell with the US Navy eventually entering the programme as a funding agency. The first hover flight of this second XV-15 was accomplished on 23 April 1979 but it was grounded on 22 August for a required 100-hour inspection. The first aircraft eventually became registered 'NASA 702' and the second 'NASA 703'.

On 30 August 1977, the name of the US Army Air Mobility Research and Development Laboratory at Ames, was changed to the US Army Research and Technology Laboratories with no change in mission task. The Ames historical records indicate that the Bell Model 301 XV-15 (c/n 1) arrived on 23 March 1978 for research flight testing, and c/n 2 arrived on 30 October 1980, to be involved in a similar programme which continues today.

Research Convertiplanes and Tilt Rotor Vehicles

Desig.	Manufacturer	Serial	Specification
VZ-1	Hiller	56-6944/45	One-man flying platform, propelled by a ducted airscrew. Two were delivered to the US Army. Original designation YHO-1E.
VZ-2	Vertol	56-6943	Model 76 tilt-wing research aircraft, powered by a YT53-L-1 engine driving two wing-mounted airscrews. One built with US Army funding, and flight tested by the US Army, the US Navy and the NASA.
VZ-3	Ryan	56-6941	Model 72 VTOL research aircraft powered by a 1,000shp Lycoming T-53-L-1 engine. One built and initially flown by the US Army with tail-wheel undercarriage. Modified in 1958 with a nosewheel, ventral fin and other modifications. Following a crash in February 1959, it was rebuilt for NASA as 'NASA 235' later 'NASA 705' with a lengthened fuselage and open cockpit, flying from Ames until it was retired and presented to the US Army Aviation Museum, Fort Rucker, Alabama.
VZ-4	Doak	56-6942	Model 16 VTOL flying platform. With a 1,000shp Lycoming T53-L-1 engine driving tilting ducted airscrew. The US Army took delivery of one VZ-4.
VZ-5	Fairchild	56-6940	Model M224-1 VTOL research aircraft. Powered by a single YT58-GE-2 engine buried in the fuselage driving four wing mounted airscrews and a small tail rotor. The US Army procured a single VZ-5.
VZ-6	Chrysler	58-5506/07	Two VZ-6 aircraxft were built for the US Army as a VTOL ducted-fan test vehicle.
VZ-7	Curtiss-Wright	58-5508/09	Two VZ-7 aircraft were delivered to the US Army powered by a Turbomeca Artouste turboshaft driving four direct-lift airscrews.
VZ-8	Piasecki (Vertol)	58-5510/11	Model 59K VTOL research aircraft powered by a Turbomeca Artouste IIB driving two ducted rotors. Two VZ-8 vehicles were delivered to the US Army.

Research Convertiplanes and tilt Rotor Vehicles — continued

Desig.	Manufacturer	Serial	Specification
VZ-9	Avro Canada	58-7055	'Flying saucer'-shaped VTOL platform powered by three J69 turbojets providing fan lift. One VZ-9 was built for the US Army. Named *Avrocar*.
VZ-10	Lockheed	62-4503/04	Became XV-4 in 1962.
VZ-11	Ryan	62-4505/06	Became XV-5 in 1962. 62-4506 to 'NASA 705'.
VZ-12	Hawker-Siddeley	62-4507/08	Became XV-6 in 1962. Serial batch 64-18262 to 64-18268 allocated. Named Kestrel.
XV-1	McDonnell	53-4016/17	Model M82 ex-XL25. Powered by a 400hp R-975-19 engine, driving a rotor and conventional wing layout. Two built.
XV-2	Sikorsky		Model S.57 radial powered convertiplane project. Not proceeded with.
XV-3	Bell	54-147/48	Model 200 powered by 450hp Pratt & Whitney R985-AN-1 radial engine driving a tilting rotor. Two were built for evaluation by the US Army. Originally ordered as XH-33.
XV-4	Lockheed	62-4503/04	Jet-lift research aircraft for the US Army. Two built initially as VZ-10 powered by two Pratt & Whitney JT12A turbojets. Second aircraft converted to XV-4B with six 3,015lb static thrust J85-GE-19 turbojets, four of which were vertically mounted in the centre of the fuselage and remaining two horizontally in the fuselage side.
XV-5	Ryan	62-4505/06	Direct-lift research aircraft for the US Army, with lifting fans buried in the wings and two 2,900lb static thrust J85-GE-5 for propulsion. Two prototypes were ordered as VZ-11 being redesignated XV-5A in 1962. The second aircraft extensively modified as XV-5B and transferred to the NASA at Ames as 'NASA 705'.
XV-6	Hawker-Siddeley	62-4507/08	The designation VZ-12 was reserved for two Hawker-Siddeley P.1127 VTOL research aircraft which were never delivered. Seven slightly modified tri-service P.1127 *Kestrals* were later procured with the designation XV-6A being test-flown and used by RAF Triparte squadron with serials XS688/696. Two additional Kestrels were to have become 64-18269/70 but were cancelled.
XV-8	Ryan	63-13003/04	Model 164 flexible-wing research vehicle, powered by a 200hp Continental pusher engine. The sail-like flexible wing was stretched between tubular frames. One XV-8A was built for US Army use. A second prototype was cancelled. Designation clashes with AV-8A *Harrier*.
XV-9	Hughes	64-15107	Model 385 high-speed research helicopter, with two auxiliary jet pods and a crew of two side by side. Only one XV-9A was built.
XV-10	Parsons	63-13070	Boundary layer-control development machine, designed by the Aerophysics Department of the Mississippi State University and built by Parsons Corporation with funding from the US Army. One XV-11A was completed with a T63-A engine driving a ducted pusher airscrew. It was initially flown with an experimental type of undercarriage which permitted operation from very soft ground.
XV-15	Bell	NASA 702/3	Model 301 tilt-rotor research aircraft designed to meet a joint US Army/NASA requirement. Two XV-15A ordered, the first of which 'NASA 702' first flew in May 1977. Side by side seating for pilot/co-pilot with cabin for up to nine passengers. Powered by two 1,550shp Avco Lycoming LTC1K-4K turbo-shaft engines mounted on the tips of the high mounted wings.
X-13	Ryan	54-1619/20	Model 69 VTOL fighter research aircraft, revolved from the XF3R-1 project and powered by a 10,000lb static thrust Rolls Royce Avon RA28. Two X-13 aircraft were built. A US Air Force fighter version designated XF-109 was not proceeded with.
X-14	Bell	56-4022	VTOL research aircraft powered by two 1,750lb static thrust Bristol Siddeley *Viper* turbo-jets and jet thrust diverters. Used Beechcraft *Bonanza* wings and undercarriage and Beech T-34A *Mentor* tail surfaces. One X-14 (later X-14A) built, and re-engined with two 2,450lb static thrust J85-GE. Transferred to NASA at Ames as 'NASA 234' later 'NASA 704'. The X-14B was a projected VTOL trainer with two J85-GE engines and the X-14C was a projected two-seat VTOL close support development version powered by three J85-GE turbojets.
X-18	Hiller	49-2883 57-3078	Tilt-wing VTOL transport aircraft, powered by two 5,850shp T40-A-4 driving three-bladed contra-props and a J34-WE in the rear fuselage to control pitch. A single YC-122C was rebuilt as X-18 being redesignated X-18A in 1962. Two Allison T40-A-14 turboprop engines came from the US Navy Lockheed XFV-1 and Convair XFY-1 *Pogo* programmes. A number of miscellaneous parts came from the US Navy Convair R3Y *Tradewind* transport programme. First flew at Ames.
X-19	Curtiss-Wright	62-12197/8	Model 200 tilt-wing VTOL light transport, powered by two 2,650shp T55-L-7 engines in tandem, each driving two airscrews at the tips of the wings. Only the first aircraft completed.
X-22	Bell-Aerospace-Textron	BuNo.51520/1	Model D.2127 tilting ducted-fan VTOL research aircraft powered by four 1,250shp YT58-GE-8B engines. Two built.
XC-142A	LTV-Hiller-Ryan	62-5921/25	Tri-service light VTOL research transport, powered by four T64-GE-1 mounted on tandem tilt-wings. Five XC-124A were built with 62-5924 eventually going to the NASA at Ames as 'NASA 522'.

CHAPTER ELEVEN
Blackbirds, Valkyrie, to the Shuttle

Between 6 August 1959 and 24 October 1968, the three North American X-15 research aircraft involved in the joint NASA, US Air Force and US Navy, aeronautical research programme made a total of no less than 199 flights. Until the first orbital flight of the Space Shuttle *Columbia* in 1981, the X-15 held the altitude and speed records for winged aircraft with flights as high as 67 miles and a maximum speed of 6.7 times the speed of sound at 4,518 miles per hour. According to John Becker the pioneering re-entry systems, their derivatives and the X-15's re-entry flight experiences led directly to the systems and techniques later employed in the Shuttle. Public relations literature surrounding the impressive success of the latter has quite rightly emphasised the development of the reusable ceramic tile heat-protection system, the enormous boosters and the automatic flight control systems but John Becker believed that too little has been said about the Shuttle's aerodynamic design features and re-entry operation modes, which were established by NACA some twenty years before the Shuttle's first orbital flight. The Shuttle's re-entry characteristics including the transition from the reaction controls used in space to aerodynamic controls, the use of high angles of attack to keep the dynamic pressures and heating problems within bounds and the need for artificial damping and other automatic stability and control devices to aid the astronaut pilot are similar in all important respects to those of the X-15 conceived by the scientific research engineers at Langley.

The X-15A-2 project as it eventually became known, was a conscious attempt to increase the performance capabilities of the standard X-15 aircraft to its absolute maximum potential.

Douglas R4D-5 BuNo.17136 c/n 12287 NASA 817 seen at Long Beach, California. It was acquired by the Flight Research Centre at Edwards AFB in January 1956. On 23 June 1969 it appeared on the US civil register as N817NA, being sold on 6 February 1981. It made the first tow of the unique M2-F1 on 16 August 1963. (Douglas CJ-006747)

The NASA, the US Air Force, the US Navy and North American had concluded, following successful completion of the X-15's performance envelope expansion programme, that the airframe was not being utilized to its fullest potential. The basic materials of which the X-15 was constructed would permit flights at speeds significantly greater than those being achieved by the aircraft in its early configuration. At the time of the decision to rebuild 56-6671 into a more advanced machine, the US aerospace industry was beginning to express strong interest in the potential of the supersonic combustion ramjet, know more popularly as the 'scramjet'. This propulsion system could be tested most effectively by the X-15.

Approval for the modification was given to North American under contract AF33(657)-11614 on 13 May 1963. Contract funding was $5 million and North American completed the project some nine month later, delivering the X-15A-2 to Edwards on 19 February 1964 to begin four months of static ground testing. Following handover to the US Air Force and NASA on 24 February, a first captive flight was made on 15 June and ten days later Robert Rushworth made the first post-modification flight, achieving a maximum speed of 3,104mph at 83,300ft. Additional envelope expansion flights continued for the rest of the year, these being flown without the X-15A-2's distinctive external fuel tanks. The first tank-equipped flight took place on 3 November 1965. The first flight with tanks attached and fully loaded with fuel occurred on the first day of July 1966. Unfortunately a fuel-flow anomaly led to a premature engine shutdown. Instruments had shown that the propellant was not transferring from one of the external tanks. An investigation found that the flow indicating system was, in fact, working as designed. Following a number of additional exploratory flights, the X-15A-2 (56-6671) was used on 18 November 1966 to establish a world absolute speed record for conventional winged aircraft of Mach 6.33 — 4,250mph.

This flight had been successfully accomplished due in part to the use of an experimental ablative coating developed by the Martin Marietta Corporation under the designation MA-25S. This material, consisting of a resin base, a catalyst and glass bead powder was designed to be sprayed on the external surface of the X-15 in order to protect it from the extreme heat generated by high-speed flight in the atmosphere. It was expendable and could be removed and discarded following a flight and replaced with fresh ablator. After the success of the 18 November flight, a decision was made to use the aircraft in a further high-speed flight to its maximum speed potential in its modified configuration.

On 3 October 1967, with Captain William Knight USAF at the controls, the X-15A-2 launched on what was to become its most memorable and historically most significant flight. The aircraft was to carry a dummy 'scramjet' out to the maximum Mach number, in the process carrying out another test on the new ablative material. Refurbishing the ablative material required approximately 700 man-hours over a two-week period. The procedures and various alternatives were well rehearsed by the pilot, who spent 35 hours in the simulator. Fortunately the flight proceeded basically according to plan. The Boeing B-52 *Stratofortress* took one hour to reach launch altitude. This was the 188th X-15 flight and an altitude of 102,100ft was reached at Mach 6.70 (4,520mph), which was an unofficial world absolute speed record. Post-flight examination of the aircraft revealed a number of serious problems with the ablative coating. The combination of high temperature and extremely high-speed airflow had charred and pitted the MA25S so badly that it was impossible to restore it for use on another flight. Additionally the airframe had suffered serious damage, particularly in the ventral fin area and in various segments of the fuselage and associated chine tunnels. A decision was made to ground 56-6671 permanently and hand it over to the US Air Force Museum for display.

With the demise of the third X-15 and the X-15A-2 making its last flight on 3 October 1967, only the first X-15 remained airworthy by early 1968. This aircraft was also reaching the end of its flying career, and though it continued to explore unknown ground for the NASA and the US Air Force, it was finally removed from active flight duty in late December. On 12 December 1968 an unsuccessful attempt was made to complete the 200th X-15 flight. Due to adverse weather conditions the flight was aborted prior to air launch and the mated X-15 returned to Edwards in the middle of a snowstorm. Although further efforts were made to consummate the flight on 13, 16, 17, 18 and 20 December, the attempt was eventually cancelled and the X-15 was demated from its B-52 for the last time. This X-15 (56-6670) was shipped to the Smithsonian National Air & Space Museum on the last day of 1968.

The North American X-15's legacy remains visible today, not only in the form of two preserved airframes but also in the many contributions it made to the aerospace community. A number of advanced X-15 configurations were proposed by North American during the programme's lengthy history. Amongst these was a delta-wing modification to permit a maximum speed of between Mach 6 and Mach 8. This was considered to be the next step in the X-15 programme. North American spent four years researching the delta X-15 configuration, gathering some 300 hours of wind tunnel test data by way of a 1/15th scale model for low speed (Mach 5) trials and a 1/50th scale model for high-speed (Mach 8) research. Finally, it is interesting to record details of the flights made by the five NASA pilots assigned to the programme — Neil A. Armstrong 7, William H. Dana 16, John B. McKay 29, Milton O. Thompson 14, Joseph A. Walker 25. The remaining pilots included five from the US Air Force, one from the US Navy and Scott Crossfield from North American. John Becker enumerated 22 accomplishments from the research and development work which produced the X-15; 28 from its actual flight research and 16

from test-bed investigation. As of May 1968, the X-15 had generated no less than 766 technical reports on research stimulated by its development, flight testing and test results.

Unique line-up at Ames in 1957 depicting four types used by the NACA — Cessna T-37A/B 54-158; Douglas F5D-1 Skylancers BuNo.142350b and BuNo.139208a; Convair F-102A Delta Dagger 56-1358 and Douglas F4D-1 Skyray BuNo.134759. In 1961 the two Skylancers were transferred to Edwards AFB as NASA 213 and NASA 212. (NACA A-23483)

Dyna-Soar

The X-20 was designed to provide a piloted manoeuvrable vehicle and associated equipment for conducting experiments in the hypersonic and orbital flight regime in order to gather research data to solve design problems associated with controlled, manoeuvring re-entry from orbital flights, it was also intended to demonstrate a pilot-controlled manoeuvring re-entry with tangential recovery at a pre-selected conventional landing site and to test vehicle equipment and explore military uses of space — primarily reconnaissance. The contract with Boeing was dated 11 December 1959, the contract number being AF33(600)-39831. US Air Force serial numbers assigned to ten production X-20s were 61-2374 to 61-2383.

Dyna-Soar was described as a World War II German Sänger-like boost-glider designed to be lifted into orbit by a Titan III booster. Its general configuration was that of a hypersonic slender delta, a flat-bottom glider using radiative cooling. Dyna-Soar was pushing the technology in many areas, including high-speed aerodynamics, high-temperature structural materials and re-entry protection concepts, A Dyna-Soar project office, in conjunction with NASA had selected a Flight Research Centre NASA pilot, Milton Thompson, as the only Agency pilot to fly the craft. NASA had complete responsibility for stipulating the X-20's instrumentation requirements. Research engineers had already prepared papers on the expected operational problems and the possibility of air-launching the machine from a Boeing B-52 or North American B-70 'mothership'. The project office now anticipated nineteen air-drop tests to begin during April 1962 with the first of eight unmanned sub-oribital flights to take place during July 1963 and the first of eight piloted, sub-orbital flights during May 1964. The first manned, global flights as to follow during August 1965. On 2 November 1959 the US Air Force Weapons Board approved the Dyna-Soar programme. Seven days later USAF awarded a contract to Boeing for the Dyna-Soar now having the designation Systems 620A. In addition to NASA's Milton Thompson, five US Air Force pilots were selected for the programme.

During early 1961 NASA at Edwards received two prototypes of the Douglas F5D-1 *Skylancer* — BuNo.139208a which became 'NASA 212' and BuNo.142350 becoming 'NASA 213'. The F5D was an experimental US Navy fighter which was not put into production. The aircraft had a wing plan form very similar to the projected Dyna-Soar, and NASA flight-test pilot Neil Armstrong recognised that the *Skylancer* could be used to study Dyna-Soar abort procedures. How to save the pilot and spacecraft in the event of a launch-pad booster explosion was a problem of great concern to the project team. The X-20A had a small escape rocket to kick it away from its booster but the separation flight path and landing approach to recover it safely to earth had yet to be evaluated. Neil Armstrong developed a suitable manoeuvre using the F5D-1. This consisted of a vertical climb to 2,100 metres, easing back the control column until the simulated X-20A was on its back, rolling the craft upright, and then setting up a low lift-to-drag ratio approach, touching down on part of the Rogers Lake marked out to simulate the 3,200-metre landing strip at Cape Canaveral, Florida.

During February 1963, the NASA at Edwards AFB purchased this Lockheed L.1329 Jetstar 6 c/n 5003. It was originally NASA 14, later NASA 814. It served as a General Purpose Airborne Simulator and went into service during November 1965. During 1981 it was involved in the flight-test of a Hamilton Standard SR-3 propfan mounted on a pylon atop the Jetstar. (NASA 68-H-1260)

Eventually questions over its utility, research potential, and safety forced cancellation of the X-20A project in December 1963, before the first aircraft could be completed. The fabulous Boeing X-20 Dyna-Soar, destined to be launched into Earth orbit atop a Martin Titan 3c booster and there to perform top-secret military missions, was just another cancelled project. It was designed to make a gliding re-entry and 'deadstick' landing just as the Space Shuttle Orbiter does today. Following the Dyna-Soar's cancellation, NASA continued to fly the F5D-1 *Skylancers* in support of the lifting-body project and SST studies with BuNo.139208a being transferred to Ames in 1963 for SST project studies. BuNo.142350 was flown until 1970, when it was retired.

Early Space Research

Support of the NASA space programme continued in 1959 when, at the request of the Space Task Group, the Flight Research Centre flew a series of Lockheed F-104 *Starfighter* trials in order to drop flight-test versions of the Project Mercury spacecraft's drogue parachute from altitudes above 15,000 metres. As a result of the flight tests, critical design problems were discovered and corrected prior to the spacecraft being launched. However the major space research activity carried out at the centre involved the 'Parasev' and the Lunar Landing Research Vehicle — LLRV — developed and flown at the Flight Research Centre in support of the Gemini and Appollo programmes.

During May 1961, when Mercury Mark II was slowly evolving, Robert R. Gilruth (director of the NASA Space Task Group) requested studies of an inflatable Rogallo-type 'Parawing' spacecraft. Several companies responded to the request, with North American producing the most acceptable concept, and development was contracted to that company. On 28/29 November 1961, at NASA headquarters a paraglider development programme was launched, with Langley conducting wind-tunnel studies and the Flight Research Centre supporting the North American test programme and with the NASA grafting the parawing scheme into the Mercury Mark II programme. In January 1962, Mercury Mark II became the well-known Gemini programme. Paraglider development involved solving major design difficulties including slowing and deploying the wing and ensuring that the crew would have adequate control and handling. Eventually, because of poor results, and rising costs and time delays, the idea was dropped from the Gemini programme in mid-1964.

The NASA research team at Edwards, supported by flights-test pilots, Neil Armstrong and Milton Thompson, decided to built and fly a Parawing. They approached director Paul Bickle, who approved the idea but recognized that both pilots had heavy Dyna-Soar commitments and so could not be spared for the proposed Parawing trials. Bikle instructed a group of centre engineers to build a single-seat Paraglider Research Vehicle (conveniently abbreviated Parasev) just prior to Christmas 1961. Seven weeks later, after expending $4,280, the team rolled out the Parasev I. It resembled a grown-up tricycle with a rudimentary seat and with the addition of an angled tripod mast supporting a 14sq-metre Rogallo-type parawing. The vehicle weighed 272kg, had a height of 3.4 metres and a length of 4.5 metres. Strapped in his seat, the pilot sat out in the open and controlled the descent rate by tilting the wing fore and aft; turns were initiated by tilting the wing from side to side. NASA registered the Parasev with the Federal Aviation Administration on 12 February 1962 and it was now ready for flight testing.

Initially the engineers tested the Parasev by towing it behind a utility vehicle driving at speeds of up to 95km per hour on the lakebed. Milton Thompson and Bruce Peterson became project pilots. The original configuration had several faults, and Milton Thompson considered it more demanding than the later lifting bodies. He made several hundred ground tows and 60 air tows. During one ground tow Peterson got out of phase with the control system and the Parasev developed a pronounced lateral rocking motion. This was followed by a wing-over into the lakebed, virtually demolishing the Parasev and slightly injuring Peterson. The engineers rebuilt the vehicle, installing a more sophisticated control system using a conventional stick and rudder system and using a Dacron-covered wing instead of linen covering. This became the Parasev I-A.

The Flight Research Centre rented a Stearman biplane (N69056) for towing, later using a Cessna L-19 *Bird Dog* acquired from the US Army Reserve. In addition to Milton Thompson and Bruce Peterson, Messrs Neil Armstrong, astronaut Gus Grisson and Langley research test pilot Bob

Champine flew the craft. Emil 'Jack' Kleuver, a US Army pilot assigned to the Flight Research Centre, also sampled it. As the Parasev, I-B, it was subsequently equipped with an inflatable Gemini-type wing. Flight testing ended in 1964 with the machine having completed well over 100 flights. Suspending the Gemini space capsule beneath a delta-shaped Rogallo-wing would have allowed landings on dry land, rather than ocean splash-downs beneath the traditional cluster of parachutes.

Flying Bedsteads

The Apollo programme became NASA's major undertaking in the 1960s and as is well known, this was an ambitious effort to place astronauts on the moon's surface by the end of the decade. Every NASA centre became involved with various level of contributions to the national space effort. One of the critical problems facing the Apollo programme was the descent to the lunar surface. The vehicle used would be operating in a gravity 1/6 that of the Earth, but the airless Moon dictated a strictly propulsion-borne, not aerodynamic descent. Grumman became the subcontractor for the landing vehicle, the Lunar Excursion Module (later known simply as the Lunar Module). An exotic simulator was required to give NASA astronauts some useful experience before they tackled the landing on the moon. Several possibilities were presented but NASA opted for three — an electronic simulator, a free-flight test vehicle and a tethered vehicle suspended beneath a framework. The most ambitious of the three was the free-flight vehicle, and this was the Flight Research Centre's contribution to Apollo.

It was the Edwards NASA engineering team which conceived the idea for a free-flight lunar landing simulator and this was approved by NASA headquarters. However Bell Aerosystems Company, unknown to the Agency, was also examining how to build a free-flight simulator. Bell and the Edwards team got together and the outcome was a $50,000 study contract to Bell, which the Flight Research Centre awarded in December 1961. NASA was thinking of the vehicle primarily for research, rather than as a training aid. At this time Langley was supporting a much less ambitous concept involving a tethered rig. When constructed, the large gantry — 120 metres long and 75 metres high — supported 5/6 of the test vehicle's weight with rockets supporting the remaining 1/6. This rig cost $3.5 million and commenced during June 1965, by which time Edwards had already amassed considerable flight experience with its own lunar landing simulator, the unique Lunar Landing Research Vehicle — LLRV. A jet engine supported 5/6 of the weight with rockets lifting the remainder, simulating the descent propulsion system of an actual lunar landing vehicle — aerodynamics played no part. This was not a new idea since aircraft companies had built and flown similar vehicles appropriately dubbed 'flying bedsteads' to obtain information required for the design of VTOL aircraft. In the United Kingdom Dr A. A. Griffith, a pioneer in British VTOL technology, had built a flying bedstead powered with a pair of Rolls-Royce *Nene* turbojets. It had an open framework supporting the pilot perched at the top of the rig, with instrumentation, fuel system, the engines, under his control. The Griffith or Rolls-Royce Flying Bedstead first flew in 1954 and was operated by RAF test pilots seconded to the Royal Aircraft Establishment, Bedford. It produced a great deal of data and pilot experience for future VTOL vehicles then being developed. NASA test engineers naturally kept this British vehicle in mind when conceiving the LLRV.

In the United States, Bell was the only company who had a mass of experience in the design and construction of VTOL aircraft using jet lift for take-off and landing. The NASA engineers conferred with Bell before drawing up the specification for their vehicle. Early in 1962, following the award of the Bell study contract, a proposal was put to Bell with reference to fabrication of such a vehicle for NASA.

Invited to the Buffalo, New York factory, the NASA personnel flew in the company helicopters on simulated lunar approaches, quickly learning that a helicopter could not simulate the descent rates and paths expected from a jet-lift lander. The NASA personnel involved included Donald Bellman, project chief, Gene Matranga and Lloyd Walsh, the latter the contracting officer from Edwards. Following the Buffalo visit, Bellman passed along their tentative finds to Walter Williams at Houston, who immediatley endorsed the concept. Out of this came full support from the Manned Spacecraft Centre and the NASA headquarters. On the first day of February 1963, NASA awarded Bell a $3,610,632 contract for the design and construction of two lunar landing research vehicles capable of taking-off and landing under their own power, attaining an altitude of 1,200 metres, hovering and horizontal flight. Bell were given fourteen months in which to build and deliver the first vehicle, with the second to follow two months later. Each LLRV would be capable of carrying 70kg of research equipment. The NASA had a full programme for the two vehicles including studies of problems envisaged during the final phase of a lunar landing and the intitial phase of take-off.

Bell unveiled the LLRV No.1 at its Wheatfield, New York, factory on 8 April 1964, with C. Wayne Ottinger from Edwards representing NASA during the ceremony. The LLRV weighted 1,680kg, was slightly higher than three metres and had four aluminium legs gving an 'undercarriage track' of four metres. Power was produced by a General Electric CF-700-2V turbofan providing 4,200lb thrust, enough to boost the vehicle to altitude. The pilot had use of two lift rockets capable of modulation between 100 and 500lb of thrust for controlling the lunar descent. These rockets burned hydrogen peroxide. Sixteen smaller rockets, arrayed in pairs, controlled pitch, yaw and roll. To permit the turbofan to

On 16 June 1961 this Douglas F5D-1 Skylancer BuNo.142350b NASA 213 was transferred from Ames to Edwards AFB. It was flown until retired in 1970 being involved in rehearsing X-20 Dyna-Soar procedures, the lifting-body project and SST studies. (Harry Gann)

maintain vertical thrust when the vehicle assumed other than a horizontal attitude, Bell gimballed the engine at the apex of the vehicle's legs. Six back-up rockets capable of 500lb thrust for emergency use in case the turbofan engine failed were also fitted. The pilot sat out in the open behind a Plexiglass shield, on an emergency zero-zero ejection seat. Endurance was fourteen minutes at full thrust, though safety factors reduced this to ten minutes. An electronic fly-by-wire control system was connected to a conventional aircraft-type centre stick for pitch and roll control and rudder pedals for yaw control. There were no aerodynamic control surfaces. The fly-by-wire also simulated the vehicle motions and control system response which an astronaut could be expected to encounter whilst piloting a descending lunar module.

Bell transported both LLRVs to Edwards partially dismantled and incomplete to assist the Agency in installation of the instrumentation. After arrival at the Flight Research Centre on 16 April 1964, personnel immediately set to work in preparation for the first flight. By September LLRV No.1 was ready for trials, mounted on a fixed-tilt table constructed by the NASA engineers. Flight-test pilot Joe Walker did the first tests prior to free-flight trials. Test operations were set up at the south base, scene of the old High-Speed Flight Station. On 30 October 1964 Joe Walker took the LLRV on its first flight completing three separate lift-offs and landings, reaching an altitude of three metres, and remaining aloft for just under one minute. By the end of 1964, Joe Walker was joined by Donald Mallick, a pilot new to the Flight Test Centre who had been transferred from Langley. He went solo on the LLRV on 9 December. The flight programme continued, and by the end of August 1966 the LLRV had completed 175 flights, flown by Walker, Mallick and Jack Kleuver from the US Army.

The Douglas F5D-1 NACA 212, later NASA 212, was fitted with a wing design concept involving a nonlinear (S-curve) leading edge which came from the United Kingdom. Known as the OGEE wing it is seen fitted on NASA 212 prior to it being transferred to Edwards AFB on 15 June 1961. It returned to Ames in 1963 for SST programme studies. (NASA A-32660-4)

In preparation for an LLRV training programme for the Apollo astronauts based at Houston, Texas, NASA flight-test research pilots Joseph Algranti and H. E. Bud Ream from the Manned Spacecraft Centre were checked out in the unique vehicle. On 11 March 1966, piloted by Don Mallick, LLRV No.1 flew with a three-axis side-arm controller, making it comparable to the actual Grumman lunar-module control system. NASA moved the control panel of the LLRV from the centre to the right-hand side of the cockpit, again matching the Grumman vehicle. During January 1967, Jack Kleuver of the US Army completed the first flight in LLRV No.2 which had an enclosed cockpit like the lunar module. By the time Edwards concluded its programme on the two vehicles in the winter of 1966/67, LLRV No.1 had completed 198 flights whilst LLRV No.2 completed five more flights. The longest flight endurance was 9½ minutes and altitudes reached 240 metres. LLRV No.1 was shipped to Houston on 12 December 1966, followed by No.2 On 17 January 1967, with Jack Kleuver flying No.1 at Ellington AFB during March. Joe Algranti and Bud Ream were appointed as instructor-pilots for the astronauts. In mid-1966 NASA ordered three more lunar-landing simulators from Bell, these being designated Lunar Landing Training Vehicles (LLTVs). Each cost about $2.5 million and incorporated modifications resulting from experience with the LLRVs. The LLTV weighed 1,860 kg and could attain an altitude of 120 metres. The cockpit display and control system was modelled on that of the lunar module, and the pilot's visibility was also restricted to match that offered by the lunar module. The first

LLTV arrived at Houston during December 1967, making its first flight on 8 October 1968. The Manned Spacecraft Centre modified the two original LLRVs as training vehicles and they became LLTVs A1 and A2. The new vehicles ordered direct from Bell became LLTV B1, B2 and B3. The NASA flight test pilots made the intial LLTV flights at Houston acting as instructor-pilots to the astronauts. The centre evolved an astronaut training programme in which potential lunar crewmen went to a helicopter flight-training school for three weeks and then to Langley's Lunar Landing Facility, followed by fifteen hours in a ground simulator. Finally they went on to the LLTVs, which they flew from Ellington AFB.

All prime and backup commanders of lunar landing missions practised on the LLTV-A and B vehicles, and a number of other astronauts also flew them. Gene Cernan completed the last LLTV flight on 13 November 1972. However the LLTV operations were not without incident, with three of the vehicles crashing. On 6 May 1968 Neil Armstrong took off in LLTV-A1 and whilst hovering ten metres above the ground, the vehicle suffered a loss of helium pressure in the propellant tanks, causing shutdown of its attitude-control rockets. It started nosing up and rolling over, so Armstrong ejected immediately. His zero-zero seat kicked him clear of the stricken craft, which fell to the ground and exploded as the astronaut descended safely on his parachute. On 8 December 1968, gusty winds forced LLTV-B1 out of control and Manned Spacecraft Centre flight-test pilot Joe Algranti ejected safely just a second before the uncontrollable simulator crashed. Finally, on 29 January 1971, LLTV-B2 suffered an electrical system failure causing loss of attitude control. Stu Present, a NASA flight-test pilot, abandoned the craft safely. Two still exist — the LLTV-A2 and the LLTV-B3. Astronaut chief Donald 'Deke' Slayton remarked that there was no other way to simulate moon landings except by flying the LLTV.

NASA and the Blackbirds

It is coincidental that not long before this volume was completed, the Lockheed SR-71A *Blackbird* long-range reconnaissance aircraft was retired from service with the US Air Force. Its programme over two decades is still shrouded in secrecy whilst its exact performance figures are still highly classified. Official sources still refer to the SR-71 as a Mach 3 vehicle capable of flying at altitudes in excess of 24,400 metres. It has been suggested that NASA will operate three of the retired aircraft, and there is no reason to doubt that the Agency already has plans for their use.

The type was conceived in the Lockheed Advanced Development Project Group, the famous 'Skunk' Works, by a team headed by Clarence L. 'Kelly' Johnson with a group of no fewer than 200 highly skilled engineers. The aircraft had a largely titanium airframe featuring a distinctive blended wing-body shape with long chines running along the fuselage sides from the wing roots. Each engine was located at mid-span, and each nacelle was surmounted by a large inwardly-canted vertical fin. For additional stability, the YF-12A variant had a folding ventral fin and two smaller fixed ventral fins as well.

NASA's involvement with the Blackbird programme commenced in 1967, when Ames opened negotiations with the US Air Force for access to the early YF-12 wind-tunnel data which had been generated at the centre under extreme secrecy. In return for NASA assistance with the flight test programme then under way at Edwards, they agreed. This arrangement dovetailed with the plans of the Office of Advanced Research & Technology, which saw the *Blackbird* as a means to advance high-speed technology, especially with reference to the future of SSTs. During the summer of 1967, the US Air Force and the Agency agreed to Flight Research Centre participation, with Gene Matranga representing NASA. The US Air Force was keen to get the SR-71A fully operational with Strategic Air Command as quickly as

The Martin Marietta X-24A 62-13551 fitted to the Starboard wing pylon on the NB-52B. Flown 1969-72. (Office of History — USAF Edwards)

Boeing NB-52B 52-008 taking off with the Pegasus air-launch space booster at 11.03 a.m. 5 April 1990. (NASA EC90-111-1)

The Wallops Flight Facility today operates this Lockheed NP-3A Orion which was transferred from Johnson on 29 September 1977. It is highly instrumentated and often involved in overseas international projects. Also at Wallops is Lockheed L-118C NASA 429 c/n 1103 ex-FAA and delivered on 27 Septemberr 1978. (NASA WI 78 312-1)

Gates Learjet 24A NASA 705 c/n 24-102 ex-NASA 365 was acquired on lease on 28 March 1973 being purchased on 10 June 1974. It is equipped with a 30cm infra-red telescope and its international studies take it overseas on occasions. The aircraft is current today at Ames. (NASA AC-76-0994)

This Grumman Gulfstream II NASA 650 fitted with a prop-fan on the port side, underwent a 150-hour assessment by Lockheed-Georgia under contract to the NASA at Lewis, and registration N650PF applied as depicted in this excellent night-shot. Today this Gulfstream is a Shuttle Training Aircraft, NASA 945, based at El Paso, Texas with the Johnson Space Flight Centre. (Gulfstream Aerospace Corportion)

Opposite: During 1978 with Lockheed producing the new TR-1 aircraft, the NASA showed interest in this upgraded U-2 model. Today Ames has two ER-2s plus a TR-1 NASA 708 ex 80-1069 on inventory. Prior to retirement in 1988, the old and new types were flown over the Golden Gate bridge. Depicted are ER-2 NASA 708 and U-2C NASA 709. (NASA AC88-0755-6)

This Lockheed F-104N Starfighter c/n L683C-4045 was one of three special versions for the NASA Flight Research Centre at Edwards AFB. It was acquired in August 1963 as NASA 011. It is seen at an open house at Norton AFB, California, on 23 October 1982 and is currently in open storage for the USAF Flight Test Centre Museum. (Frank Hudson)

Seen parked at Edwards AFB is Northrop T-38A-60-NA 65-1035, NASA 821, c/n N5772. It is used for continuation training for NASA test pilots, and as a chase aircraft. It came from Johnson ex-NASA 911, on 28 September 1972 and is current today. Johnson has no less than 28 Talons on strength, which are used by the NASA astronauts for space flight training. (Ralph Peterson)

possible, whilst NASA wanted an instrumented SR-71A for its own research — failing that, they were willing to install an instrument package in a US Air Force SR-71A stability and control test aircraft. The US Air Force declined, but did offer the Agency use of the Lockheed YF-12As then in storage at Edwards. Using funds made available by termination of the North American X-15 and XB-70 programme NASA took the unusual step of paying the operational expenses of the aircraft. A test team from the Air Defense Command supplied maintenance and logistic support. A memorandum of understanding was signed on 5 June 1969 and a public announcement followed on 18 July. Matranga and his team from Edwards set to work instrumenting the two YF-12A aircraft (60-6935 and 60-6936). The installed strain gauges and thermocouples instrumented the wing and fuselage for aerodynamic loads. They also instrumented the left side of the aircraft for temperature measurements with a view to obtaining a better understanding of the aircraft's thermal environment.

It was 29 February 1964 when President Lyndon B. Johnson revealed to the world that the United States was flying a new and highly advanced aircraft, using photos of a Lockheed YF-12A to back up his statement. On 24 July, President Johnson revealed that Lockheed had developed a new aircraft designated the SR-71, a strategic reconnaissance machine with an outstanding performance. The President made a slight error in his speech — the correct designation for the new aircraft was RS-71. No one likes to correct the President, so the new design retained the designation SR-71. However, there was a third Lockheed aircraft in the family. This was the single-seat A-12, bult for the Central Intelligence Agency (CIA) although apparently funded by the US Air Force. At least one A-12 carried USAF insignia and national markings. CIA-operated A-12s covered the world, several of twelve built being lost in training and operational accidents. The A-12 first flew from Groom Lake, Nevada, 26 April 1962, being withdrawn from operations in 1968.

Other US companies were also working on advanced aircraft. The General Dynamics Convair Division proposed using a ramjet-powered Mach 4 aircraft carried aloft by a B-58 *Hustler* bomber and dropped when the desired speed and height were achieved. US Navy concept was to altitude by a balloon and then rocket-boosted to a speed where the ramjets could produce thrust. Studies proved the concept to be totally unfeasible; the carrying balloon would have had to be a mile in diameter to lift the unit, which had a proposed wing area of $\frac{1}{17}$th of an acre!

The Lockheed trio — A-12, YF-12, SR-71 — were rumoured to have a top speed of Mach 3+ at heights in excess of 92,000ft. The YF-12 was to carry the advanced Hughes ASG-18/GAR-19 fire control and missile system and had a two-man crew. Also assembled at Groom Lake, Nevada, the first US Air Force YF-12 first flew on 7 August 1963. Only three were built. Two were passed on to NASA, whilst the third was extensively modified to create the dual control SR-71C (60-6937), this also going to NASA.

Both Ames and the Flight Research Centre at Edwards had a great interest in the YF-12A aircraft. Its airframe, propulsion system and related equipment were what most NASA engineers anticipated they would see on a future Mach 3 aircraft. It was an ideal vehicle for assessing the state of the art of wind-tunnel prediction, aerodynamics, propulsion and structural design. Other NASA centres were also interested; Langley engineers were keen to run fundamental aerodynamic experiments and tests of advanced structures, whilst Lewis was interested in propulsion research. Ames was interested in inlet internal aerodynamics and the correlation of wind-tunnel and flight data. NASA and US Air Force technicians took three months preparing YF-12A 60-6935 for flight, and a US Air Force crew accomplished the first sortie on 11 December 1969. To predict loads and structural response, NASA had developed two computer programmes using a technique known as finite element analysis. Both programmes, FLEXSTAB and NASTRAN, were applied to the YF-12A.

The North American XB-70A Valkyrie 62-0001 seen on take-off from Edwards AFB with a B-58 Hustler as chase during March 1968. It was involved in a joint NASA/USAF programme in support of the US National Supersonic Transport Programme. This aircraft is today preserved in the USAF Museum at Wright-Patterson AFB. (NASA 68-H-191)

The first flight of 60-6935 with a NASA flight engineer, Victor Horton, took place on 26 March 1970, and Donald Mallick became the first NASA pilot on the first day of April. Whilst the programme with 60-6935 progressed smoothly, the programme on 60-6936 resulted in near-tragedy for the US Air Force crew. This aircraft had completed 62 flights, primarily with US Air Force flight-test crews, before being lost in an inflight fire on 24 June 1971, the crew ejecting safely. Fatigue failure of a fuel line and fire in the right engine was the cause, and the crew — Lieutenant Colonel Ronald J. Layton and Major Billy A. Curtis debated whether or not they could land the aircraft. Wisely, they elected to eject. This loss did not seriously affect the NASA structure programme since it was almost complete, but it did delay plans for the propulsion research programme. A month after the crash, the US Air Force made available 60-6937, which was designated YF-12C although having SR-71A features — it was a real oddball. As the SR-71A programme was shrouded in the highest-ever security classification, the US Air Force restricted the Agency to using the aircraft solely for propulsion testing with YF-12A-model inlets and engines in place of the more sophisticated inlets and engines of the SR-71A. For the NASA programme, the Flight Research Centre designated pilots Fitz Fulton and Don Mallick together with flight-test engineers Victor Horton and Ray Young. Both pilots spent time at Beale AFB receiving indoctrination in the two-seat SR-71B, which had a second raised cockpit in place of the navigation/systems-operator position.

On 24 May 1972 Fulton and Horton crewed the new YF-12C on its first NASA flight. By this time some 53 flights had been accumulated on the remaining YF-12A and on 23 February 1972 — after its 53rd flight, which included being refuelled by a US Air Force Boeing KC-135Q *Stratotanker* over El Paso, Texas — it flew maximum time at Mach 3 on a 2hr 30min flight. Following this sortie the aircraft was grounded for more than a year for studies in the Flight Research Centre heat loads laboratory. It did not fly again until 12 July 1973. The YF-12 programme was ambitious; the aircraft flew an average of once a week unless grounded for extended maintenance or modification. Programme expenses averaged $3.1 million per year just to run the flight tests.

The preparation required for a NASA YF-12 flight was enormous. The crew suited-up 1½ hours prior to take-off, using a special US Air Force aeromedical van to drive them out to the flight line to join the aircraft. Engine and system checks all took time, and often the two aircraft would fly as a pair. Other NASA personnel were involved in preparing chase aircraft, either a Lockheed NF-104 *Starfighter* or a slower Northrop T-38 *Talon* to follow the YF-12s on take-off. To the north at Beale AFB, a Boeing KC-135Q *Stratotanker* loaded with special JP-7 *Blackbird* fuel would already be airborne climbing to the rendezvous point. The exercise would be monitored from take-off to touch-down by the US Air Force and Federal Aviation Administration radar units. On the 28th YF-12C flight on 2 May 1973, a stuck inlet spike caused high fuel consumption so an emergency landing was made at Fallon Naval Air Station in Nevada. The crew was Mallick and Young, who flew the aircraft in subsonic mode back to Edwards the following day.

Flight test of a proposed 'Coldwall' experimental package (a Langley-supported heat transfer experiment) caused anxious moments. A stainless steel tube equipped with thermocouples and pressure-sensing equipment was carried beneath the YF-12A. A special insulation coating covered the tube, which was chilled with liquid nitrogen. At Mach 3 the scientists hoped that the insulation would be blown away from the tube, instantly exposing it to the thermal environment. Its data could be compared with results taken from testing a similar tube using ground-based wind-tunnel facilities caused numerous in-flight difficulties. On the 'Coldwall' flight on 21 July 1977, the YF-12A experienced a phenomenon similar to

engine surge, known as 'inlet unstart', which was followed by decidedly poor engine performance. As it made a descent, followed by the YF-12C chase aircraft, the latter also experienced multiple unstarts. Both aircraft limped back to base under reduced power. On 13 October the final and successful 'Coldwall' flight was made.

By the beginning of 1977 the YF-12 aircraft had completed over 175 flights, much of the time above Mach 3. However, the two aircraft were becoming increasingly expensive to maintain and more difficult to justify. Other research, programmes such as the new McDonnell Douglas F-15 *Eagle* research programme, could lay greater claim to the NASA funding. In the Spring of 1977 at an OART centre director's management council meeting it was decided to retire the YF-12 aircraft, but residual funding enabled the YF-12C to fly through to 27 October 1978, when 60-6937 was handed back to the US Air Force. The last NASA flight, the 88th, was flown by Mallick and Young on 28 September. The other YF-12A, 60-6935, ended its research programme on the last day of October 1979, the 145th flight with Fulton and Horton as the crew. On 7 November 1979, a US Air Force crew delivered 60-6935 to the USAF Museum at Wright-Patterson AFB, Dayton, Ohio. The Flight Test Centre at Edwards had hoped (and even planned) to operate the type into the 1980s.

SST

Despite NASA's priority mission of landing a man on the moon before the end of the 1960s — aeronautics taking second place — several major developments in the free world suggested a need for greater aeronautical research and development. There was a protracted war raging in Southeast Asia which was revealing surprising problems with US aircraft and airpower doctrine. At times the victory-loss ratio slightly favoured the North Vietnamese. Foreign military aircraft technology was moving forward rapidly, especially in the USSR. During 1971 the NASA OART study concluded that the United States' traditional pre-eminence in military airpower had been lost in recent years. Whilst progress in foreign airpower had been rapid during the last decade, very few truly advanced aircraft had been developed in the United States. Supersonic jet transports (SSTs) were flying in the USSR and Europe, but only paper development was taking place in the USA. There was a need to strengthen America's traditional position of leadership in the field of civil air transport.

In this period the Langley Research Centre remained prominent as the agencies' principal aeronautical research resource. It was leader of the national SST programme, then a joint effort between the NASA, the Federal Aviation Administration and the industry. The Flight Research Centre at Edwards remained heavily committed to space-related efforts but the centre became involved in four SST-related aeronautical research programmes. These involved the North American XB-70A, the Lockheed YF-12, the Ling-Temco-Vought F-8 Supercritical Wing and the General Dynamics F-111 Transonic Aircraft Technology efforts. The first two involved deep research into sustained M2.5 M3+ flight, whilst the latter two were concerned with transonic aircraft design.

By 1963 there were three military aircraft flying at Edwards on SST research studies. The Douglas F5D-1 *Skylancer* had a modified delta-wing platform similar to a wing configuration proposed for a Mach 3 SST. The NASA flight test pilots flew the F5D-1 on SST landing studies and accumulated data on sink rates and approach characteristics. A North American JF-100C *Super Sabre* (53-1709) with variable-stability characteristics was acquired from Ames and flown to generate information on predicted SST handling qualities. NASA at Edwards acquired a North American A-5 *Vigilante* attack bomber from the Naval Air Test Centre located at Patuxent River in Maryland, and it was flown to determine the let-down and approach conditions of a SST flying into a dense air traffic network. NASA pilots Milt Thompson and Bill Dana flew the *Vigilante* over remote areas on expected SST profiles, even flying supersonic approaches into the terminal approach control zone to Los Angeles International Airport controlled from the ATC centre located at Palmdale. At the end of the year the A-5A (BuNo.147858) was returned to the US Navy.

During February 1963, the Agency at Edwards purchased a Lockheed L.1329 *JetStar* which became (NASA 814). This four-engined jet was instrumented with an analogue computer and could simulate the handling characteristics of a wide range of aircraft, including a SST. The aircraft cost NASA $1,325,000 (and installation of the required equipment cost a further $1.3 million) and it was prepared at the Cornell Aeronautical Laboratory at Buffalo, New York. The *JetStar* went into service during November 1965, and was known as the General Purpose Airborne Simulator — GPAS. Flight test research engineers believed that any SST would require a moveable droop nose such as employed on the *Concorde*, for adequate pilot visibility in the high angle of attack required on take-off and landing. Other engineers thought visibility could be provided by an extendable periscope binocular system. Binocular optics were installed in the NASA two-seat Lockheed F-104B *Starfighter* (57-1303; NASA 819) and Bill Dana evaluated it in flight. The research staff planned to provide funds and eventually did contribute approximately $2 million to instrument the experimental North American XB-70A *Valkyrie* Mach 3+ bomber in order to obtain supersonic-cruise research data.

XB-70A Valkyrie

By far the most controversial of all North American aircraft was the XB-70 supersonic bomber. Launched as a project to replace the Boeing B-52 *Stratofortress* as America's primary deterrent force, the B-70 was overtaken by changing defence policies and escalating costs and only two prototypes were built in what was possibly the most expensive aerospace programme ever in relation to units flown. The project began

formally on 14 October 1954, when HQ US Air Force issued its General Operational Requirement GOR-38 for an inter-continental piloted bomber. The aircraft was to be all-chemical-fuelled and was expected to be in the active Strategic Air Command inventory in the period 1965–75, with a target introduction date for the first thirty aircraft by 1963. Prior to the issue of the GOR, the US Air Force, the NASA and industry had completed two years of studies of the feasibility of such a weapon system, which was initially designed for an aircraft with a Mach 0.9 cruise and supersonic dash capability for a 1,000-mile penetration of enemy airspace. By the time the prototypes were launched, however, the aircraft was designed to maintain a supersonic cruise at all times with a top speed of Mach 3.

Both Boeing and North American competed for the contract, the latter being announced winner in December 1957. The initial design studies were launched by North American as NA-329 on 12 September 1955, but constant changes by US Air Force directives took the project through the company charge numbers NA-259, NA-264, NA-267 and NA-274 to NA-278 which covered three prototypes ordered on 4 October 1961. The bomber became the B-70 and was named 'Valkyrie' by the company. The third prototype was cancelled on 3 May 1964, and construction of two XB-70As was allowed to proceed only as a research programme.

It was an advanced aerodynamic design of delta-wing layout with a foreplane and hinged wing tips which could be folded down to a 65-degree angle to enhance stability at supersonic speed. The design made use of compression lift theories to allow the B-70 to ride on its own shock wave at Mach 3. Advanced structural design and many new techniques were required to allow the aircraft to tolerate the high temperatures derived from skin friction at high speeds. The power plants were planned for six General Electric J93-GE-5 turbojets with afterburner, using JP-6 high-energy fuel. It was

The XB-70A was designed as a Mach 3 strategic bomber with delta wing and large canard foreplane, intended as a Boeing B-52 replacement, but was cancelled in defence cuts. Power was by six 3100lbst YJ93-GE-3 engines. The unique XB-70A 62-0001 is seen in landing configuration at Edwards AFB with its three landing chutes fully deployed. (Office of History — USAF Edwards)

eventually calculated that mission performance could be achieved without the latter and J93-GE-3 engines were substituted rated at 30,000lb thrust each and were located behind a complex air intake control system with fixed and variable ramps. It was an expensive programme rumoured to have cost $2,000 million to design and build the two prototypes, with each test flight costing $800,000.

The maiden flight was on 21 September 1964, and the XB-70A had a wing-span of 105 feet length and a height of 30 feet. It had two large vertical fins, a canard horizontal control surface mounted on the fuselage and a sharply swept delta wing constructed of titanium and brazed stainless-steel honeycomb materials. The aircraft could withstand sustained temperatures olf 332°C as it cruised at high altitude at Mach 3. The first XB-70A (62-001) had a NASA Flight Research Centre-funded package of test instrumentation capable of telemetering some thirty-six measurements of aircraft performance and conditions to ground stations. A further 900 measurements were recorded by digital pulse-code-modulation and analogue frequency-modulation recording systems on magnetic tape at the rate of 20,000 samples per second. This was a far cry from the scratchy oscillograph film used on the earlier Bell X-1.

Flown by North American and US Air Force flight-test pilots during the first phase of its flight-test programme, the two XB-70A aircraft (62-001 and 62-207) were routinely flying above Mach 3 by early June 1966. It was soon discovered that an SST would not be able to use conventional airway routes since turns required flight corridors hundreds of kilometres wide. The first aircraft proved to have poor stability characteristics above Mach 2.5. On the basis of Ames wind-tunnel studies, North America added 5° dihedral to the wing of the second XB-70A which gave it better stability characteristics; for this reason it became the prime Mach 3 research aircraft.

Depicted in flight and being towed by Douglas C-47 NASA 817, is the first lifting body, a home-built, hence the US civil registration N86652. It was unveiled to the press on 3 September 1963. The landing gear came from a Cessna 150, later a Cessna 180, Milton Thompson was the pioneer test pilot on the first tows both ground and air. (NASA 66-00057)

The complex systems of the aircraft posed maintenance problems, and poor bonding of the stainless steel skin on the wing allowed whole sections to peel off in flight. The complex landing gear plagued the XB-70A on retraction, and on one occasion partial gear failure caused the aircraft to veer almost a kilometre off the runway centre line. However, despite all the difficulties, the two aircraft produced a great deal of useful information for the SST designers — on noise, operational problems, control system requirements, validation of tunnel-test techniques by comparison with actual flight-test data, and high level clear-air turbulence.

The NASA Office of Advanced Research & Technology — OART — had allocated $10 million for support of the XB-70A programme, primarily for flight-test instrumentation on the two aircraft. In the spring of 1966, the US Air Force and the Agency announced a joint $50 million programme to be controlled by the Flight Test Research Centre and the US Air Force Aeronautical Systems Division forecast to begin in mid-June 1966, at the conclusion of the North American airworthiness programme. This joint programme was planned to study problems of sonic booms and evaluate the aircraft during typical SST flight profiles. NASA flight test pilot Joe Walker was designated project pilot for the programme.

On 8 June 1966, 62-207 was airborne from Edwards at 7.15am piloted by North American test pilot Al White and new co-pilot Major Carl Cross making his first flight in the XB-70. The programme was to make a series of passes over the control tower at various airpseeds to calibrate its onboard airspeed system, then make a single pass at Mach 1.4 and 9,450 metres to acquire sonic boom information over the Edwards instrumentated test range. There was another item on the flight programme. General Electric had approached the US Air Force for the XB-70A to lead a formation of aircraft equipped with their power plants. For this a briefing had been held the day before; John Fritz, a General Electric test pilot who would fly a Northrop YF-5A, advised all the pilots to fly a loose formation, with about one wing-span clearance between each other.

A US Navy McDonnell F-4B *Phantom* flew from Point Mugu Naval Air Station to join the formation and Joe Walker flew in the NASA Lockheed F-104N *Starfighter* NASA 813. A US Air Force T-38A *Talon* and the YF-5A completed the formation, whilst a *Learjet* was the photographic aircraft. The rendezvous point was Lake Isabella and by 8.43am all had joined up with the large white delta XB-70A which was formation leader on a racetrack pattern between Mojave and Mt. Whitney at 6,100 metres. Clouds forced some modification of the original plan, so the racetrack was changed to northeast of Rogers — a much shorter track with the straight leg covered in just over a minute, after which the formation made a 180° three-minute turn. The T-38 and F-4B were on the left wing of the XB-70A, with Joe Walker in the F-104N and the YF-5A on the right. A two-seat F-104D of the US Air Force returning from a test mission joined the formation, the cameraman in the back seat only too pleased to use up his movie film on such a unique group of aircraft. It was suggested that a tighter formation would improve the looks, after which the F-104D would break off and return to Edwards. Joe Walker in the F-104N complied and edged closer to the *Valkyrie*, then for reasons unknown he closed with the giant, his horizontal stabilizer touching the downturned tip of the delta. Passing through the leading edge of the XB-70A wing, the *Starfighter* rapidly rolled over the top, hooking its left wing tank on the *Valkyrie* wing. The tip tank broke up, and the right-hand tip tank automatically jettisoned. Still rolling over the F-104N smashed into the right and left vertical fins of the huge delta and exploded in flames, making its final impact on top of the delta's left wing. Joe Walker was killed instantly. Apparently White and Cross in the XB-70A were initially unaware they had been involved in a collision until the aircraft yawed right and rolled right, tumbling over and over violently. A parachute appeared as Al White got out but he suffered severe internal injuries in the struggle to leave in the escape capsule. Carl Cross died in the wreckage of the *Valkyrie*. The co-pilot's ejection capsule never left the aircraft, and the XB-70A spun into the ground and exploded six kilometres northwest of Barstow.

NASA Deputy Administrator Robert Seamans cited Joe Walker for his many contributions to flight research, and President Lyndon Johnson issued a statement of tribute from the White House. The Flight Research Centre had lost a valued colleague and the XB-70A programme had received a serious setback, which drastically altered plans for NASA's joint SST research programme with the US Air Force. The first XB-70A (62-001) was grounded at the time for maintenance and modification and did not resume flying again until November 1966.

Subsequently, the US Air Force transferred the SST programme and funding responsibility to NASA. Expenditure to this point had amounted to approximately $2 million per month and the Agency opted to reduce this figure to $800,000 per month, so cutting back the flight programme. $10 million in FY1968 funding was requested, sufficient to take the programme through 1968. The Flight Research Centre awarded North American an $8.9 million contract for maintenance and support of the one XB-70A, and a $1.9 million contract to General Electric for engine maintenance.

On 3 November 1966, Joe Cotton and the veteran NASA flight-test pilot Fitz Fulton took 62-001 over the instrumented test range for sonic boom assessment at Mach 2.1, making ten more flights by the end of January 1967. These eleven flights also supported the National Sonic Boom Programme. Begun in June 1966, it involved a number of military aircraft flights over selected cities in the United States although the XB-70A was restricted to the instrumented test range at Edwards for its sonic boom contribution. By the end of March 1968 the delta had completed twelve more flights. Pilots included Fulton, Cotton, Van Shepard, Lieutenant Colonel Emil Ted Sturmthal and NASA pilot Donald Mallick. Following the 73rd flight on 21 March 1968, the NASA grounded the aircraft for installation of a structural dynamics research package dubbed ILAF (identically located acceleration and force). So equipped it made its first flight on 11 June 1968, the new system reducing the buffeting in clear-air turbulence and rapidly fluctuating atmospheric temperature. From then on until the end of the programme in 1969, the XB-70A acquired a great deal of information applicable to the design of future SST or large supersonic military aircraft.

By the end of 1968, operating costs and maintenance problems had caught up with the XB-70A and on 4 February 1969, 62-001 was flown to the USAF Museum at Wright-Patterson AFB, Dayton, Ohio. The two aircraft had completed 129 flights covering 252 hours 38 minutes of which 22 hours had been above Mach 2.5.

Unfortunately the American SST fell further and further behind its competitors, the Russian Tu-144 and the Anglo-French *Concorde*. Although the US SST had the full support of three successive presidents — Kennedy, Johnson and Nixon — it had numerous critics, ranging from thoughtful spokesmen who questioned its economics to others who had a anti-technological bias. There was an attempt to keep the Boeing SST project alive but to no avail. On 24 March 1971, the US Senate declined to appropriate $289 million for SST prototype fabrication.

F-4 Phantom

Of all the aircraft operated by NASA, the aircraft records reveal that only one McDonnell F-4 *Phantom* was used by the agency. In December 1965 the Flight Research Centre received an ex-US Navy F-4A fighter, BuNo.145313. Apart from using the *Starfighter* as a workhorse for X-15 mission support and as chase aircraft for the lifting body, the NASA used them in a number of short programmes such as base drag measurements, sonic boom measurements in support of Langley research, and tests of balloon-parachute deceleration devices known as ballute. In the early 1960s Flight Research Centre flew a brief military-inspired programme to determine whether an aircraft's sonic boom could be directed; if so, it could possibly be used as a weapon of sorts, or at least an annoyance. This was to be the flight test programme of the *Phantom*.

The aircraft was a very early-model F-4A delivered to the US Navy at Point Mugu Naval Air Station, California, on 14 July 1960, remaining there until 3 December 1965 when it was

assigned to the Flight Research Centre at Edwards. Here, NASA discovered that the aircraft was in poor condition, requiring considerable effort in removal and repair of portions of the electrical wiring system. Up to July 1967 the total airframe hours were 630, and only fifty hours had been flown on it by NASA. This was due, in part, to its poor condition upon receipt from the US Navy.

On 15 July 1967, flown by NASA flight test pilot Hugh Jackson, the F-4A departed Edwards at 9.13am with no external stores and a normal afterburner take-off was accomplished. Approximately two and a half minutes after brake release, the afterburners were de-selected. A short time later, while passing through 13,000 feet at 350–400 knots, the pilot felt a thump and immediately requested another aircraft to look him over. A large hole in the forward centre section of the right wing was observed. The pilot dumped the fuel in the left wing and a low-speed handling check was carried out. With the aid of a flapless approach, the aircraft was landed safely at Edwards with no difficulty. Hugh Jackson stated that the only abnormal flight characteristics due to the loss of the upper and lower torque-box wing skins were a slight in-flight buffet and a slight right roll just prior to touch down. The *Phantom* was not repaired for flight, being turned over to the US Air Force to be used in tests for the development of short field arresting gear.

In early August 1964 it was agreed to incorporate XLF-111 rocket engines in the M2-F2 and HL-10 lifting bodies. This rear view of NASA 804 shows to good effect the jet outlet cluster of the rocket motors. (Vic Seeley via Peter M. Bowers)

The Lifting Bodies

By the 1960s the concept of ballistic re-entry from space travel had become familiar, there being a group of prominent scientific engineers who were in favour of lifting re-entry. The idea was to build a space-craft with aerodynamic characteristics enabling a crew to fly back through the Earth's atmosphere and make a conventional landing at an airfield. The Boeing X-20A DynaSoar project of the US Air Force was a premature attempt to develop such a craft. But the DynaSoar never flew, a victim of budget constraints and new technology. It was, however, not the only lifting re-entry approach to orbital flight; there were also wingless shapes known as 'lifting bodies'. At Ames, a series of exploratory studies during the 1950s by H. Julian Harvey Allen, conceived the blunt-body theory in 1951. A design known as the M2 was a modified half-cone — flat on top — with a rounded nose to reduce heating. The NASA engineers at Edwards kept up with much of the theoretical ideas percolating out of Ames, and Robert D. Reed, a Flight Research Centre Research Division engineer fond of building flying models, became interested in the M2 — by now called *Cadillac* in reference to the two small fins emerging at the blunt tail. Reed built a successful flying model, which led to authorization for a manned glider. It was suggested that a piloted M2 shape for low-speed stability and control tests could be built and launched from a Boeing B-52. At Langley, the NASA engineers favoured a more traditional approach over that of sawing a cone in half. They opted eventually for modified delta configurations, as a result of the work of Eugene S. Love. Langley developed the shape into the HL-10 — the HL standing for 'horizontal lander'. This first appeared on Langley drawing boards in 1962 as a manned lifting re-entry vehicle. The US Air Force, though still working on DynaSoar, considered other lifting re-entry schemes and in the early 1960s commissioned a series of studies which eventually spawned the Martin SV-5D — a configuration which resembled a cross between the cone-like M2 and the modified-delta HL-10.

In February 1962 Robert Reed built a 60-centimetre model of the M2, which he launched from a larger radio-controlled mothership having a 150cm wing span. Ames promised the use of wind-tunnels and Paul Bikle authorised a six-month feasibility study of a cheap, manned lightweight M2 glider. This was followed in September by authorization for design and construction of a manned M2 glider — the authorization was local as Paul Bikle feared he could not secure NASA headquarters support quickly enough to permit an immediate development programme. One aircraft company later estimated it would have cost at least $150,000 to build the M2, but the Edwards team did it for less than $50,000. A nearby sailplane company built the laminated wooden shell, with a considerable amount of fabrication work being done by NASA personnel, most of who were practised hobbyists in the art of home-built aircraft. The landing gear was scrounged from a Cessna 150. By 1963 the M2-F1, as it was now called, was completed. It was six metres long, three metres high and had a width of four metres. It had two vertical fins on which were mounted stubby elevons and weighed 516kg. The pilot sat under a large bubble cockpit, and NASA later fitted a lightweight Weber rocket zero-zero ejector seat. Later on, a solid-fuel rocket of 24lb thrust, developed by the US Naval Ordnance Test Centre at nearby China Lake, was fitted to assist in the pre-landing flare manoeuvre if this became necessary.

The weird craft went to Ames for low-speed tests in the 40×80ft tunnel. Completed in March 1963, these were very encouraging and the NASA project pilot, Milton Thompson, often sat in the cockpit of the M2-F1 during the wind-tunnel tests.

Initial flight required a vehicle to tow the M2-F1 above the dry lake bed, but none of the NASA fleet of trucks and vans was fast enough. The Edwards team therefore procured a stripped-down Pontiac convertible with a large and powerful engine having a 4-barrel carburettor, and 4-speed stick shift, capable of towing the lifting body to 177km/hr in 30 seconds. A custom-car dealer in Long Beach fitted rollbars, radio equipment and special adapted seats for observers. Fearful lest a critic hastily concluded that here was a private toy paid for by US government funds, the team wisely painted 'National Aeronautics and Space Administration' on the side and sprayed the hood and trunk high-visibility yellow to comply with the paint scheme of other flight-line vehicles. During road tests in the Nevada desert to calibrate the speedometer, this vehicle attracted great interest from the highway patrolmen as it roared along the highway driven by Walter Whiteside at high speed; it did six kilometres to the gallon. Finally by the Spring of 1963 all was ready.

First ground tows were completed on 5 April 1963, and by the end of the month the M2-F1 had made forty-five others. From then until the first air tows, the aircraft made over 100 tows, accumulating four hours in the air. Results were good, so the next step involved aerial tows behind the Edwards-based Douglas C-47 (an ex-US Navy R4D-5, BuNo.17136 and 'NASA 817'). One of the team, Vic Horton, scrounged a C-47 tow mechanism from a junk yard. On 16 August 1963, Milton Thompson piloted the M2-F1 behind the *Skytrain* transport. On this and other flights the ensemble climbed up

to 3,000 metres at 190km per hour to release the lifting body above its intended landing spot on Rogers Lake, normally two minutes after release. Landing speed was 137-145km/hr. On 3 September 1963 NASA unveiled the M2-F1 to aviation news reporters. There was great interest shown by Capitol Hill as word got back to Washington. By mid-April 1963, many congressmen were quizzing NASA headquarters on the M2 flight programme, and causing consternation among some Department of Defence officials who apparently had no idea that the M2 was flying. Some congressmen feared the low-budget M2 might soar overnight to a major multi-billion post-Apollo development programme. Others suspected the programme represented a way for NASA to circumvent the decision to cancel DynaSoar. Dr Hugh Dryden defended the Agency effort, and the M2 programme continued.

At Edwards, seven other pilots checked out in the M2-F1. It was an unusual craft in many ways including its civil US registration, N86652. This was way out of context with the usual NASA numbering system, so one can only assume that it was treated as a home-built craft. It carried the NASA 'meatball' insignia just aft of the cockpit. The M2-F1 pilots included NASA test pilots Bruce Peterson, Donald Mallick and Bill Dana; US Air Force pilots Chuck Yeager, Gerauld Gentry and Donald Sorlie. Eventually the M2-F1 completed approximately 100 flights and 40 ground tows prior to retirement for the National Air & Space Museum. There were various incidents, one being when Bruce Peterson landed with sufficient force to shear off the landing wheels. The NASA replaced the landing gear with more rugged items from a Cessna 180. On two other flights, Gerauld Gentry became involved in some extremely hazardous rolling manoeuvres. Not wishing to take further chances, Paul Bikle closed down the programme. It had served its purpose, proving that the lifting-body shape could fly and encouraging further research with supersonic rocket-powered lifting bodies to determine if the shapes so desirable for hypersonic flight could safely fly from supersonic speeds down to landing through the still tricky area of transonic trim changes.

When the M2-F1 completed its last air-tows in August 1964, work was already well advanced on the heavyweight aluminium lifting-body follow ons — the M2-F2 and the HL-10, both products of the Northrop Corporation.

Northrop became the prime contractor for the new family of lifting bodies sponsored by NASA. The M2-F2 was a refined version of the M2-F1. The company also delivered the HL-10, which had a very short angled-delta wing and a different fuselage shape.

US Air Force interest resulted in the formation of a joint NASA-USAF lifting body programme and a Memorandum of Understanding was issued in April 1965. It was February 1964 when the Flight Research Centre solicited proposals from twenty-six companies for two heavyweight, low-speed lifting-body gliders. The Agency would test them in the full-scale Ames wind tunnel and also air-launch them from a Boeing B-52 flying at 13,700 metres. The companies were given five weeks in which to submit proposals; the Office of Advanced Research & Technology would supervise the programme, with Ames, Langley and FRC at Edwards participating. Only five companies responded, NASA selecting the Norair Division of the Northrop Corporation to build the vehicles. On 2 June 1964 a fixed-price contract was awarded to Northrop for fabrication of the M2 and HL-10 heavyweight gliders for $1.2 million each. Delivery of the M2-F2 was planned for Spring 1965, the HL-10 following six months later. In early August 1964 it was agreed to incorporate provision for XLR-111 rocket engines in the two new craft.

A memo created a Joint Flight Research Centre/Air Force Flight Test Centre Lifting Body Flight Test Committee composed of ten mixed members. The FRC had responsibility for maintenance, instrumentation, and ground support of the vehicles, whilst the AFFTC assumed responsibility for the launch and support aircraft, medical support, the rocket power plant and pilot's personal equipment. Joint responsibility was assumed for planning the research flights, analyzing flight data, test piloting, range support and overall flight operations.

The M2-F2 and the HL-10 were not the only lifting bodies under construction. Eighteen months later, on 11 October 1966, the AFFTC and NASA amended the memo to cover the agency's participation in a US Air Force-sponsored lifting-body programme involving the Martin SV-5P, which had a complex genesis. During 1960 the US Air Force had commenced examining manned manoeuvrable lifting-body spacecraft as alternatives to the ballistic-type orbital re-entry concepts then in favour. In May 1961, the US Air Force Flight Dynamics Laboratory awarded the McDonnell aircraft company a contract for a suborbital lifting body re-entry vehicle called ASSET (Aerothermodynamic/Elastic Structural Systems Environmental Tests). Measuring over 1.5 metres long, it generally resembled the cancelled A-20A. Six were built and were launched down the Eastern Test Range from Thor-Delta and Thor boosters between September 1963 and March 1965. These shapes reached speeds between 16,000 and 21,000 km per hour while making lifting re-entries from 60,000 metres over the South Atlantic. All the vehicles survived re-entry, though some were lost at sea before recovery crews could pick them up.

The next step, PRIME (Precision Recovery Including Manoeuvring Entry) began in November 1964 when the Space Systems Division of the US Air Force Systems Command awarded the Martin Marietta Company a contract to design, fabricate and test a manoeuvring re-entry vehicle to demonstrate whether a lifting-body could be guided. Martin had been in the DynaSoar competition and was therefore conversant with lifting re-entry vehicles, having put more than two million man-hours into studies. From Martin came the SV-5 bodyshape; they built a 1.5 metre-span radio-controlled model and flew it at their Middle River, Maryland, plant. It was raised to altitude under two balloons, then dropped and guided it down to land. The design was refined into the SV-5D, a 894lb aluminium vehicle with ablative heat shield. The US Air Force ordered four of the SV-5D PRIME vehicles, designating them X-23A and launching three of them between December 1966 and mid-April 1967 over the Western Test Range using Atlas boosters which blasted them at 24,000 km per hour toward Kwajalein Island in the Pacific. The vehicles performed so well that the US Air Force cancelled the fourth launch to save money. From PRIME the US Air Force and Martin derived PILOT — Piloted Lowspeed Tests — a proposed Mach 2 research vehicle which the service could test to determine its supersonic, transonic and subsonic landing behaviour. It was designated SV-5P and a contract for one was awarded in May 1966; it was completed in a little over a year and rolled out on 11 July 1967. The US Air Force designated it the X-24A and it was taken to Ames for comprehensive wind-tunnel testing. From there it went to Edwards, where the other lifting bodies (the M2-F2 and HL-10) had already flown. The SV-5P had the USAF serial 66-13551.

The Uglies

'Lifting bodies fly a lot better than they look', remarked one flight-test pilot, but generally the vehicles were definitely the ugliest of postwar research aircraft. The M2-F2 appeared at the Northrop plant at Hawthorne California on 15 June 1965, being transported to Edwards the next day. It lacked the XLR11 rocket engine since NASA would fly it initially as a glider. Constructed of aluminium, it weighted 9,400lb and measured 22 ft 2 in in length, with a span of 9 ft 7 in. It had two vertical fins but lacked the horizontal control surfaces of the earlier M2-F1. It also had a retractable main landing gear taken from a Northrop T-38 *Talon* and the nose gear was borrowed from a North American T-39 *Sabreliner*. A modified zero-zero ejector seat came from a Convair F-106 *Delta Dart*. The Flight Research Centre technicians checked

out the aircraft, added research instrumentation and then transported it to Ames for two weeks in the full-scale wind tunnel, completing 100 hours of testing during August 1965. It returned to Edwards for initial flight trials and Northrop fitted a special adapter so that the M2-F2 could be launched from the Boeing B-52 mothership using the X-15 launch pylon.

The Flight Research Centre launched a co-operative pilot training and simulation programme with the Cornell Aeronautical Laboratory. Earlier, NASA had flown the laboratories' highly modified variable-stability Lockheed T-33A to simulate the low lift-to-drag re-entry characteristics of the X-15. In the spring and summer of 1965 they again flew Cornell's T-33A, this time on lifting body studies, using the M2-F2 as the reference. The variable-stability T-33A had drag petals installed on its wingtip tanks. Extended in flight, these varied the lift-to-drag ratio of the aircraft from the normal 12–14 to as low as two — the approximate L/D ratio of the M2 lifting body. Approaches were flown by Cornell test pilot Robert Harper and the NASA flight test pilots Milt Thompson, Bruce Peterson, Bill Dana and Fred Haise. The NASA pilots also simulated lifting-body approaches and landings using the centre's F-104s and the Douglas F5D.

Northrop HL-10 NASA 804 was acquired by the NASA Flight Research Centre in January 1966. It had a very short start, angled delta wing and a different fuselage shape. With flat bottom and rounded top fuselage it was capable of Mach 1.86 and 90,000 ft. The first unpowered landing was made on December 22, 1966, piloted by Bruce Peterson. (NASA 74-03422)

The M2-F2 completed its maiden flight on 12 July 1966, piloted by the NASA pilot Milt Thompson. It was launched from the B-52 at 13,700 metres altitude, flying at 725 km/hr. The flight lasted for four minutes, and touch-down was 'spot-on' the planned aiming point on Rogers Lake. By November 1966 the craft had completed an additional thirteen flights piloted by Thompson, Bruce Peterson and the USAF pilots Jerauld Gentry and Donald Sorlie. Following flight fourteen on 21 November, NASA grounded the M2-F2 for installation of the XLR11 rocket engine. It made its first flight with — but not using — the rocket engine on 2 May 1967, piloted by Jerauld Gentry. On 10 May 1967, Peterson launched from the B-52; all initially went well until the M2-F2 began a dutch roll from side to side at over 200° per second. Recovery carried the craft away from its intended flight path and Peterson, realizing he was too low to reach the planned landing site and rapidly sinking toward a section of the lakebed without the usual visual runway reference markings, accurately needed to estimate height above the lakebed. A rescue helicopter appearing in front of the M2-F2 did not help the situation, but Peterson was reassured that the approach path was clear by chase pilot John Manke, flying a Douglas F5D. Peterson fired the landing rockets and lowered the undercarriage, but he was too late — before the landing gear locked, the M2-F2 hit the lakebed, shearing off its telemetry antennas and rolling over at more than 400 km/hr. It turned over no less than six times prior to coming to a halt on its flat back, minus canopy, main gear and right vertical fin. Bruce Peterson was badly injured; he was rushed to the hospital at Edwards AFB for emergency surgery, then to March AFB hospital. He pulled through but lost the sight of one eye.

The remains of the M2-F2 were returned by NASA to Northrop where it was given a thorough inspection. In March 1968, the NASA Office of Advanced Research & Technology authorized Northrop to restore the primary structure and return the vehicle to Edwards. In the light of its poor landing characteristics, the M2-F2 obviously required modification. On 28 January 1969, NASA headquarters announced that the Agency would repair, modify and return the M2-F2 to service as the M2-F3. The lifting body was fitted with an additional central vertical fin to improve yaw control. On 2 June 1970, the M2-F3 — still NASA 803 — made its first unpowered flight; it was released from the B-52 over Edwards with Wilton H. Dana as the pilot. This key event was followed by first powered flight on 25 November 1970, with the M2-F3 achieving Mach 0.8 at 53,000 ft after Dana ignited three of the four XLR11 rocket engines. Later in the programme the M2-F3 reached a maximum altitude of nearly 90,000 ft and a speed of Mach 1.7. By the completion of the programme the M2-F3 had logged 27 flights, with the basic M2 airframe recording a total of 43 flights. Retired at the end of December 1972, the craft subsequently joined the collection of the National Air & Space Museum in Washington, DC and is on display.

Lt Col Donald Sorlie USAF, seen in the cockpit of the Northrop M2-F2. He made his first flight in this lifting body on 20 September 1966, being launched from the NB-52. He achieved 45,000 feet and Mach .64 in a flight lasting 211 seconds. The zero zero ejector seat was from a Convair F-106 Delta Dart. (NASA E-15781)

Following a series of gliding flights by Edwards, the Langley-designed HL-10 made its first rocket-powered flight, from the Boeing B-52 on 13 November 1968. John A. Manke was in the cockpit. In contrast to the accident-marred M2 flight test programme, testing of the HL-10 (NASA 804) progressed quite smoothly. Externally, the M2-F2/3 and HL-10 had dissimilar aerodynamic lines; internally they were alike, with identical systems and accessories. Both had a riveted aluminium alloy semi-monocoque forebody. The aft structure was basically an aluminium box with side fairings.

Two full-depth keels extended from the cabin to the base of each vehicle; this allowed for the provision of non-structural equipment with its access doors on the outside and acted also as an isolation bay for the rocket fuel.

The HL-10 was the most successful of the lifting bodies and, like the M2-F3, measured 22 ft 2 in in length but was wider — 15 ft 1 in — and higher — 11 ft 5 in. It used many of the off-the-shelf components from the T-38, T-39 and F-106. Roll out from the Hawthorne facility was on 18 January 1966. The HL-10 initially went to Ames for tests in the 40 × 80 ft full-scale tunnel, on on 22 December 1966 NASA flight test pilot Bruce Peterson completed the first glide flight — which was anything but routine. Peterson discovered that he had minimal lateral control over the lifting body; flow separation was much worse than anticipated. He managed to set the HL-10 down safely on Rogers Dry Lake, no small tribute to his piloting skills. The Agency immediately grounded it for study, also taking the opportunity to install the rocket engine. Langley undertook a series of tunnel tests, resulting in some modifications. When the HL-10 took to the air again on 23 October 1968 it handled well. After John Manke reached Mach 0.84 using two of the four thrust chambers, NASA incrementally worked towards the maximum performance. The HL-10 went supersonic for the first time on 9 May 1969 — the first supersonic flight of any manned lifting body and a major milestone in the entire lifting-body programme. The craft exhibited acceptable transonic and supersonic handling characteristics. On 18 February 1970, US Air Force test pilot Peter C. Hoag reached Mach 1.86, the fastest any lifting body had ever been, and on 27 February Bill Dana reached an altitude of 27,524 metres — another record for the programme. The HL-10 became the fastest and highest-flying piloted lifting body ever built.

By 1970 the Space Shuttle was an item under discussion. A series of powered landing trials with the HL-10 was embarked on by NASA, since the flight-test engineers were not certain whether a Space Shuttle should make an unpowered landing approach, or like a conventional aircraft, fly a powered approach for landing. During February 1970, following the record altitude and speed flights, NASA grounded the HL-10 and replaced its XLR11 rocket engine with three 500 lb thrust Bell Aerosystems hydrogen peroxide rocket engines. It was planned to launch from the B-52 in the vicinity of Palmdale, the HL-10 pilot igniting the rocket engines as the craft passed through an altitude of 2,000 metres. The rockets would reduce the approach angle from the customary 18° to 6° and give an airspeed in excess of 560 km/hr. At 60 metres above the lakebed, the pilot could shut down the rockets, extend the landing gear, to make a routine landing. Two of these flights were completed by Peter Hoag on 11 and 17 June 1970, the latter flight being the final mission for the HL-10. During its brief flying career, the HL-10 completed 37 flights, and it contributed to the decision to design the Space Shuttle without landing engines.

Initially NASA had no role to play in the US Air Force X-24A (SV-5P) programme. Completed by Martin Marietta at its Middle River, Maryland, plant in the summer of 1967, its body shape differed greatly from that of the M2 and HL-10. After rollout on 11 July 1967, the X-24A went to Ames for full-scale tunnel testing and in early 1969 it went to Edwards for flight trials with Jerauld Gentry making the maiden glide flight on 17 April. The new craft completed nine more such flights prior to powered flight. It was 19 March 1970 when Gentry flew the X-24A on its first powered flight, reaching Mach 0.87 — well into the transonic region. Following this initial powered flight, NASA pilot John Manke and US Air Force test pilot Major Cecil Powell undertook a series of envelope expansion missions that included — on 14 October 1970 — a flight to a speed of Mach 1.19 at 68,000 ft. Less than two weeks later Manke flew the aircraft to 71,000 ft and simulated, for the first time in an aircraft with similar performance and handling characteristics, a

Space Shuttle-type approach and landing. This was followed, several flights later, by a speed run to Mach 1.60 on 29 March 1971. The latter would be the fastest speed recorded by the X-24A configuration. The twenty-eighth and final X-24A flight took place on 4 June and ending prematurely when two of the XLR11 powerplant's four chambers refused to ignite.

During the flight-test programme the X-24A accumulated a total flight time of 2 hours 54 mins 28 secs and achieved a maximum speed of Mach 1.60 and altitude of 71,400ft. In general, pilots considered the X-24A to be relatively easy to fly with few vices. It demonstrated decisively that lifting-body configurations could successfully complete spot landings on a consistent basis. Tests at Edwards utilizing the Air Force Base's Runway 18 resulted in a 250ft average longitudinal miss distance from the intended touchdown spot. Over 20 hours of simulator time were commonly utilized by the pilot in preparation for a flight. In-flight practice in a NASA F-104 included approaches to as many as five possible landing runways. It has been estimated that the pilots performed as many as 60 simulated landing approaches during the two-week period prior to a flight in the X-24A.

The Martin Marietta X-24B was the last post-war rocket aircraft produced in the United States. Following the last flight of the X-24A on 4 June 1971, the aircraft was stored for some six months while final modification negotiations were being completed. On the first day of January 1972, the US Air Force awarded the Martin Marietta Corporation the contract. Modifying the existing craft secured for $1.1 million a research vehicle that could have cost $5 million if built from scratch. The programme was evolved around a family of re-entry shapes, the FDL-5,6,7 all having a reasonable lift-to-drag ratio at hypersonic speeds and a large internal volume. The X-24B represented the FDL-8 shape. The aircraft was airlifted to Denver, Colorado and work by Martin Marietta commenced on 7 April 1972. It was returned as the X-24B to Edwards on 24 October after roll-out by the company on 11 October. Like the earlier lifting bodies it contained several off-the-shelf components, with portions of its landing gear, control system, and ejection system coming from the Northrop T-38, Lockheed F-104, Martin B-57, Grumman F11F, Convair F-106 and the North American X-15. It had an XLR11 rocket engine and Bell Aerosystem landing rockets. Once the X-24B was back at Edwards, the NASA technicians installed a research intrumentation package. Its length was increased from 24ft to 37ft and the span was also increased, with gross weight increased to 13,800lbs from 11,450lbs.

On 19 July 1973 the X-24B was carried to near-launch altitude on its first post-modification captive flight on the B-52. On 1 August, with NASA test pilot John Manke at the controls, the X-24B successfully completed its first unpowered glide flight. A first powered flight by John Manke was completed on 15 November 1973, and on the sixteenth flight — on 24 October 1974 — pilot Mike Love accelerated out to Mach 1.76, recording the highest velocity achieved during the X-24B programme. Another flight, on 22 May 1975, resulted in the highest approach for landing, the aircraft touching down on the dry lake bed at Edwards after having reached an altitude of 74,130ft. The pilot was John Manke.

NASA's interest in the ability of low lift-over-drag aircraft like the X-24B, with their similarities to the future Space Shuttle, to land at predetermined spots utilising conventional runways with their fixed geogaphical and heading constraints led, during mid-1975, to a short X-24B landing-test programme. Utilizing the 15,000ft-long 04/22 runway at Edwards, the tests commenced on 5 August 1975 when John Manke touched down within a prescribed 5,000ft-long target segment after an approach from 60,000ft. Two weeks later, just to prove the repeatability of Manke's approach, US Air Force test pilot Mike Love duplicated the landing.

On 9 September 1975 Bill Dana completed the X-24B's last powered flight, bringing to an end not only the successful X-24B programme but also the era of manned rocket-propelled research aircraft at Edwards. The X-24B had completed 36 air-launched flights over a 26-month period with a total accumulated flight time of 3 hours 46 mins and 43.6 secs. The basic research flight programme had been completed during six glide and 24 powered flights over 24 months by three pilots — two NASA and one US Air Force. A six-flight pilot check-out programme was accomplished after completion of the basic research programme. This increased the experience level of the pilots for future programmes, and increased the database of handling information on the X-24B. The craft is today on display at the USAF Museum at Wright-Patterson AFB, Dayton, Ohio, to where it was delivered on 19 November 1976. All the lifting bodies gave great confidence to advocates of landing an unpowered Space shuttle on a conventional runway after its return from space. This plan was ultimately followed for the Space Shuttle.

In 1972, the Martin Marietta X-24A 66-13551 was rebuilt as the X-24B, was delta shaped and had twice the lifting surface of the X-24A. It explored handling qualities of wingless configuration for extended near-earth flights, and for conventional runway approaches and landings. Its last flight ws on 9 September 1975 and today is preserved in the USAF Museum. (USAF Museum)

CHAPTER TWELVE
The New NASA

During 1955 both the United States and the USSR had announced plans for launching satellites, but the appearance of *Sputnik I* on 4 October and of *Sputnik II* on 3 November 1957 shocked much of the civilized world. It was evident that the Russian techniques for the launching and control of large rocket vehicles were far advanced. Confronted with this combination of spiritually inspiring and competitive prospects, the American people required no further convincing that a major space-research effort, costly though it would be, should promptly be undertaken. But by whom should it be administered? The space research operation was a tempting programme for any scientific, military or political organization. One of the major contenders was the Department of Defense, with each of the three armed services appearing willing to take on the task individually, and each competing with the other two armed services as well as with civilian interests.

The US Navy was already involved in its *Vanguard* satellite project, whilst the US Army had possibly given earlier and more thought to the design of satellites than had either of the other two armed services. The US Air Force was busy contructing the rocket motors which were most likely to be used in any space programme. It commanded the huge Atlantic Missile Range, and so was perhaps the most aggressive and powerful — and certainly not the least ambitious — the three armed serves. There was great interest from civilian agencies including the National Academy of Sciences, the American Rocket Society, the Atomic Energy Commission and NACA. In January 1958 Dr Hugh Dryden, speaking for the Agency, proposed that the US space programme be jointly undertaken by the DoD, NACA, the National Academy of Sciences, the National Science Foundation and universities, research institutions and the nation's industry.

Early in 1958 the DoD formed the Advanced Research Projects Agency — ARPA — to manage the military (and hopefully the US) space research programme. It was headed by civilians, but the programme agencies were the three armed services. The ARPA received the endorsement of President Eisenhower. Space matters had reached a climax and naturally there was a feeling of uneasiness in civilian circles, especially with the new DoD organization moving ahead so quickly. It was time for decision-making action, and on 5 March 1958 President Eisenhower approved the recommendation of his Advisory Committee on Government Organisation that the '. . . leadership of the civil space effort be lodged in a strengthened and redesigned National Advisory Committee for Aeronautics.' On 2 April a draft legislation establishing a new National Aeronautics and Space Administration, using NACA as a nucleus, was sent to Congress. The act establishing NASA, known as the National Aeronautic and Space Act of 1958, was passed by Congress and signed by President Eisenhower on 29 July 1958. The conversion of NACA to NASA was to take place in ninety days or sooner if appropriately proclaimed by NASA's appointed Administrator.

This early production model of the Douglas F4D-1 Skyray BuNo.134759 arrived at Ames on 4 April 1956 remaining until 16 October 1959, being used on various research and development projects. It was a US Navy single-seat delta-wing jet fighter powered by a 14500lbst J-57-P-8 engine.
(Douglas HG83-1701 and NACA A-21303)

Throughout all these proceedings, NACA personnel were naturally very enthusiastic about the prospects of undertaking space research, and very keen to have the new organization take the lead in the space effort. Although they were aware that changes were inevitable, it was well known that NACA had taken action to assume its expected space role well before the Space Act was passed. In November 1957, it had authorized the establishment of a new space technology committee to plan a research programme. It had also initiated plans for a revision of its headquarters organization to accommodate space-research requirements. So by the time the Space Act was passed, action on both these matters was very well advanced. There was naturally a bitter-sweet atmosphere of both sadness and elation at the NACA laboratories, yet future prospects — though uncertain — appeared most exciting. The general excitement of the times was heightened with the launching of the US *Explorer I* satellite on the last day of January 1958, and *Explorer III* on 26 March 1958, and the space race, though not yet officially recognized and acknowledged, became nevertheless a matter of pressing reality.

Once the general outlines of the new organization were clear, both a space programme and a new organization had to be charted. The new National Aeronautics and Space Administration came into being on the first day of October 1958. In April, Dr Hugh L. Dryden assigned Abe Silverstein, assistant director of the Lewis Laboratory at Cleveland, Ohio, to Washington to head the programme planning. Ira Abbott, the NACA assistant director for aerodynamic research, headed a committee to plan the new organization. During August President Eisenhower nominated T. Keith Glennan, president of Case Institute of Technology and former commissioner of the Atomic Energy Commission, to be the first administrator of the new organization, NASA, and Hugh Dryden to be deputy administrator. Quickly confirmed in office by the Senate, they were both sworn in on 19 August. Glennan made a review of the existing planning efforts and approved most. Talks with the Advanced Research Projects Agency — ARPA — identified the military space programmes which were space-science oriented and therefore obvious candidates for transfer to the new agency. Plans were quickly formulated for building a new centre for space science research, satellite development, flight operations and tracking. A site was chosen, covering nearly 500 acres of the US Department of Agriculture's research centre in Beltsville, Maryland, and the Robert H. Goddard Space Flight Centre — named after America's rocket pioneer — was officially dedicated in March 1961.

As early as 1958, plans had been laid to establish a new space-projects centre near Washington, DC. The resulting organizational entity, at first largely composed of Naval Research Laboratory groups occupying scattered temporary facilities, was named the NASA Goddard Space Flight Centre in May 1959, and in 1960 it was moved into new facilities built for it under the new NASA reorganization near Beltsville, Maryland. The assigned functions of the new centre included the planning and construction of vehicles and payloads for scientific applications and manned-space-flight programmes and the conducting of flight operations relating to them. The centre also has today major responsibilities in the establishment and operation of a global tracking and data acquisition network. It consists of facilities in Greenbelt, Maryland; Wallops Island, Virginia; the Goddard Institute for Space Studies in New York City, and sixteen tracking stations around the world. Personnel are situated around the globe as part of the Space Tracking and Data Network — STADN — team and at facilities of the NASA Communication Network which links these networks together. The centre conducts, and is responsible for, automated spacecraft and sounding-rocket experiments in support of basic and applied research. Satellite and sounding rocket projects at Goddard provide data about the earth's environment, sun-earth relation-

ships and the universe. These projects also advance technology in such areas as communications, meteorology, navigation and the detection and monitoring of our natural resources. The centre has been assigned a lead role in the management of the International Search & Rescue Satellite Aided Tracking — SARSAT — project.

Six Douglas DC-4 Skymaster transports were employed by the Goddard Space Centre including NASA 432, a C-54G 45-637 c/n 36090 ex-NASA 232. They were operated by the Bendix Radio Division on behalf of the NASA. Acquired by the NASA in April 1960, the Goddard fleet travelled worldwide during the early days of the space programme. On 26 January 1977 it went into storage at Davis-Monthan AFB, going to the USAF Museum on 16 October 1979.
(Harry Gann)

The aircraft fleet which has supported the Goddard Space Flight Centre since 1960 initially consisted of a mixture of leased and US Air Force aircraft, used mainly in support of space programmes. The bases for these aircraft have included Friendship International Airport at Baltimore and the Martin Company's Strawberry Point Hangar and Middle River, both in Maryland. The first aircraft listed as being used by Goddard was a Grumman S2F-1 *Tracker* (BuNo.129151) which was used during the Vanguard project to perform mini-track calibrations for the centre. In February 1960, a Douglas DC-3 transport (N2733A) was leased from the Bendix Field Engineering Corporation for programme support until April 1962. This was during Project Mercury. In June 1960 a Piper *Tri-Pacer* (N3659P) owned privately by Charles D. Mason was also used for support of Project Mercury to check-out a quad-helix antenna at Wallops Island. Six Douglas DC-4 four-engined *Skymaster* transports followed, together with three US Air Force Lockheed C-121 *Constellation* and a single Douglas DC-6/C-118A *Liftmaster*. In 1967 no less than nine Boeing EC-135N *Stratotanker* aircraft, designated Apollo Range Instrumented Aircraft — ARIA — were designed by Goddard to be used as airborne tracking stations. The nose section was modified to house an S-band tracking antenna, and the interior was configured to contain all the necessary electronic support equipment. The airframe modifications were completed at McDonnell Douglas at Tulsa, Oklahoma, and the Bendix Field Engineering Corporation performed the installation of the electronics. Four aircraft — 61-0327, 61-0326, 61-0329 and 61-0328 — were also modified to use an Airborne Lightweight Opitical Tracking System (ALOTS) pod which was mounted in place of the standard cargo door. Unfortunately EC-135N 61-0328, acquired on 27 December 1967, was lost in a fatal accident near Frederick, Maryland on 6 May 1981. One Boeing EC-135N (61-0331) was modified with fan-jet engines. The Goddard Space Flight Centre paid for the overall modification for these eight aircraft at an approximate cost of $60 million. The home base of the ARIA aircraft was Patrick AFB, Florida until November 1975, after which this operation was reassigned to Wright-Patterson AFB, Dayton, Ohio. These aircraft are mainly used by NASA for coverage of launch support, for which Goddard pays an hourly rate. Since they no longer support Apollo missions, their new role is as Advanced Range Instrumented Aircraft — which still allows use of the acronym ARIA.

Space Centre

The huge Johnson Space Centre — formally the Lyndon B. Johnson Space Center — was established in September 1961, and is located on NASA Highway 1, adjacent to Clear Lake, two miles east of the town of Webster and approximately twenty miles south-east of Houston, Texas. Aircraft used by the centre are located at nearby Ellington AFB, which is approximately seven miles to the north. Johnson is NASA's primary centre for design, development and manufacture of manned spacecraft, selection and training of space flight crews, ground control of manned flights from 'Mission Control' and many of the medical, engineering and scientific experiments carried aboard the flights. It is the lead NASA centre for management of the Space Shuttle.

The Mission Control Centre has monitored all the NASA manned space flights starting with Gemini IV and the Apollo and Skylab series and continuing into the current missions of the huge Space Shuttle programme. The centre is also responsible for the direction of operations at the White Sands Test facility — WSTF — located on the western edge of the US Army White Sands Missile Range at Las Cruces, New Mexico. The WSTF supports the Space Shuttle propulsion system, power system and materials testing. The facility served as the alternate landing site for the second test flight of the orbiter *Columbia*.

The Goddard Space Centre operated a large fleet of aircraft from airports in Maryland. This included this Lockheed R7V-1 Super Constellation ex-US Navy BuNo.131642 which was also C-121G 54-4065 c/n 4143. It was used for Mercury, Gemini and Agena evaluation and tracking tests at stations around the world including Australia. It was transferred to the US Army on 23 January 1973. (Harry Gann)

The aircraft fleet listing for Johnson will be found as an Appendix, and it is unfortunate that space restricts an adequate description of the many unique aircraft types operated over the years and their NASA tasks. One type used by Johnson was the unusual Martin/General Dynamics RB-57F aircraft obtained from the US Air Force in the early 1970s, with whom they had flown on a series of clandestine operations. One of the main prerequisites of a successful and effective airborne reconnaissance platform is that it is capable of attaining extremely high altitudes. For many decades the objective has been pursued with the utmost urgency, ultimately leading to the production of reconnaissance systems which are capable of operating at altitudes approaching the absolute physical limits of manned air-breathing aircraft. In the United States the requirements of various airborne reconnaissance systems has, over the past three decades, led to the birth of some of the world's most extraordinary flying machines.

On 27 March 1962, the US Air Force awarded General Dynamics a study contract calling for the configuration and modernization of the Martin B-57-series aircraft. The contract called for a rather unique altitude capability and a payload capacity of just over two tons. Intelligence-gathering equipment was rapidly growing in size, weight and sophistica-

tion. The reconnaissance cameras alone, in consideration of their immense 16- to 18-inch apertures and the amount of glass associated with such large openings, were approaching weights of well over one and three quarter tons. On the second day of October 1962, the US Air Force authorized General Dynamics to commence construction of two RB-57 reconnaissance aircraft. It was anticipated that the first would be delivered by August 1963, followed by the second a month later. The aircraft were obtained under the US Air Force 'Big Safari' procurement policy in order to get the product into the air in absolutely minimal time. Following completion, the prototype was rolled from General Dynamics' special projects division hangar on 16 May 1963 — only eight and a half months after initial contract signing. One of the related sensor systems was a very high-altitude, high-resolution reconnaissance camera. Referred to by the acronym of HIAC — High Altitude Camera — it was optimized for side-looking at extreme ranges. At a slight angle of some 40 degrees, and a range of some 60 miles, resolution was expected to be no less than 30 inches. Improved resolution could, of course, be achieved at shorter ranges and more obtuse angles. The first models, costing over $1 million each, weighed in excess of 3,500lbs.

The original RB-57F contract was for $9,314,350 (later increased to $16,500,000) which made the company responsible for all spares, personnel and sensor costs for the two aircraft. Eventually twenty-one RB-57F aircraft were built by General Dynamics. During the spring of 1972, the US Air Force made a decision to transfer nine aircraft to the huge Military Aircraft Storage and Disposition Centre at Davis-Monthan AFB in Arizona for inviolate storage. Two of the aircraft were transferred to NASA's jurisdiction and formed the basis for what was quickly to become the new high-altitude airborne instrumentation research programme. The WB-57Fs of the USAF 58th Weather Reconnaissance Squadron supported NASA with launch and recovery area weather reconnaissance and took part in 'Operation Cold Flare' — a NASA/FAA project to study high-altitude solar flare activity in preparation for the establishment of intercontinental polar routing or supersonic flights. The aircraft operated out of Eielson AFB in Alaska. 'Operation Cold Hunt' was in support of NASA/MSFC Airborne Visual Laser Optical Communication (ALOC) project. 'Operation Cold Rap' was conducted in conjunction with NASA for earth resources studies. Sensors included metric and multi-spectral cameras, infra-red instruments and active and passive radar systems. 'Operation Secede III' was a NASA experiment, in which a rocket was fired from Alaska into near space where it released a barium cloud. Again the aircraft operated from Eielson AFB, Alsaka.

Eventually three WB-57F aircraft operated under the auspices of NASA, though technically one remained in the possession of the US Air Force. They were used on a number of environmentally related programmes and flown and maintained for the NASA Airborne Instrumentation Research Programme — AIRP. Two retained stock RB-57F configuration whilst the third was substantially modified in order to carry the 'Universal Pallet System'. This concept consisted of a series of interchangeable pallets which, when sensor and related systems were added, comprised the aircraft payload. Use of the NASA WB-57F aircraft was open to legitimate organizations societies, companies and individuals although charges were made for their use. Johnson Space Center had been engaged in the AIRP since 1964. Photographic facilities at the centre gave a full capability of acquiring and processing large and small-format photographs including both colour and black and white. Sensitometry, density and other photo-science capabilities were also available. The NASA Universal Pallet System — UPS — offered metric, high resolution and/or multiband camera capability, a five-channel multispectral scanner, and X-band side-looking radar, a central data recording system and various other playload capabilities up to 4,000lbs.

NASA at Langley did the nose modification optimized for gust research on RB-57F (63-13503 and 'NASA 926', known as 'Earth Survey 6') with crews using full-pressure suits — A/P22S-2s valued at $13,000 each. Later an A/P22S-6 suit, a modified Gemini space garment much like the type worn on many Apollo missions, was used. It gave complete protection to the crew member whilst allowing him the freedom of movement necessary to perform normal flight tasks. RB-57F 63-13298 was passed to NASA on 24 June 1974, becoming 'NASA 928'. Aircraft 63-13501 went to NASA on 19 July 1972, becoming 'NASA 925' and christened 'Earth Survey 3'; it was retired and stored at Davis-Monthan AFB on 15 September 1982. The third RB-57F (63-13503) was delivered to the Agency on 21 July 1972, becoming 'NASA 926' and 'Earth Survey 6'. The aircraft had the ability to carry a Los Alamos Scientific Laboratory — LASL — wing slipper tank.

Power for the RB-57F was provided by two Pratt & Whitney TF33-P-11A turbofan engines and two Pratt & Whitney J60-P-9 turbojet engines. The wing span was increased to 122ft 5ins and the service ceiling was listed as 81,000ft with an absolute maximum altitude of 88,300ft.

The Johnson Space Centre Earth Observations Aircraft Programme conducted in the early 1970s used five aircraft types equipped with a variety of remote sensors for recording earth resources data. Users of data acquired by the programme were usually government agencies and universities designated to work with the NASA ain development of remote sensing applications. The US government agencies included the US Department of Agriculture, US Geological Survey, National Oceanographic & Atmospheric Administration, US Department of the Interior and the US Corps of Engineers. The programme supported user agency requirements such as co-operative research projects and earth resources investigations connected with data acquired by Earth Resources Technology Satellite — ERTS — and Earth Resources Experiment Package — EREP — flown on board Skylab. In addition the programme supported sensor research development testing and evaluation, applications technique development and other investigations in support of state and local governments.

The Lockheed NP-3A *Orion* used by Johnson was a real hybrid, being the prototype P-3A (BuNo.148276) designated YP3V-1 and converted from an L-188A *Electra* N1883 original constructor's number was 003. Designated by the NASA as 'Earth Survey 1' and 'NASA 927', it was acquired on 20 December 1965. Powered by four Allison 501-D-13A turbo-prop engines, it was modified to carry remote sensors. The NP-3A operated from sea level to approximately 30,000ft at an airspeed of 150 to 330 knots. The maximum weight at take-off was 113,000lbs and it could remain at operating altitude for about seven hours. At sea level it required a 6,000ft runway. Equipment included a Litton LTN-51 inertial navigation system for long range tactical, airway and terminal area navigational capabilities; a Loran APN-70 long-range navigation system for use primarily over water, and a PB20-N automatic pilot for manually flying the aircraft during routine and operational flight. Remote-sensor capability of the NP-3A was provide by photographic cameras, infra-red and multispectral scanners, thermal radiometers and microwave scatterometers. An onboard photographic darkroom provided a film-magazine reloading capability whilst in flight. Six windows were installed in the lower fuselage for use with the camera systems. Provisions were available in the *Orion* for installation of a passive microwave imaging system, a side-looking airborne radar (SLAR) and a Geodolite 3 laser profiler and other environmental sensors. The aircraft was considered so valuable by NASA that another *Electra* was dismantled for a set of replacement wings for 'NASA 927' rather than obtain an entirely new aircraft. It was transferred to Wallops Island on 29 September 1977 and was still in use in 1990.

The Lockheed NC-130B *Hercules* used by Johnson was formerly the Boundary Layer Control (BLC) test-bed aircraft

(58-0712, c/n 3507) and was delivered to NASA on 18 July 1968, after a period of storage at Davis-Monthan AFB, Arizona. It became 'NASA 929', 'Earth Survey 2'. The aircraft was equipped with an LTN-51 inertial navigation system an AN/APN-159 radar altimeter and an autopilot. Modifications to the NC-130B were carried out by the Agency to provide for the installation and support of various types of remote sensing equipment. Four windows in the lower fuselage and two right hand windows were provided for use with the camera and television systems. Also installed in the aircraft was a side-looking airborne radar (SLAR) infra-red and multispectral scanners, thermal radiometers and microwave scatterometers. A non-standard nose was fitted, this coming from the *Orion* 'NASA 927', whilst the airframe had been modified to accept a Geodolite 3 laser profiler. NASA also provided for the installation of environmental sensors indicating liquid water content, total air temperature and a dew point hygrometer. An onboard photographic darkroom provided a film-magazine reloading capability while in flight. On 9 October 1981, the *Hercules* was transferred to the Ames laboratory and became 'NASA 707'. It was still in service there in 1990.

The only Douglas C-133A-1-DL Cargomaster, 54-136, with NASA 928 was delivered to the Johnson Space Centre from the US Air Force at Dover AFB, Delaware, for programme support on 13 April 1966. It remained until 19 August 1969, completing over 200 flying hours with the NASA. The four turboprop strategic transport is seen parked at Long Beach, California. (Douglas C 105651)

As mentioned earlier, Johnson operated the unique R/WB-57F aircraft, two of the three used being involved in the Earth Observations Aircraft Programme. These were 'NASA 925' ('Earth Survey 3') and 'NASA 926' ('Earth Survey 6'). The aircraft's navigation system included a G-12 gyrosyn compass, LTN-51 inertial navigation system and a A/A 24G-9 true air speed computer. This system measured the aircraft's attitude, linear acceleration and ground speed and computed the current latitude and longitude, ground speed, drift angle, track error and the distance and time remaining. The aircraft was also equipped with an auto-pilot system.

On 5 December 1967, Johnson took delivery of a Bell 47G-3BI helicopter (c/n 6663) which became 'NASA 946'. Sensors carried on this helicopter included a field spectrometer system and the S-191 infra-red spectrometer. The latter is the same instrument as that carried on Skylab with the exception that the field of view is larger on the helicopters' unit. Liquid nitrogen was used for cooling the S-191 detector. Unfortunately 'NASA 946' was destroyed in an accident on 2 August 1969, the flight test pilot Gibson suffering no injury. It is of interest to note than Johnson used two more Bell 47G-3BI helicopters, one of which was 'NASA 947' (c/n 6665) acquired on 11 December 1967. This was also destroyed, in an accident on 23 January 1971 in Florida, with no injuries to the NASA flight test pilot Gene Cernan. The third helicopter was 'NASA 948' (c/n 6670) acquired on 10 January 1968, which was disposed of to the Dryden Flight Research Centre at Edwards AFB on 1 November 1973. A single Bell 206B *Jet Ranger* (c/n 508) was delivered to Johnson from the Jet Propulsion Laboratory at Pasadena, California, where it had been registered 'NASA 12' on the first day of October 1973. It later became 'NASA 950' and bore the civil registration N950NS on the fuselage.

The NASA Jet Propulsion Laboratory — JPL — is located in Pasadena, California, some twenty miles north-east of Los Angeles. It is a government-owned facility which is staffed and managed by the California Institute of Technology; it occupies 177 acres of land, of which 155 are NASA-owned and 22 acres on lease. The laboratory is funded under a NASA contract and also operates the Deep Space Communications Complex, a station in the Worldwide Deep Space Network located at Goldstone, California, and sited on 40,000 acres of land occupied under permit from the US Army. The Jet Propulsion Laboratory is engaged in activities associated with deep-space automated scientific missions tracking, data acquisition, data reduction and analysis required by deep-space flight, advanced solid propellant and liquid propellant spacecraft engines, advanced spacecraft guidance and control systems, and integration of advanced propulsion systems into spacecraft. The laboratory designs and tests flight systems, including complete spacecraft, and also provides technical direction to contractor organizations. It operates the worldwide deep-space tracking and data acquisition network, known as the Deep Space Network, and maintains a substantial programme to support present and future NASA flight projects and to increase capabilities of the laboratory.

Used as an administrative aircraft this Grumman G-159 Gulfstream I c/n 92 ex-N710G, was acquired by Huntsville on 29 March 1963 becoming NASA 3. A Beech Model 200 Super King Air c/n 1092 arrived on 3 November 1982, becoming NASA 9, whilst a second Gulfstream I c/n 125 NASA 10 was procured on 28 April 1971, it having a cargo door modification. (Harry Gann)

Aircraft assigned to the Jet Propulsion Laboratory are mainly administrative fixed-wing and helicopter types. On 17 October 1960, a Beech C-45H *Expeditor* was acquired (52-10887, c/n AF-817) and became registered as 'NASA 6' and N9525Z. On the first day of May 1966, it was transferred to the Kennedy Space Centre. On 1 July 1965 Beech Model 65-80 *Queen Air* c/n LD-77 ('NASA 7') was acquired, initially as a lease aircraft and purchased by NASA in March 1968. It was traded-in for a Beech Model 200 *Super King Air* (c/n BB-997) which was acquired on 19 January 1982 and in turn became 'NASA 7'. A Beech Model 65-80 *Queen Air* (c/n LD-79) was obtained on lease and then purchased by NASA in March 1968, becoming 'NASA 8'; it was transferred to the Wallops Flight Facility on 8 July 1968.

The first helicopter used by the Jet Propulsion Laboratory was a Bell Model 206B *Jet Ranger* (c/n 508) which replaced two leased Bell 47J helicopters. It became 'NASA 12' and was later transferred to the Johnson Space Centre. Bell Model 47J-2A N8550F had been leased for use in the Los Angeles area and was involved in a minor accident on the last day of March 1969, with no one injured. The second leased

Bell 47J-2A was N8581F, which was involved in a minor landing accident in the Goldstone area on 29 April 1969 — again with no injuries. A third Bell 47J-2A is listed but unidentified; it was leased for use in the Goldstone area with a lease termination date of 30 June 1968.

Between 16 February and 10 March 1973, a US Air Force NC-135A (60-370) was operated on research and development work involving atmospheric impact tests, and an unidentified Beech Model 60 *Duke* was used on hydrogen engine tests at the Beech facility during 1976.

The New Centres

The Kennedy Space Centre — KSC — is located on the east coast of Florida, 150 miles south of Jacksonville and approximately 50 miles east of Orlando. It is immediately north and west of Cape Canaveral. Its full title is the 'John F. Kennedy Space Centre', named after President John F. Kennedy. The centre is about 34 miles long and varies in width from five to ten miles, and the total land and water area occupied by the installation covers 140,393 acres. This large area, with adjoining bodies consisting of water, provides sufficient space to afford adequate safety to the surrounding civilian community for planned vehicle launchings. The centre serves as the primary focus within the huge NASA organisation for the test, checkout and launch of space vehicles. This presently includes launch of manned and unmanned vehicles both at the Kennedy Space Center and the US Air Force Eastern Space & Missile Centre in Florida, together with the US Air Force Western Space & Missile Centre located at Vandenberg AFB, in California. Kennedy is currently (1990) concentrating on the assembly, checkout and launch of the huge Space Shuttle vehicle programme and their various payloads, landing operations and the turn around of Space Shuttle orbiters between missions, as well as research and development involving operational unmanned vehicles.

Patrick AFB, some two miles south of Cocoa Beach, Florida, supports NASA at the Kennedy Space Centre, and it is here that a small fleet of aircraft both fixed-wing and helicopters, are based. The first aircraft in the fleet was a Lockheed L-1329 *JetStar-6* (c/n 5015, ex-N172L and N130KC) which was on a 1,400 hours lease, used for administration flights and operated as 'NASA 4'. It was acquired in January 1963 and used until June 1965. Its replacement on 8 June 1965 was a Grumman G-159 *Gulfstream I* (c/n 151) which became 'NASA 4' in turn. On the first day of May 1966, a Beech C-45H *Expeditor* (52-10887, c/n AF-817) was received from the Jet Propulsion Laboratory, becoming 'NASA 6' and being used until April 1982 in research and development work. It was a 'Dumod' conversion.

Three Bell UH-1B *Iroquois* were acquired in the early 1980s. Army 62-1908 ('NASA 414', ex-'NASA 732') arrived from Ames on 15 February 1980. Army 63-8695 arrived from Fort Rucker, Alabama, on 18 October 1981 to become 'NASA 415' and Army 62-2064 arrived also from Fort Rucker on 27 October 1981 and became 'NASA 416'. In 1990, all were based at the Wallops Flight Facility, with 'NASA 414' modified to UH-1D standard and 'NASA 415' operating with floats.

During the summer of 1975, in conjunction with the National Oceanographic & Atmospheric Administration (NOAA) the Agency conducted a research and development programme involving a Grumman S-2D *Tracker* BuNo.149240 which was loaned by the US Navy's Naval Research Laboratory.

The George C. Marshall Space Flight Centre — MSFC — is located on the US Army's Redstone Arsenal, just outside Huntsville, Alabama, and has deep-water access to the Tennessee River. Located on 1,840 acres of land, the centre is one of the United States' pioneering space centres; it was established in 1960 by a team of former US Army rocket experts headed by Dr Werner Von Braun. Marshall provides project management as well as scientific and engineering support for many of NASA's prime space programmes and scientific endeavours. It has a wide spectrum of technical facilities and laboratories. Originally the primary propulsion development centre for the NASA, the centre has diversified into an establishment for the development of payloads and space science activities. The Marshall centre is responsible for managing the development of the Space Shuttle main engines, solid rocket boosters and external propellant tank. It is also responsible for the Space Telescope, the Spacelab orbital research facility and for many other key research and development programmes.

Located some ten miles east of downtown New Orleans, Louisiana, is the Michoud Assembly Facility — MAF — managed by the Marshall Space Flight Centre occupying 833 acres of land encompassing 32 buildings with an area of about 3.5 million square feet. The primary mission of the facility is the systems engineering design, manufacture, fabrication, assembly, testing and checkout of the Space Shuttle External Tank. The Slidell Computer Complex — SCC — is also managed by the Marshall Space Flight Centre, and is located at Slidell, Louisiana, approximately 22 miles north-east of New Orleans. The complex was transferred from the Federal Aviation Administration to NASA and became operational in 1962. The facility consists of a large office building and several smaller support buildings which are well suited for computer operations. This computer complex is primarily responsible for fulfilling the computional requirements of NASA at the contractor-operated plant at Michoud, and also provides computing support for the National Space Technology Laboratories. These requirements are in the areas of scientific, management and engineering automated data processing and in static and flight test data reduction and evaluation.

The Marshall Space Flight Centre is the location of the world's largest space museum and the state-operated Alabama Space & Rocket Centre, featuring some of the most exciting space exhibits to be found anywhere. A tour of the centre includes the modest test stand where the original Huntsville team tested the Redstone rocket which carried astronauts Alan Shepard and Gus Grissom into space. Located here is the world's largest indoor swimming pool; actually not a swimming pool at all but a huge Neutral Buoyancy Simulator which allows astronauts and engineers to work in an environment closely duplicating the weightlessness of space. The first aircraft used by the Marshall Space Flight Centre was a Martin 404, N442E, on contract to the NASA in 1960

with East Coast Flying Service of Washington DC to provide a shuttle service between Marshall and the Kennedy Space Centre. This aircraft was still in use for a period after the first Grumman G-159 *Gulfstream I* (c/n 92, 'NASA 3') was acquired on 29 March 1963. This aircraft was previously N710G. A contract shuttle service with Lockheed *Electra* airliners was established between NASA at Houston and Marshall. Other aircraft operated under contract to the Agency at Marshall included a Douglas/Tempo A-26 *Invader* (N214A) in 1961, whilst Rocket City Air Activity contracted two *Aero Commanders*.

On 29 May 1965, a Beech Model 65-80 *Queen Air* (c/n LD-49) was acquired at Marshall, becoming 'NASA 9' and used, like the *Gulfstream*, for administrative purposes. On 3 November 1982 the *Queen Air* was traded-in for a new Beech Model 200 *Super King Air* (c/n 1092) which then became 'NASA 9'. A second Grumman G-159 *Gulfstream I* (c/n 125, ex-N738G and N205S) was procured for programme support on 28 April 1971, becoming 'NASA 10' and having a cargo-door modification during October 1971. On 15 September 1976 it was transferred to Lewis as 'NASA 5'. A Beech C-45H *Expeditor* (c/n AF-534, becoming 'NASA 650') arrived at Marshall on 28 May 1972 for use in research and development as well as environmental duties. It was transferred to the Wallops Flight Facility on 1 July 1976.

The ubiqutous Douglas DC-3 is unique in that it has served at all the NACA and NASA establishments over the years. On 19 January 1965 a Douglas C-47J (c/n 25771, ex-BuNo.17268 and US Air Force 43-48510) arrived from Lewis where it had been 'NASA 268' — it became 'NASA 423' and then finally 'NASA 11' before being transferred to Auburn University, Alabama, on 17 July 1972. The transport was used for programme support.

Perhaps the most unusual NASA aircraft to use the Marshall space Centre are the Boeing 377PG ('Pregnant Guppy') and 377SG ('Super Guppy'). The first was N1024V (c/n 15924) which was leased and used in programme support and also utilized to transport oversize cargoes such as rocket stages, spacecraft. The 337SG Super Guppy (ex-YC-97J N1038V, c/n 15938) was also contracted, NASA today has its own YC-97J Super Guppy (ex-52-2693, 'NASA 940') which was acquired by the Johnson Space Centre on 13 July 1979.

Boeing 377-SG Super Guppy N1038V c/n 15938 powered by four Pratt & Whitney YT-34-P engines first flew on 31 August 1965 at Santa Barbara, California. A new centre section added 15 ft to the wing span, and its fuselage was 31 ft longer than the model 377. Now NASA 940 it was acquired by the NASA on 13 July 1979 and is today based at El Paso, Texas. (NASA 79-H-658)

The many contracts issued by the NASA over the years would themselves fill a large volume; however it is felt that at least one is worthy of mention. Late in 1983 when the auxiliary power units from the Space Shuttle Orbiter *Challenger*, had to be airlifted back to the manufacturer for tests, the job was accomplished by the oldest operational Douglas DC-3 in the world — N133D, c/n 1499, a Douglas Sleeper Transport delivered to American Airlines on 12 July 1936. The aircraft, operated by Academy Airlines of Griffin, Georgia, was the sixth DST to roll off the production line at Santa Monica, California, and as of 1990 it was still registered to the company. This veteran had the proud distinction of being the oldest aircraft ever to use the Shuttle Landing Facility at the Kennedy Space Centre. To that date in 1983, the DC-3 had logged 71,640 flying hours.

The John C. Stennis Space Centre is located in Hancock County, near Bay St Louis, Missouri, on the east Pearl River, and situated midway between New Orleans, Louisiana, and the Mississippi Gulf coast. Initially the Mississippi Test Facility, it was given full field installation status by the NASA in 1974 and was renamed the John C. Stennis Space Centre by an Executive Order signed by President Reagan on 20 May 1988. In October 1961, the federal government announced the selection of a site in Hancock County to locate the nation's test facility for static firing of the Saturn V rocket engines. The Saturn rocket test stands have been modified for Marshall Space Flight Center main engine testing and for orbiter main propulsion testing for the Space Shuttle programme. The mission of the centre is support of Space Shuttle main engine and main orbiter propulsion-system testing. Static test firing is conducted on the same huge test towers used between 1965 and 1970 to captive-fire all first and second stages of the Saturn V used in the Apollo manned lunar landing and Skylab programmes. Shuttle main engine testing has been under way at the centre since 1975. The main engine test programme is expected to continue into the 1990s and beyond, supporting the Shuttle mission programme and the planned space station.

The front section of the Sikorsky RSRA — Rotor Systems Research Aircraft seen protruding from the huge cavity of the NASA Super Guppy 940 during 1979. Location was Wallops, the RSRA was from Ames, and the Guppy from Johnson. The RSRA had just completed 23 hours of flight testing at Wallops and was named 'Heathcliff' by Sikorsky. (NASA 79-H-659)

Historical Update

Many of the new NASA facilities described were previously occupied by other government agencies. In the case of the Johnson Space Centre, this was established on 1.020 acres of land originally the property of the Humble Oil & Refinery Company property which had been donated to Rice University and then transferred to NASA by the university. The centre has participated with other NASA installations in the Mercury, Gemini, and Apollo space programmes, which culminated in the first manned lunar landing in July 1969. During 1975 the Johnston Space Centre participated in the joint US-USSR Apollo Soyuz space mission, which has highlighted international co-operation in space to date.

The Jet Propulsion Laboratory in Pasadena, California, was transferred from US Army jurisdiction to that of NASA on 3 December 1958, just two months after the new agency was created by Congress. Every rocket that flies today shares the common heritage of pioneering research by Professor Theodore von Karmán of the Californian Institute of Technology — Caltech. It is a pre-eminent national laboratory under contract to NASA with a budget of $892 million and a workforce of 5,465 personnel. Today, the laboratory's principal responsibility is exploration of the solar system with automated spacecraft. JPL was selected to launch the nation's first satellite; on 31 January 1958, 66 days after receiving project approval, the laboratory launched the 14kg (31lb) Explorer I. It transmitted until 23 May 1958, and discovered the Van Allen Radiation Belts that surround the earth. JPL went on to lead the United States mission to explore the Moon and planets. It designed, built and operated the Deep Space Network — DSN — of antennas for NASA. These are clustered at sites located at Goldstone, California, near Madrid in Spain and near Canberra in Australia such locations allow communications with spacecraft anywhere in the solar system.

Langley's Research Centre, called the Langley Memorial Aeronautical Laboratory after Samuel P. Langley (a contemporary of the Wright brothers) has today expanded to cover 787 acres, and is considered one of the world's premier research facilities. Since those vintage days in 1917, when the centre was established, it has received no less than five coveted Collier Trophies — an annual award given for the greatest accomplishment in aeronautics and astronautics. Today, under NASA, the primary work at Langley is basic research in the fields of aeronautics and space technology including aerodynamics, materials structures, flight systems, information systems, acoustics, aerolasticity, and atmospheric sciences. Approximately 60 per cent of Langley's work is aeronautical research to improve aircraft of the future. This research includes investigation of the full flight range, from low-speed general-aviation and transport aircraft up to high-speed hypersonic vehicles. Approximately 40 per cent of the work supports the US national space programme. Prior to the formation of the Johnson Space Centre, Langley, was the home of the Mercury and Gemini manned spacecraft programmes. Five successful lunar orbiter mission to photograph candidate Apollo landing sites were managed by Langley, as were two unprecedented planetary landing Mars missions by the Viking spacecraft. Research scientists at Langley did extensive work on the Space Shuttle structure, aerodynamics and thermal protection system. The centre currently employs some 2,800 personnel and handles some 1,950 contracts.

Carved out of virgin savannah and marsh in the early 1960's, the John F. Kennedy Space Centre is the departure point for the Space Shuttle Orbiter and its variety of payloads. So far the Shuttle Orbiter has always returned to California, but it has the capability of returning to the Kennedy Space Centre. In any case it is transported atop a suitably modified Boeing 747, one of two converted for NASA, from California to Florida prior to the next launch from the huge 'Vehicle Assembly Building' in Complex 39. This covers eight full acres and stands 525ft high, one of the largest buildings in the world.

The Lewis Research Centre, established in 1941 at Cleveland, Ohio, became the nucleus of NASA in October 1958. Today Lewis scientists, engineers, technicans and support personnel number about 2,700 and occupy 100 buildings and 500 research and development facilities spread out over 360 acres. The centre is responsible for the 'Work Package 4' portion of the 'Space Station Freedom' Project. It has a large number of facilities which can simulate the operating environment for a complete system; these include altitude chambers for aircraft engines, large supersonic wind tunnels, space simulation chambers for electric rockets or spacecraft, and a 420ft deep

Unique air-to-air photo depicting the three purpose-built Lockheed Model L683C F-104N Starfighter aircraft for NASA delivered to Edwards AFB during 1963. 'NASA 013' was lost with its pilot Joe Walker in the horrific XB-70 collision on 8 June 1966. The remaining two carried on as research workhorses for the agency. Today 'NASA 011' is in open storage for the US Air Force Flight Test Centre Museum at Edwards. (Lockheed)

Depicted is 'NASA 011' delivered on 19 July 1963, and seen in the colourful markings initially adopted for the three Lockheed F-104N Starfighter research aircraft. They later received the current NASA livery system. Today the agency retains two Starfighters, one of which is a two-seater. (Lockheed)

Latest acquisition by the NASA at the Ames-Dryden facility at Edwards AFB early in 1990 was three ex-USAF Lockheed SR-71A Blackbird reconnaissance aircraft. Seen on arrival on 19 March 1990 is SR-71A 64-17971 on the NASA ramp. They are currently in flyable storage until involved in a high speed research programme later in 1990. (NASA EC 90 0096-2)

Latest research aircraft to appear on the NASA inventory are three ex-USAF Lockheed SR-71A Blackbirds. These arrived at Edwards AFB during Spring 1990, going into temporary storage pending a flight test programme. Depicted is the arrival of a SR-71A during February deploying its drag chute on landing. (NASA 90-HC-173)

Opposite: Three Lockheed SR-71 Blackbird aircraft, similar to the one depicted in this high altitude photo, were acquired by NASA from the US Air Force because of their unique capabilities for flight research. One of the allocated tasks by the agency is the study of sonic booms for NASP — National Aero Space Plane. Problems associated with engine inlet stability and hypersonic propulsion will also be studied using the SR-71. (NASA)

Full-scale wind tunnel model of the Grumman design 698, NASA-US Navy —
VSTOL demonstrator aircraft shown on the test rig at Ames during 1980, prior
to testing in the 40 × 80 ft low-speed tunnel to check out the two General Electric
TF34-GE-100 engines. Lack of funding cancelled this Special Electronic Mission
aircraft — SEMA-X. (Grumman — Bruce Montgomery)

The NASA Ames QSRA — Quiet Short-Haul Research Aircraft — NASA 715,
seen during carrier landing tests aboard USS Kitty Hawk off San Diego during
August 1986. During the trials a total of 37 touch-and-go landings, and 16 full-
stop landings were made. Able to land at 65 knots no arresting gear was needed.
(NASA HC-336)

zero-gravity facility. Some problems are amenable to detection and solution only in the complete system and at essentially full scale.

Although adjacent to Hopkins Airport in Cuyahoga County, Ohio, the high-speed research aircraft operated by the Lewis Research Centre are normally based and operated out of Selfridge AFB, Michigan, which is located three miles north-east of Mount Clemens. In the 1970s Lewis operated a Douglas C-47D *Skytrain* transport (43-49526, 'NASA 636') for transportation of flight-test pilots and personnel to Selfridge AFB. It arrived at Lewis on 28 October 1971, and in addition tgo programme support was used for water- and land-quality evaluation. The transport came from Langley where it was 'NASA 501' and was transferred from Lewis to Dryden on 14 August 1978, where it became 'NASA 817'. A full listing of the Lewis Research Centre aircraft appears as an Appendix. Lewis also has management responsibility for the Plum Brook Station, located in central Erie County, approximately three miles south of Sandusky, Ohio and about 50 miles from Lewis. It is an adjunct facility to Lewis, providing very large-scale specialized research test installations.

The Goddard Space Flight Centre, established in 1959, was the NASA's first major scientific laboratory devoted entirely to the exploration of space. Named after Dr Robert Hutchings Goddard (the man recognized as the 'Father of American Rocketry') the centre continues its commitment to scientific research and development for space exploration. It employs more scientists than any other NASA centre. It is also the only national laboratory that can develop, design, fabricate test-launch and analize space-science missions using all its own resources. Goddard also provides a launch range and research airport at the Wallops Flight Facility on Virginia's eastern shore for suborbital rocket, balloon and aeronautical missions. The NASA teams consist of 3,700 employees and 8,100 contracted personnel, and the centre is involved in the Space Station Freedom project.

Founded in 1939 as an aircraft research laboratory by NACA, Ames was named after Dr Joseph S. Ames, who was chairman of the Agency from 1927 to 1939. During 1981 NASA merged the Dryden Flight Research Centre with Ames, the two installations now being referred to as Ames-Moffett and Ames-Dryden. Currently (1990) some 2,000 NASA employees and almost 2,000 contractor employees are on site. The centre has a wide assortment of wind tunnels, one of which is the largest in the world, whilst the Ames fleet of airborne laboratories supports the Airborne Sciences and Applications Programme. These aircraft serve as flying instrument platforms for the use of scientists from all over the world in studies of both space and earth. In spacecraft technology, Ames supports the NASA Space Shuttle programme by providing research on heat protection systems and wind tunnel investigations of the stability and heating of the various configurations and modifications. In co-operation with Federal, state and local agencies, Ames conducts pilot programmes and prototype investigations of applications of space technology to earth bound problems.

Ames-Dryden occupies about 520 acres adjacent to Edwards AFB on the edge of the Mojave Desert some 80 miles north of Los Angeles. In 1959, the station became the NASA Flight Research Centre, and in 1976 was renamed the Dryden Flight Research Centre in honour of Dr Hugh Dryden, chairman of NACA from 1947 to 1958 and deputy administrator of the NASA from 1958 to 1965. It was established as the major NASA facility for high-speed flight test and the primary research tools are its aircraft. It has operated a unique and wide range of types from the X-series and the Century series of fighters to advanced supersonic and hypersonic aircraft and aerospace flight research vehicles such as wingless lifting bodies. There are many special ground-based facilities, including a fully instrumented Flight Test Range with tracking stations, a high-temperature loads calibration laboratory, and a remotely-piloted research-vehicle facility. The facility also conducted studies into terminal area operations of the Space Shuttle vehicle and flight investigations involving the flight test of the orbiter.

The Wallops Flight Facility is now part of the Goddard Space Flight Centre, and is located on the Delmarva Peninsula on the Atlantic coastline of Virginia. It is approximately 40 miles south-east of Salisbury, Maryland and 72 miles north of the Chesapeake Bay Bridge Tunnel. The facility consists of three separate areas which include the main base, the Wallops Island launching site and 1,140 acres or marshland. Wallops Island is about seven miles south-east of the main base and is five miles long and a half mile across at the widest point. The facility is responsible for managing the NASA Suborbital Sounding Rocket Projects from mission and flight planning to landing and recovery, including payload and payload carrier design development, fabrication, and testing — experiment management support — launch operations and tracking and data acquisition. The NASA balloon programme is managed by Wallops and it is responsible for managing the National Scientific Balloon Facility located at Palestine, Texas.

The XC-142A tri-service light VSTOL research aircraft NASA 522 ex-62-5924 seen at the USAF Museum. It served at Langley between 13 October 1968 and 5 May 1970 going to Wright-Patterson AFB in October 1974. It was powered by four TA64-GE-1 engines mounted on tandem tilt-wings.
(Coombs via Peter M. Bowers)

CHAPTER THIRTEEN
Diversity at the NASA Centres

At first NASA depended on the US military for launch operations, but in 1960 the Agency established a Launch Operations Directorate to assume general responsibility for NASA space launchings at both the Atlantic and the Pacific Missile Ranges. Originally connected with the Marshall Space Flight Centre, Huntsville, Alabama, the directorate became a separate entity known as the NASA Launch Operations Centre in 1962. It was located at Cape Canaveral, Florida — today known as the John F. Kennedy Space Centre — and soon employed a staff of well over 1,000 personnel and was participating in the design of some immense assembly check-out and launching facilities which NASA was planning to build at the Cape for future manned-flight operations. It was in 1961 when, in view of the growing magnitude and importance of the manned space flight programme, plans were laid for the construction of a Manned Spacecraft Centre in order to conduct research and development in manned spacecraft and to plan and carry out manned spaceflight missions. Although not the subject of this volume, space travel must be briefly mentioned. The huge Apollo programme lasted 11½ years, cost $23.5 billion, landed twelve men on the Moon and produced an enormous amount of evidence and knowledge. Who will ever forget those historic words transmitted by astronaut Neil Armstrong on 20 July 1969, 'Houston — Tranquility Base here — The Eagle has landed'. It was not only a giant leap for mankind, it was also a giant leap for NASA.

During 1964 Langley leased the prototype Boeing 707 for three months for $800,000 for flight tests. It was equipped with large wing flaps, a boundary layer control system, a thrust modulating system and a mass of instrumentation. During 1967 it was under contract to Ames equipped with aircraft response sensing system installed in the 15 ft nose spike. (NASA 67-H-538)

Supercritical Wing

With investigation of the supersonic regime continuing, a major breakthrough at the transonic level occurred — the supercritical wing. The transonic region had beguiled aerodynamicists for years. At transonic speeds, both subsonic and supersonic flow patterns encased an aircraft. As flow patterns went supersonic, shock waves formed across the wings, resulting in a sharp rise in drag. During the 1960s Richard Whitcomb at Langley, committed himself to a programme intended to resolve the transonic problem. Over several years, Whitcomb analysed what came to be called the 'supercritical Mach number', spending four years on a wind-tunnel study programme at Langley. The shape finally selected had a flattened top surface, with a downward curve at the trailing edge — this helping to restore lift lost from the flattened top. By 1967 Whitcomb was convinced he had a major breakthrough and the search began for a suitable aircraft to serve as a testbed for a supercritical wing.

The aircraft chosen by NASA was the Vought F-8U *Crusader* a single-seat, single-engined US Navy jet fighter. The structure of the aircraft's shoulder-mounted wing made it easy to remove and replace with the new supercritical wing design. The F-8U was built with landing gear which retracted into the fuselage, leaving the new wing with no outstanding encumbrances. The US Navy had surplus *Crusaders*, and the aircraft's speed of Mach 1.7 made it ideal for transonic flight tests. The wing programme was principally aimed at civil applications; the airlines as well as the civil aircraft manufacturers closely followed development of the new airfoil. NASA acquired three surplus F-8U *Crusaders* including BuNo.141353 which became 'NASA 810' and BuNo.145546 ('NASA 802').

A decision to use the F-8U was decided by the Langley NASA team on 21 March 1967. In February 1969, the Agency announced that Whitcomb's supercritical wing concept would be flight-tested at the Edwards Flight Research Centre. NASA Administrator Thomas O. Paine testified before a congressional committee that the tests would probably commence in late 1970. The Langley 8ft wind tunnel was employed in testing a model F-8 with the new planform.

The Vought F-8U (BuNo.141353) arrived at Edwards on 25 May 1969 and NASA flight test pilots Thomas C. McMurtry and Gary Krier commenced flying to gain operational experience prior to modification. The Flight Research Centre contracted North American (Rockwell Division) to fabricate the supercritical wing, at a cost of $1.8 million. Meanwhile North American Rockwell fitted a supercritical airfoil on the wing of a US Navy T-2C *Buckeye* jet trainer at the company's Columbus, Ohio, plant to gain preliminary experience with such wings. This made its first flight on 24 November 1969, and three weeks earlier the new supercritical wing was delivered to Edwards. By this time the two NASA pilots had completed 32 flights in the *Crusader*, which received the designation TF-8A.

By early 1971, the NASA technicians had installed the shapely wing on the TF-8A. Tom McMurtry was the lead project pilot and engineer John McTigue, who had played a part with the lifting bodies, was the programme manager. The technicians had built a supercritical wing simulator which the two flight test pilots used. On 9 March 1971, McMurtry took-off on the first supercritical wing flight, evaluating the aircraft's low-speed handling qualities and stability augmentation system and reaching an altitude of 10,000ft and a maximum speed of 300 miles per hour. It was described as the most graceful aircraft flown by NACA/NASA at Edwards. Flight testing went smoothly as the Agency gradually expanded the flight envelope to higher altitudes and speeds. On its fourth flight on 13 April 1971, McMurtry reached Mach 0.9 at 33,000ft. The first data-gathering flight came on 18 August, following installation of special instrumentation including a network of 250 pressure sensors on the wing's upper surface to locate and measure shock-wave formation.

The programme gave sufficient encouragement for NASA and the US Air Force Flight Dynamic Laboratory to commence another supercritical wing research programme, the military-orientated TACT (Transonic Aircraft Technology). Whitcomb envisaged the ideal transonic transport as having both a supercritical wing and transonic area ruling, and at a later date winglets. In May 1972 NASA reworked the TF-8's instrumentation an installed new fuselage fairings which gave it pronounced area ruling. It first flew on 28 July 1972, and by the end of the year the research utility of the *Crusader* was nearing an end. Commencing in January 1973, the aircraft began flying pilot familiarization flights. The author's good friend Ron Gerdes from Ames had the honour of making the last flight on 23 May 1973. As if sensing the end, the TF-8 chose this flight to develop a serious problem, its primary hydraulic system failing, but Ron landed the aircraft safely at Edwards. The aircraft is today preserved at Dryden. Industry rapidly applied the results of supercritical wing technology to new designs such as the Boeing YC-14 and the Douglas YC-15, the Rockwell *Sabreliner 65* and the Canadair *Challenger*. At NASA headquarters on 4 June 1974, the NASA Administrator James C. Fletcher conferred on Whitcomb the maximum $25,000 prize for invention of the supercritical wing. The National Aeronaut Association awarded him the 1974 Wright Brothers Memorial Trophy.

The TACT (Transonic Aircraft Technology) programme involved modifying a General Dynamics F-111A 63-9778 to explore how supercritical wing technology could benefit new military aircraft designs. The Langley Research Centre had undertaken a great deal of wind-tunnel work on the aircraft type. Whitcomb had devised a supercritical wing for a transonic manoeuvring military aircraft. The F-111 was chosen because of its variable-sweep wings, and the new wings could be easily installed with a minimum of extra modification. By mid-1971 the Agency and General Dynamics had expended over 1600 hours of wind-tunnel test time on development of a suitable wing for the F-111. Whitcomb determined its shape, twist and airfoil co-ordinates. General Dynamic built the wing, and the US Air Force Flight Dynamic Laboratory provided the money. On 16 June 1971, the Agency and the US Air Force signed a joint TACT agreement, with the F-111 to be flown at the Flight Research Centre and development of the advanced configuration of the wing to be undertaken at Ames Research Centre. The F-111A was ideal for a supercritical wing capable of supersonic speeds above Mach 2, the aircraft having a large volume for both fuel and instrumentation with the wings easily removable, and the variable-sweep provision enabled supercritical wing testing over a wide range of wing sweep angles and aspect ratios. The aircraft made available — 63-9778 — was the 13th F-111A and NASA signed a loan agreement for the aircraft with the US Air Force on 3 February 1972; fifteen days later NASA pilot Einar Enevoldson and US Air Force pilot Major Stu Boyd checked out at Edwards in the aircraft. The modified F-111A was ready by the autumn of 1973 and on the first day of November the two pilots made the first TACT flight, reaching Mach 0.85 at 28,000ft. On the sixth flight on 20 March 1974 they exceeded Mach 1, and on the twelfth flight Mach 2 was reached.

The Dryden Flight Research Centre's Lockheed L1329 Jetstar 6 NASA 814 c/n 5003, acquired in May 1963, was used for a wide variety of general-purpose airborne simulator — GPAS — studies, general aviation research, plus support of the Space Shuttle approach and landing tests. During 1981 it had a Hamilton Standard SR-3 propfan mounted on a pylon atop the aircraft. (NASA 71-H-102)

This F-111A flew frequently, with a mixed NASA and US Air Force crew, and soon became a workhorse, flying with variety of aerodynamic experiments including special shapes to evaluate base drag around the tail, experimental test instrumentation and equipment destined for use with other aircraft. The experiment with TACT encouraged the US Air Force Flight Dynamic Laboratory to proceed with further research efforts. Advanced Fighter Technology Integration — AFTI — was another joint effort consisting of various technology sets and involved the F-111. It was one of many programmes for the future.

RPRVs

Radio controlled aircraft, or remotely piloted vehicles, appeared as early as World War I, and by the end of World War II, the major powers had made extensive use of remotely controlled guided weapons. Today, most of the major military forces of the world employ RPVs as target-drones and they were employed extensively in the Vietnam conflict and during the Middle East hostilities in 1973. In support of the M2 lifting-body programme in the early 1960s, Dale Reed built a number of small lifting-body shapes and launched them from a twin-engined radio-controlled model called 'Mother'. By late 1968 Mother had achieved over 120 launch drops. More sophisticated equipment became available and Dale Reed selected the Langley *Hyper III* configuration, a slender re-entry shape having a flat bottom and sides having a lift-to-drag ratio of about three. NASA shop personnel built the vehicle at a cost of $6,500. It weighed 450lb, measured 30ft in length and had a span of 17ft. By December 1969 the Flight Research Centre was ready for the initial trials; the *Hyper III* was launched from a helicopter at 10,000ft, at which height control was handed over to a ground-control cockpit. The *Hyper III* was not tested further since it had a much lower L/D ratio than predicted.

Basically a Ryan BQM-34 Firebee 2 RPV, acquired by NASA in 1979 this DAST — Drones for Aerodynamic and Structural Testing — aircraft performs its first free flight over the Mojave Desert on 17 November 1982. It took 2½ years to develop and was launched from the NASA B-52. Recovery was by helicopter. It was a flight load alleviation test-bed for experimental wing shapes, being destroyed during one of its flights. (NASA 82-H-746)

Dale Reed and his RPV team decided to attempt to control an actual manned aircraft by means of a ground pilot, with a safety pilot in the aircraft. The centre selected a Piper PA-30 *Twin Comanche* (c/n 30-1498, registered 'NASA 808'), a light twin-engine aircraft already configured as a test bed for general-aviation flight trials. As flown by the Agency, it had dual controls, one side having an electronic fly-by-wire system, the other having conventional controls. During October 1971 NASA commenced flight trials, with Einar Enevoldson flying the PA-30 from the ground and Tom McMurtry as safety pilot. Eventually, Enevoldson flew the aircraft unassisted from take-off to landing, making precise instrument-landing-system approaches, stalls and stall recoveries. The next step was to apply the RPV to some meaningful research project.

During April 1971, the US Air Force issued a memorandum calling for a national programme to investigate stall and spin phenomena, this area having become critical with many US Air Force fighter aircraft being lost in spinning accidents. A steering commitee including NASA representatives recommended expanding existing programmes, using radio-controlled free-flight models to evaluate spin entry and post-stall gyrations. Langley had made stall-spin studies using small-scale models dropped from helicopters, but it was recommended that larger models be used. During November 1971 the Flight Research Centre put a proposal to the NASA

headquarters for stall and spin testing a ⅜ scale model of the McDonnell Douglas F-15 *Eagle* configuration. In April 1972 the Agency awarded McDonnel Douglas a $762,000 contract for the construction of three ⅜ scale F-15 models. NASA also place contracts with other firms for support equipment including parachute recovery units.

The first F-15 RPV arrived at Edwards on 4 December 1972, the centre adding the avionics, hydraulics and other sub-systems. It cost $250,000 compared to $6.8 million for a full-scale piloted F-15 Mach 2 aircraft and was launched from a Boeing B-52 at about 48,000ft, after which a NASA pilot put the drone through its planned research programme. On recovery and reaching 16,000ft the RPV streamed a spin-recovery parachute having a diameter of 13ft. This chute extracted two other parachutes — a 26ft engagement and an 80ft diameter main chute. As the F-15 model descended, a helicopter snatched the engagement chute with grappling hooks. The helicopter reeled in the RPV with a winch prior to returning to base. Eventually NASA planned to land the F-15 RPV on the Edwards lakebed using skids. On 12 October 1973 the first F-15 RPV was flown under the Boeing B-52 for a flawless nine-minute flight, remotely piloted by Einar Eneldson.

Since the F-15 RPV programme, work has continued at Edwards. During June 1973 it was revealed that the NASA was hoping to acquire its first Teledyne-Ryan BQM-34E *Firebee 2* drone during FY 1974, having already had experience with US Navy versions of the vehicle to obtain data. This, and a second drone planned for acquisition a year later, would be operated under Project DAST — Drones for Aerodynamic & Structure Testing. The US Navy drone used had the NASA instrumentation fitted as a pack on top of the fuselage, whilst a second BQM-34 was modified to include 30 pressure orfices on one wing semi-span. It flew on normal US Navy drone sorties, with both agencies sharing the data obtained. By July 1979, a BQM-34 *Firebee 2* at the NASA Flight Research Centre had been fitted with an aeroelastic research wing in the DAST programme. The wing was designed to provide a better understanding of transonic flutter and improve prediction techniques. First flight was scheduled for August 1979. Milton O. Thompson, director of the NASA Research Projects Division at Edwards, indicated that the first step would be to learn how to fly the *Firebee 2* using the ground control loop developed with the Piper PA-30 'NASA 808' and proven with the F-15 model tests. Techniques for landing the *Firebee* horizontally on skids instead of with a parachute recovery system had to be worked out.

Fly-by-Wire

In the pioneer days of aviation, pilots controlled aircraft by moving a stick or pushing a rudder pedal connected to cables which, in turn, pivoted a control surface. In those days an on-off switch provided full engine power or none at all. In time, sets of throttles and fuel mixture controls regulated engine power. With increases in speed and control loads, flying reached a point where pilots could no longer exert the brute strength to control the aircraft at high speed. The came hydraulicly boosted controls, and by the 1960s jet aircraft were operating with boosted hydro-mechanical controls. These were very vulnerable to damage — loss of hydraulic pressure in the control system could mean the end of an aircraft even if all other systems were functioning smoothly. The necessity for redundant backup systems complicated aircraft design. The US Air Force lost many Republic F-105 *Thunderchief* aircraft over North Vietnam to anti-aircraft fire which damaged the hydraulics. The prototype Grumman F-14 *Tomcat* was lost on its maiden flight as a result of hydraulic failure — an accident which delayed the programme at a critical time.

Electronic controls, where the commands go to a digital computer which sends a signal flashing through a wire to move the controls electronically, appeared to be the answer.

Electronic fly-by-wire controls are much less vulnerable to damage than conventional hydro-mechanical controls. Several wire bundles could be routed through an aircraft with greater flexibility than a maze of pushrods, pulleys, and cables. The advantages of electronic controls are numerous — they are simpler, smaller and lighter, and translate directly into improved performance, reliability, payload and fuel consumption.

The Flight Research Centre flight-test engineers required a true fly-by-wire testbed having strictly electronic controls. The possibility of reconfiguring a conventional fighter, such as the Lockheed F-104 or a Vought F-8, was discussed. This would have fly-by-wire controls and revised flight control surfaces, perhaps reducing tail proportions or incorporating a canard layout. NASA engineer Melvin Burke was particularly interested in flying a digital fly-by-wire testbed. NASA headquarters showed little interest in the project until Neil Armstrong became NASA's deputy associate administrator for aeronautics within the Office of Advanced Research & Technology. During the Apollo programme Neil had become acquainted with fly-by-wire technology at the controls of the lunar module. That vehicle had a digital computer and inertial measuring unit. Armstrong believed this off-the-shelf system could be readily applied to a testbed aircraft, and supported Melvin Burke and his project.

A Vought F-8C *Crusader* was acquired from the US Navy and NASA engineers replaced all cables, push rods and bell cranks with the Apollo-derived digital flight computer and inertial sensing unit; sets of wire bundles were routed from the pilot's control stick to the computer, and so to the control surfaces. The Massachusetts Institute of Technology's Charles Stark Draper Laboratory supported the effort of re-programming the Raytheon computer from the lunar module, and the Sperry Flight systems Division supplied a backup fly-by-wire system for the *Crusader*. On 25 May 1972, the NASA flight-test research pilot Gary Krier flew the F-8 testbed — 'NASA 802' — in the first flight of an aircraft completely dependent upon an electronic control system.

By early 1973, after fifteen flights without incident, Krier testified before the House Committee on Science & Astronautics on the benefits already demonstrated by the programme. The F-8 flew 42 times without incident, and it was never necessary to resort to the emergency backup flight-control system. Before completion of the test programme, a prototype version of the electronic sidestick planned for the General dynamics F-16 fighter was tested on the F-8, including formation flights and landings.

The fly-by-wire research programme was only one of the electronic control programmes which influenced development of this new technology. The Flight Research Centre became involved in the Integrated Propulsion Control system — IPCS — evaluated on a US Air Force General Dynamic F-111E (67-0115). The programme was conducted between March 1973 and February 1976 and was a co-operative effort involving the Lewis Research Centre and NASA at Edwards, the US Air Force Flight Propulsion Laboratory, The Boeing Airplane Company, Honeywell and Pratt & Whitney. The US Air Force was willing to fund an experimental effort using a suitable aircraft. The General Dynamics F-111 was a large, two-seat twin-engined aircraft with a complex propulsion system. It had variable-position inlet and after burning fanjet engines, as well as an internal weapons bay which test engineers could use to house the necessary electronic controls. The aircraft loaned by the US Air Force was the first prototype of the F-111E series. Contracts were awarded for the IPCS programme in March 1973. For various reasons, including flight safety, the programme could not have been conducted at the Lewis Research Centre, so it was decided to carry out flight trials at Edwards.

The F-111E was received at Edwards in mid-1974 and embarked on a series of thirteen flights prior to modification. These flights acquired baseline data for comparison with results of later IPCS tests. Installation commenced in March 1975. The system comprised an instrumentation package, power supply, digital computer and interface equipment installed in the fuselage weapons bay. The hydromechanical inlet and afterburner controls were replaced by new electronic controls. Two software programmes supported the IPCS evaluation including a digital representation of a TF30-P-9 afterburning turbofan engine used for assessing the ability of the IPCS system to duplicate the hydromechanical control functions. The other (called the 'IPCS control mode') integrated the inlet and engine control functions into one operation. All the software and hardware were rigorously bench-tested, installed on a Pratt & Whitney TF90-P-9 engine, run on a test stand and then installed in the altitude test chamber at the Lewis Research Centre.

This North American Rockwell OV-10A Bronco NASA 718 BuNo.152881 was delivered to Ames on 8 April 1968 remaining until 7 October 1976. It was equipped with experimental rotating cylinder flaps, one of a number of STOL concepts then being investigated for possible application to both civil and military aircraft. It was fitted with new engines with props interconnected. (NASA 71-H-1606)

On 4 September 1975 the F-111E completed its first IPCS flight, piloted by NASA flight-test pilot Gary Krier and US Air Force pilot Stan Boyd. A further fourteen IPCS exercises were carried out before the programme was complete, with the last flight on 27 February 1976. The F-111E was returned to the US Air Force, restored to its original condition and ultimately served as chase aircraft for the new Rockwell B-1 strategic bomber. NASA interest in electronic controls has continued, however, with the similar but more advanced Digital Electronic Engine Controls — DEEC — research programme using a Flight Research Centre McDonnell Douglas F-15 *Eagle*.

This General Dynamics F-16A Falcon 75-750 NASA 750 AFTI — advanced fighter technology integrates — aircraft programme is devoted primarily to investigate flight control and system concepts that could be incorporated in advanced aircraft. Flight testing began in June 1982 and the F-16 is current with a new three year research programme planned. (Office of History — USAF Edwards)

Aviation Safety

Relatively little work in the field of aviation safety had been conducted by the old NACA. The High-Speed Flight Station undertook virtually no aviation safety projects related to air transportation, the closest being the Boeing KC-135 studies supporting the introduction of the Boeing 707 generation of jetliners into service. for the most part, NACA had left aviation safety to the Civil Aeronautics Administration — the forerunner of the Federal Aviation Agency, later the Federal Aviation Administration together with organizations as the Flight Safety Foundation and the Cornell-Guggenheim Aviation Safety Centre. All this changed in the 1960s and early 1970s. Between 1964 and 1966, the Flight Research Centre conducted a number of flight evaluations of general-aviation aircraft. Similar work was also conducted at Langley. Though general aviation posed some major research problems, two other areas attracted particular attention — wake vortex and clear air turbulence (known as CAT) which still present difficulties today. Turbulent vortices trailing behind an aircraft can affect other aircraft which pass through them. In simple terms, one vortex streams from the right wingtip, rotating anti-clockwise, whilst the vortex from the left wingtip, rotating anti-clockwise, whilst the vortex from the left wingtip rotates clockwise. The intensity of these vortices is directly related to the size and weight of the aircraft generating them. The wake vortex of a light aircraft such as a Cessna 150 is negligible, whereas that of a Boeing 747 can exceed 120mph in rotational velocity and can occasionally persist for distances of 10–15 miles.

The problems engendered by what is now usually referred to as 'wake turbulence' first became serious following the introduction of large jetliners; when the wide-bodied jets such as the Boeing 747, McDonnell Douglas DC-10 and Lockheed 1011 entered service, wake vortices became a major hazard. These aircraft trailed vortices powerful enough to roll over business jets and even other airliners. In response Air Traffic Control agencies throughout the world imposed separation distances, which automatically reduced the number of aircraft able to approach and land at an airport in a particular time. The Flight Research Centre became involved once NASA viewed vortex research both from the safety aspect and as a matter of aerodynamics. A wingtip vortex seriously reduces efficiency, causing drag to rise with a consequent penalty in fuel consumption and performance. The desire for efficiency and minimization of the wake vortex prompted Richard Whitcomb at Langley to develop the winglet concept; small, nearly vertical wing-like surfaces mounted on the wing tips. They reduced drag and offered fuel savings. The NASA centre subsequently testing a Boeing NKC-135A (55-3129) equipped with winglets in a proof-of-concept demonstration. The Ames Research Centre engineers experimented with small fins mounted above or below a wing. Langley tunnel-tested a $^3/_{100}$ scale model of a Boeing 747, and following this the Flight Research Centre flew a Boeing 747 on wake-vortex alleviation studies.

During November 1973 the Flight Research Centre studied wake vortices with a Boeing 727, equipping the airliner with smoke generators to trace the patterns and following it with instrumented Piper PA-30 and Lockheed F-104 chase aircraft to measure the force and effects. The Agency had purchased a Boeing 747-100 from American Airlines on 22 June 1974, for use as the Rockwell Space Shuttle carrier (ex-N9668, c/n 20107) and this became 'NASA 905'. NASA headquarters were petitioned for use of this aircraft, which was assigned to the Johnson Space Centre. On 16 August 1974 the request was granted and the Boeing 747 made some 30 flights in a wake-vortex research programme. Test crews varied the position of the spoilers and used various spoiler segments in an attempt to determine the optimum method of alleviating wake vortices. Chase aircraft included a Gates *Learjet* ('NASA 701') and a Cessna T-37 trainer, both from Ames, representing business and small jet aircraft. They probed the vortices to measure their strength, with surprising results. In July 1977 a Lockheed 1011 *TriStar* was used in a brief series of tests for comparison with the Boeing 747.

Another research area involving the Flight Research Centre at Edwards were studies of pollution of the upper atmosphere. NASA sponsored Lockheed U-2 and Martin WB-57F high-altitude sampling flights.

Clear-air turbulence was also a problem which was investigated. Both Langley and Edwards engineers joined in research to provide a limited amount of highly accurate measurements associated with mountain waves. Jet streams, convective turbulence, and CAT near thunderstorms. At a joint meeting on 3/4 June 1969 the team agreed to use a NASA Martin B-57B aircraft, which operated out of Edwards although normally based at the Johnson Space Centre. The B-57B supported three atmospheric science programmes; measurement of atmospheric turbulence sponsored by Langley; aerosol-sampling sponsored by the University of Wyoming; and detection of clear-air turbulence sponsored by the US Department of Transportation.

New Aircraft

Due to commitments to advanced research programmes such as the X-15, the Flight Research Centre lacked manpower to participate in any new military aircraft programme such as the McDonnell F-4 *Phantom*. Paul Bikle would have preferred to have been involved but, under new procedure policies, if NASA flew an aircraft on loan from a US military service, it had to pay its operational costs. The Agency at Edwards had close personal ties with the resident US Air Force Flight Test Centre, but NASA headquarters refused several requests to acquire service aircraft on budgeting grounds. Bikle needed new aircraft, since, in addition to research, pilots needed to stay current with the latest technology. In 1963 the Flight Research Centre acquired three Lockheed F-104N *Starfighters* especially ordered, these being initially registered 'NASA 011', 'NASA 012' and 'NASA 013'. Bikle also obtained a Northrop T-38 *Talon* supersonic trainer. This useful and reliable jet could perform a variety of mission-support chores, as well as simulating lifting-body landing approaches. Paul Bikle's managerial philosophy stressed diversity, which helped to save the Flight Research Centre from the criticism of those who sought to close it down during the 1960s.

The centre became involved in programmes with a wide variety of service aircraft, whilst other progammes such as the YF-12A, XB-70A and TACT F-111 were military-related. Aircraft were also acquired from abandoned projects, such as the Northrop A-9A, but no programme was planned. During the 1960-70s the Agency flew the Lockheed F-104, not only as a workhorse but as various forms of flying testbeds. In addition to using them for X-15 mission support and chase and in support of the lifting-body programme, they were used in a number of short projects — base drag measurements, sonic boom measurements in support of Langley research and tests of ballute (balloon-parachute) deceleration devices. The centre also received two early General Dynamics F-111A aircraft — 63-9771 and 63-9777. Historians will record and remember the Tactical Fighter Experiment — TFX — programme involving the US Navy and the US Air Force, both committed much against their will. This called for the development of a single aircraft — the F-111 — to fulfil a US Navy fleet-defence interceptor requirement. The naval version, the F-111B, was never put into production. The US Air Force produced a variety of models, and it was only for F-111F which actually fulfilled the original TFX design specification.

The early F-111A had extremely bad engine problems and suffered from compressor surge and stall. In January 1967 the US Air Force sent the sixth production F-111A (63-9771) to NASA at Edwards for flight testing. Eventually as a result of co-operation between the Agency, the US Air Force and General Dynamics, the engine problems were solved by a major inlet redesign. During April 1969 the twelfth F-111A (63-9777) arrived and was flown in a handling qualities investigation programme. Both aircraft were retired to the Davis-Monthan AFB boneyard in Arizona during 1971. The later F-111 aircraft operated in the TACT and IPCS programmes were far superior in many respects.

NASA also flew the Lockheed T-33 jet trainer on a human-factors study to evaluate the effects of visibility restriction upon a pilot's performance during landing. Advanced types tended to have restricted visibility forward and laterally during a landing approach. The centre also undertook a comprehensive study of high-lift flaps as aids to transonic manoeuvrability with a series of tests on the F-104, Northrop F-5A *Freedom Fighter* and Vought F-8C *Crusader* aircraft during 1970/71. Wind-tunnel tests were apparently not reliable for this purpose, and the flight-test data was useful for the development of new military types.

Sponsored by the Department of Defense, the Flight Research Centre's work in this area led to derivation of agility criteria for fighter turn rates, buffet, maximum lift, and handling qualities. This assisted in the development of the McDonnell Douglas F-15A *Eagle*, the General Dynamics F-16A, the Northrop YF-16 *Cobra* and the F-18 *Hornet*. The NASA used a Lockheed T-33 *T-Bird* for evaluating a self-contained liquid-cooled flight garment providing the pilot with heating, cooling and an anti-G function.

New shapes soon appeared at the Flight Research Centre in the form of the McDonnell Douglas F-15A *Eagle* and the Northrop YF-17 and F-18. Involvement with the F-15 programme came out of earlier work with the F-15 RPV model, and the need to have a representative of the latest US Air Force fighter. The *Eagle* was a turning-point in the US Air Force's doctrine, a return to an aircraft designed primarily for agility and air-to-air ability — the first since the North American F-86 *Sabre*. The NASA engineers and flight test pilots were looking forward to renewing the service-testing policies and NASA headquarters approved transfer of two aircraft from the US Air Force F-15 Joint Test Force, whose activities were winding down and which had two specialised prototypes available. The second F-15A (71-0281) had been used for propulsion tests, and the eighth F-15A (71-0287) had been used for spin tests. The Agency acquired both aircraft on indefinite loan, and they were utilised on a variety of research missions. In 1990 they were still on the inventory at Edwards.

The Light Weight Fighter — LWF — programme in the early 1970s produced two technology demonstrators, the single-engine General Dynamics YF-16 and the twin-engine Northrop YF-17 *Cobra*. The programme later became a Department of Defense competition for a new fighter for both the US Navy and the US Air Force. The YF-16 was judged superior, going to the US Air Force as the F-16A. The US Navy unhappy with the outcome, proceeded with a derivative of the YF-17 which evolved into the Northrop F-18 *Hornet* fighter programme. Remaining briefly in storage, the two YF-17 prototypes flew again as development aircraft for the new proposed F-18. NASA flew the first YF-17 (70-1569) in 1976 for base drag studies and to evaluate the manoeuvring capability and limitations of the aircraft. All the NASA flight test pilots flew the aircraft whilst engineers examined the buffet, stability and control, handling qualities and acceleration characteristics. The flight testing of US service aircraft continues today.

The Lockheed XH-51N rigid-rotor helicopter arrived at RAF Bedford in the United Kingdom on 12 October 1970 by RAF C-130K completing 20 hours of test flying. Photo taken on 14 October 1970 depicts NASA 531 flying over the Aero Flight apron at Bedford. There has always existed a close relationship between RAF-NASA and the Hunting 126 research aircraft was flown to the USA for NASA evaluation. (Crown Copyright Reserved — B2858C)

Space Shuttle

In July 1970 NASA awarded study contracts to North American-Rockwell and McDonnell Douglas for the Space Shuttle. Various designs were submitted, including vehicles launched from the backs of other winged re-entry vehicles, launched on top of boosters and vehicles attached to large fuel tanks and solid-fuel boosters — the 'parallel burn' configuration, in which both liquid-fuel engine and solid-fuel booster would burn during ascent. In March 1972 the Agency selected the parallel-burn approach and on 16 July selected the Rockwell proposal for development. The construction of the first Space Shuttle orbiter, vehicle OV-101, commenced at Rockwell's Downey, California, plant on 4 June 1974. Components were delivered to Palmdale, where final assembly began in August 1975. Roll-out was on 17 September 1976. The craft, the size of a Douglas DC-9 airliner, was christened *Enterprise*. During January 1977 it was moved to Edwards.

The Boeing 747 purchased by NASA was at the same time being modified by Boeing at their Wichita plant so that the Shuttle could be mounted on top of the fuselage. It could then be ferried from one site to another as required. In the Fall of 1974 the US Air Force and NASA executed a joint agreement to establish Space Shuttle facilities at Edwards, the site already designated for the Space Shuttle approach and landing tests and as the prime landing site for the return from the first orbital flights. Overall control of the Shuttle would be held by the Johnson Space Centre, with the Flight Research Centre operating in a supporting role.

The de Havilland (Canada) C-8A Augmented Wing Jet STOL aircraft, NASA 716, seen during its first flight on 12 February 1973 from Ames. The conversion was a joint NASA — Canadian Department of Industry Trade & Commerce programme to flight list the wing system for powered lift. (NASA 73-H-106)

The flight-test programme had three phases — captive, captive-active and free flights. The unmanned captive flights demonstrated whether the combination could fly safely together. In the captive-active trials, an astronaut crew would ride in the Shuttle. In the final phase the Shuttle would be launched from the back of the Boeing 747 and flown down to a landing. A series of high-speed taxi tests by the mated 747 and *Enterprise* in mid-February 1977 were completed without a hitch. On 15 February 1977 the first Shuttle-747 flight took place with Fitz Fulton, Tom McMurtry, Vic Horton and Skip Guidry in the Boeing 747 ('NASA 905'). Its was a success and was followed by four more. NASA had selected four astronauts for the two Shuttle landing tests — Fred W. Haise and Charles G. Fullerton on one and Joe H. Engle and Richard H. Truly on the other. They had prepared for the shuttle programme by use of the ground simulator and by fly use of the ground simulator and by flying the much-modified Grumman *Gulfstream II*. Other pilots flew the NASA Lockheed *Jetstar* 'NASA 814' to test the Shuttle's microwave scanning-beam landing system. On 18 June 1977, the 747 and *Enterprise* went aloft on the first captive-active tests with Fred Haise and Gorden Fullerton in the Shuttle, and during the hour all objectives were achieved. The test data indicated that the Space Shuttle was buffet- and flutter-free up to the maximum speed of over 180mph. On 28 June, with Engle and Truly in the Shuttle, high-speed flutter tests up to 280mph were carried out. The NASA cut down the captive-active flights to three and the final test took place on 26 July.

On 12 August 1977 the air launch of *Enterprise* took place. At an altitude of 28,100ft, Fitz Fulton nosed the combination into a shallow dive and Fred Haise radioed Fulton — 'The *Enterprise* is set; thanks for the lift'. — prior to pressing the separation button. Seven explosive bolts detonated and the Shuttle was on its own at 24,250ft, briefly pitching up to the right while the 747 pitched down slightly rolling into a diving left turn as briefed. The large delta Shuttle handled well, and because of the low lift-to-drag ratio it would remain airborne for only five minutes. The *Enterprise* descended over Leuhman Ridge, passed over Highway 58 at Boron, turned

west towards Peerless Valley, swung round over North Edwards and lined up on Runway 17. Houston Mission Control radioed Haise to tell him the Shuttle had a lower lift-to-drag ratio than predicted by tunnel tests. High and hot on its final approach, Haise deployed the landing gear, the Shuttle landing long by about 3,000ft at 180mph nearly 5½mins after launch. It coasted for over 1½nm before halting on the south lakebed. The first Shuttle flight had been a success.

The Space Shuttle flew better than the Grumman *Gulfstream II* simulator, and NASA were now wondering how it would behave without its tailcone. Would buffet from disturbed air caused by the removal of the cone cause structural problems for the 747's vertical fin during the climb, and would the Shuttle's low lift-to-drag ratio, (made even lower by the removal of the tailcone) present any serious piloting problems? In theory, the descent rate of the Shuttle would approximately double, reducing the flight time from over five to just over two minutes. Pending a decision to fly with the tailcone off, testing continued with the blunt end (still with pointed tailcone) attached. A variety of delays, including rain on the lakebed, forced deferral of the next free flight until 13 September, when Joe Engle and Dick Truly completed a successful flight. Data was recorded on the Shuttle's longitudinal, lateral and directional stability and L/D ratios, together with an examination of flutter characteristics during approach and landing. Fred Haise and Gordon Fullerton completed the third free flight on 23 September, events progressing so smoothly that NASA decided to commence tailcone-off testing. In preparation for the flight, Rockwell and the NASA technicians removed the tailcone, replacing it with a configuration identical to that which it would have during re-entry from space. This included the three main Shuttle engine nozzles and the much smaller nozzles of the orbital manoeuvring subsystem.

Fitz Fulton and Tom McMurtry were briefed carefully in respect of possible severe buffeting in the Boeing 747 cabin. The loads imposed on the 747 tail would be carefully monitored, and if they proved to be excessive the flight would be aborted. The flight on 12 October 1977 went without difficulty. After being airborne for forty minutes, Fulton pushed the combination into a shallow dive at 25,500ft above the desert. Thirty-eight seconds later Joe Engle triggered the explosive bolts and separation occurred over Peerless Valley. *Enterprise* nosed down sharply and descended over North Edwards on final approach to Runway 17, followed by the Northrop T-38 chase aircraft. Removing the tailcone had made a difference; Engle pulled out of the descent into the landing flare, and deployed the landing gear and touched down 2mins 34secs after the launch. The Shuttle had flown well, but could it be landed on a confined runway? This was a critical issue since NASA planned landing the Shuttle on the 15,000ft runways located at Vandenberg, California and Kennedy, Florida. For the next flight (with tailcone off) it was planned to land on the 15,000ft runway at Edwards. So far the Shuttle had little difficulty in landing on a chosen spot on the lakebed runways.

The fifth Shuttle free flight was 26 October, with air launch at 19,000ft for a straight-in approach. Fred Haise flew a 450mph approach profile down to flare manoeuvre, the *Enterprise* losing speed very slowly, but on passing over the runway threshold the Shuttle was over 30mph faster than planned. Haise used the split-rudder speed brake and nosed down the runway at the planned impact point. The craft entered a roll to the left, which Fred corrected, touched down and bounced back into the air. Pilot-induced oscillation had occurred. It touched again, bounced again, then made a final landing and stopped. HRH the Prince of Wales witnessed this flight with great interest. Despite the landing, NASA engineers and astronauts were confident that the Shuttle had the ability to land at Kennedy and Vandenberg. Flight testing was complete.

The Flight Research Centre now prepared the combination for ferrying *Enterprise* to the NASA Marshall Space Flight Centre at Huntsville, Alabama, for a series of ground-vibration tests for which NASA technicians reinstated the tailcone aerodynamic fairing. During mid-November Fitz Fulton and the Boeing 747 crew completed a series of test flights with the Shuttle in ferry position, with its front attachment strut lowered slightly to improve cruise performance. On 10 March 1978 *Enterprise* left the Edwards runway for the last time; it was ferried to Ellington AFB, Houston, Texas, where it was displayed to 240,000 visitors over the weekend. Whilst at Houston, the Boeing 747 crew and the Shuttle project officers received the NASA Exceptional Service Medal. Nine other Johnson and Kennedy Centre personnel also received the medal. Donald K. Slayton, project director for the approach and landing tests, received the NASA Outstanding Leadership Medal.

On 13 March 1978, the 747-*Enterprise* combine departed Ellington AFB on the short flight to Huntsville, and many NASA and US Army Redstone Arsenal personnel witnessed the arrival of the strange pair. The following day, huge cranes removed the Shuttle from the Boeing 747 preparatory to installing it in a special test rig at Marshall for a series of vibration tests, simulating the loads a Shuttle would experience in flight. It was towed nearly four miles from the Redstone airfield, past the NASA Marshall headquarters building to a hangar facility where it was displayed to the public for two days, attracting a crowd of 85,000. *Enterprise* was later hoisted up the side of a 430ft-tall dynamic test stand.

Variety

By 1976 the NASA Flight Research Centre was renamed the Dryden Flight Research Centre in honour of Dr Hugh L. Dryden. After the Boeing 747-Shuttle programme, normality was resumed at the centre. The Lockheed YF-12s were still active, the F-111A TACT continued its investigation, whilst the Vought F-8 DFBW ('NASA 802') continued its flight testing. It would be April 1981 before the Space Shuttle

Columbus would land at Dryden. The centre had borrowed a US Air Force NKC-135A (55-3129) and fitted the tanker-transport with Richard Whitcomb's winglets, validating the concept of the energy-saving wingtip devices. A Grumman F-14 *Tomcat* (BuNo.157991) arrived on a joint US Navy-NASA programme following preliminary testing by NASA personnel at the Grumman New York test facility.

A new concept, originally developed by Robert T. Jones to provide more efficient transonic flight, had undergone extensive wind-tunnel tests at Ames; this culminated in a joint Ames-Dryden AD-1 flight research project aircraft, registered 'NASA 805'. In essence, this was a specially-made twin-engine research aircraft with an oblique or 'scissors' variable-sweep wing. The AD-1 made its maiden flight at Dryden on 21 December 1979, and in a thirty-eight minute flight the aircraft reached an altitude of 1,000ft and a speed of 140 knots. The wing of the AD-1, which could be moved back as much as 60° whilst the opposite wing moved forward an equal amount, remained fixed during the flight. After several more sorties flown by flight test project pilots Tom McMurtry and Fitz Fulton, to demonstrate air-worthiness and to gain familiarity with the aircraft's basic flight behaviour, NASA was to use the AD-1 to investigate the handling qualities and control characteristics general to oblique-wing aircraft.

At Ames on 15 July 1979, the Tilt Rotor Research Aircraft Project Office was abolished and the Tilt Rotor Aircraft Office established. The new office would be responsible for all tilt-rotor technology development and demonstration programmes, including completion of the Bell XV-15 research aircraft proof-of-concept flight demonstration and concept evaluation, management of advanced flight experiments, and conduct of the Tilt Rotor Systems Technology Programme.

The Sikorsky S.72 Rotor Systems Research Aircraft — RSRA — NASA 545 72-001 'Heathcliff' first flew on 4 April 1978. It was based at Langley along with NASA 546 'Gertrude' being later transferred to Ames. They are designed to fly as pure helicopters or as a fixed wing aircraft and are equipped with auxiliary jet engines. (NASA 78-H-189)

David Few was appointed manager of the new office. During June 1980, the tie-down test facility for the Bell XV-15 tilt rotor research aircraft — which would permit ground operation of the rotors in all flight configurations — was completed. The helicopter-sized blades were so large that the craft had to be raised before being ground-tested in the aircraft mode. The facility used the hydraulic lift of the existing V/STOL hover-test stand with two tie-down towers which were moved into place after the aircraft was elevated. By July the Bell Helicopter Company completed the first phase of the XV-15 tilt-rotor research aircraft flight-test programme in Texas. Subsequent flights at the Dryden Flight Research Centre would expand the manoeuvring envelope and investigate operational aspects of the tilt rotor for military and civil applications. The second Bell XV-15 was being tested at the Ames tie-down facility, with flight testing planned for late autumn 1980.

On 30 October 1980, the Agency and US Army officials of the Tilt Rotor Research Aircraft Project accepted the first Bell XV-15 aircraft 'NASA 702' in ceremonies at the Dryden centre. Government flight testing, to be conducted by NASA, the US Army, and the US Navy and Bell Helicopter Textron, would follow to demonstrate and evaluate the tilt-rotor concept. A second aircraft (NASA 703) was undergoing ground testing at Ames. For take-off and landing the engines remained vertical, with the large rotors providing lift. Once in the air the engines and rotors tilted to the horizontal, propelling the XV-15 forward.

Aircraft crash tested at Langley during August 1979. Airframes used were damaged in a flood, condemned as unairworthy, and became useful research tools. Photo depicts a twin-engined pressurised aircraft during the FAA-NASA general aviation crash test programme. Four Falcon air-to-air rocket engines give the aircraft the desired velocity. (NASA 79-H-670)

Taking advantage of US Congressional support for aeronautical research, the director of Ames, Hans Mark, guided the centre into research on short-haul aircraft, including new V/STOL designs. Since the mid-1960s, Ames had been working closely with the US Army on helicopter research, relying on the large low-speed tunnels at Ames along with its excellent simulator equipment and other facilities. By the 1970s, both the Federal Aviation Administration and the US Air Force were working with Ames on a new generation of STO transports. In 1976, to the chagrin of Langley, Ames officially became the NASA lead centre in helicopter research. Although the Pioneer project and future planetary missions were moved to the Jet Propulsion Laboratory at Pasadena in 1980, the new aircraft programmes enlivened activities at Ames.

In keeping with rising energy concerns of th 1970s, NASA committed considerable resources to new engine and aircraft technologies to increase flight efficiency as a means of conserving fuel. The Aircraft Energy Efficiency programme commenced in 1975 to develop fuel-saving techniques which would be applicable to current aircraft as well as future designs. Several areas of investigation were covered by the project — more efficient wings and propellers, composite materials that were lighter and more economical than metal, improved fuel efficiency in jet engines, and new engine technologies for aircraft in the future. Other research efforts were carried out through the Engine Component Improvement Programme. Research results were so positive (and so rapidly adaptable) that the new airliners of the early 1980s such as the Boeing 767 and the McDonnell Douglas MD-80 series used engines incorporating many such innovations. For business jets, the NASA rebuilt an experimental turbofan incorporating newly engineered components designed to reduce noise. Completed by 1980, this project successfully developed engines which generated 50–60 per cent less noise than current models. For larger transports, the Lewis Research Centre commenced tests on two research engines which reduced noise levels by 60–75 per cent and reduced emissions of carbon monoxide and unburned hydrocarbons as well.

In a different context, the Agency became engaged in procedures for flight operations in increasingly congested air space. Among the many issues requiring assessment were aircraft noise during landing and take-off over populated areas, safe approach and landing procedures in bad weather, and methods for controlling high-density traffic patterns. By October 1976 Langley had obtained a Boeing 737-13 airliner (ex-N73700, c/n 19437) which was heavily modified and became registered 'NASA 515'. In the aircraft passenger area, NASA technicans put together a second cockpit equipped with the latest innovations in instrumentation. The second cockpit became the flight centre for research operations, the crew occupying the standard cockpit and functioning as backup. In addition to precision descent and approach procedures on instruments, the aircraft played a key role in demonstrating the Microwave Landing System — MLS — in 1979. The International Civil Aviation Organisation eventually adopted the MLS over a competing Eruopean design, to be used today as the standard landing system around the world.

Overseas Co-operation

On 10 December 1964 a new helicopter was delivered to Langley in the form of the Lockheed XH-51N rigid-rotor helicopter; this was BuNo.151263, c/n 1003, soon registered 'NASA 30' and later 'NASA 531'. Very little has been publicized in reference to the close co-operation which exists between the United Kingdom's Royal Aerospace Establishment and NACA/NASA. However, co-operation is very close with a wide exchange of data, etc. NASA flight-test pilots Fred Drinkwater and Ron Gerdes from Ames are no strangers to the RAE Bedford facility, Fred having flown many of the research vehicles produced for V/STOL and supersonic research whilst Ron has flown the two-seat Harrier fly-by-wire aircraft at the establishment. Even Neil Armstrong on a visit to Bedford flew the one-off research Handley Page HP115.

During May 1970 the Deputy Director (Air) of the RAE, Mr P. A. Hufton, visited NASA at Langley and was offered the loan of the Lockheed XH-51N helicopter in a six-month collaborative research programme lasting six months, ending in April 1971. The XH-51N was interesting in two respects. Firstly it possessed a semi-rigid or hingeless rotor, a feature it had in common with the Anglo-French Westland WG13 helicopter. Secondly, it was fitted with the Lockheed integral gyro control system later incorporated in the Lockheed *Cheyenne* helicopter. At Langley the XH-51N had been involved in handling studies and investigation of rotor-blade stresses during low-flying manoeuvres.

The Royal Air Force Air Support Command, with its large fleet of Lockheed C-130 *Hercules* transports made regular training flights to North America so it was no problem diverting one into Langley to pick up the XH-51N. Despite the C-130 having undercarriage trouble at Kennedy Airport, New York, resulting in a replacement transport, the helicopter was safely delivered to RAE Bedford on 30 September 1971. It was assigned to Aero Flight and project test pilot was Flight Lieutenant Ken Robertson, with flight test observer C. A. James. Principal Scientific Officer Phil Brotherhood was in charge of the research project. A NASA Langley research engineer, Bob Houston, remained with Aero Flight for the first month to discuss the collaborative programme planned by the Full Scale Research Section of the Helicopter Division of Aerodynamics Department of the Royal Aircraft Establishment — now the Royal Aerospace Establishment. Bob Huston was the author of many NASA reports, including work on hingeless rotors.

The XH-51N flight test programme at RAE Bedford was restricted to twenty flying hours during which the effects on stability and control of varying the inertia of the gyro and associated springs of the control system were studied. Rotor blade stresses for the Lockheed mechanical gyro control system were claimed to be less than in more orthodox control system. In addition to Ken Robertson, short handling assessments were flown by test pilots from the RAE Avionics Department, the Aircraft & Armament Experimental Establishment at Boscombe Down, Wiltshire, and the Westland Helicopter company at Yeovil, Somerset. In April 1971 the helicopter was returned to Langley to complete the remaining twenty-five hours of useful life and evaluation of an active cockpit vibration oscillator. The final results of the RAE Bedford flight tests were reported and made freely available to the international aeronautical press, with due acknowledgement to NASA.

More of NASA's aircraft fleet were employed on tasks which took them overseas. On 2 March 1977, Ames announced that its Gates *Learjet* 'NASA 705' (ex-N365NA) acquired in October 1976 and equipped with a 30cm infra-red telescope, would participate in Project Porcupine — an international study directed by the Max Planck Institut für Physik and Astrophysik in a coupling between the magnetosphere and the ionosphere. The experiment called for the launch of an Aries sounding rocket from Sweden. After the rocket ejected a barium shaped charge at 450km altitude, the *Learjet* would follow the barium trail along the Earth's magnetic lines of force. The instruments required were furnished by the University of Alaska and the flight path took the Ames aircraft high over the Arctic. Collectively, these researches by aircraft on a global scale enhanced professional contacts for Agency personnel and generated favourable foreign press coverage for NASA as well as for the United States.

On 3 March 1977, the Ames-based Lockheed C-141 Kuiper Airborne Observatory — 'NASA 714', ex-N4141A, c/n 300-6110 delivered to NASA after purchase on 23 December 1971 — left California on its first international expedition. From bases in Australia it was to observe the planet Uranus during unique astronomical conditions. On 10/11 March, Uranus would move between Earth and a star. The resulting occultation, or blacking-out, of starlight would enable the international team of scientists to learn more about Uranus' atmosphere, composition, shape and size. Investigators included research scientists from both American and Australia universities. The aircraft was fitted with a 91cm (36inch) fluting telescope, computer banks, TV monitors, control consoles, Tektronix print-out machines, etc, requiring some 90kVa of electrical power, with stations for a mission director, astronomes, scientists and technicians. The Lockheed 300-50A-01 *Starlifter* has twenty-six seats and a bunk on board.

On 25 October 1977, the NASA Convair 990 'NASA 712' named 'Galileo II' surveyed archaeological sites in Guatemala in an effort to learn more about the Mayan civilization which flourished there centuries ago. Three different types of radar were used to penetrate the dense tropical foliage to different depths, allowing identification of features not readily distinguishable by other means. Signs of roads, stone walls, agricultural terraces, and other man-made structures were sought. The aircraft also carried a scanning infra-red sensor to detect differences in vegetation, seeking clues to the extent and type of farming done by the Mayans. The flight was a co-operative effort among Ames, the Jet Propulsion Laboratory and researchers at the University of Texas at San Antonio.

This same Convair 990A research aircraft left Ames on 2 May 1979 to participate in a summer-long international study of the summer monsoon which annually brings torrential rains to the Asian subcontinent. Named MONEX (monsoon experiment) it was to explore the origin of the monsoon winds in order to improve short-range prediction and understanding of the monsoon's role in global weather patterns. The NASA aircraft operated from air bases in Saudi Arabia and elsewhere in the region, in co-ordination with several other aircraft, ships and a variety of ground-based facilities. The mission was part of the large-scale atmospheric research programme being conducted by the World Meteorological Organisation of the United Nations.

This RPRV is the NASA HIMAT — Highly Manoeuvrable Aircraft Technology vehicle — making its eighth flight at Edwards AFB on 18 December 1980. Two were built by Rockwell under a $11.9 million contract, powered by a J-85-21 turbojet. Photo shows clearly the sharply swept wing canard configuration necessary to provide the technology base of an advanced fighter system. It was launched from the NASA B-52. (NASA 81-HO-140)

This Convair 990A had replaced the earlier 'NASA 711', lost on 13 April 1972 when it was involved in a mid-air collision over Sunnyvale, near Ames, with a US Navy Lockheed P-3 *Orion* (BuNo.157332). 'Galileo' II, c/n 30-10-37, had been delivered to Garuda Indonesian Airways in September 1963 as PK-GJC, returning to the US during 1973 as N7878 with California Airmotive. It went to NASA at Ames the same year. On 17 July 1985 it was unfortunately destroyed by fire during take-off from March AFB, California. It is interesting to record that a third Convair 990A was operated by Ames; this was 'NASA 710', c/n 30-10-29, which originally went to American Airlines as N5617 on 11 May 1962. It was sold to Modern Air Transport on 2 February 1968. On 8 May 1975 it was registered N713NA, becoming N710NA in 1978. It was withdrawn from service and flown to storage at Marana, Arizona on 26 October 1983, with total flying hours of 24,800.

Advanced Research Aircraft

The Ames Flight Research Centre programme in the 1970s included flight testing of an Augmented Wing Jet STOL research aircraft involving a modified DHC-8A *Buffalo* ('NASA 716'), a Grumman OV-10A *Bronco* ('NASA 718'), a Rotating Cylinder Flap STOL aircraft (ex BuNo.152881), the extensive Bell XV-15 Tilt Rotor and the Quiet Short-Haul Research Aircraft project (QSRA). The latter project commenced in January 1974, after industry design studies and a design competition. This was awarded to the Boeing Airplane Company during February 1976 and included a contract to modify a twin-turboprop de Havilland (Canada) DHC-8A *Buffalo* aircraft into the present QSRA configuration. The primary modification consisted of a completely new wing design, with four prototype Avco Lycoming ALF-502 fan jet engines installed over the new wing. It was developed as a minimum-cost research aircraft to investigate the low-speed end of the flight envelope, including take-off and landing conditions. To make the research investigation significant in the application to some future aircraft, the wing of the QSRA uses a super-critical airfoil having efficient cruise characteristics at Mach 0.74. In order to minimize costs, the landing gear on the QSRA does not retract, and the wing slotted leading-edge flap.

Developed and constructed for NASA to serve as a national research facility, the QSRA is strictly a high-performance but low-speed research aircraft with the mission of developing data on powered-lift aircraft operations in the low-speed regime for the US aerospace industry and various US government agencies. This includes technical data on upper-surface blown-flap aerodynamics, short field take-off and landing (STOL) aircraft performance, aircraft handling qualities, STOL aircraft certification criteria, terminal area guidance and control, and aircraft noise reduction.

The 'propulsive lift' wing designed by Boeing is representative of one capable of efficient cruise at Mach 0.74 — 500mph at altitude. Prior to the QSRA being designed and built, extensive wind tunnel tests and flight simulation studies were performed at Ames. A four-engine 0.55 scale model of the QSRA was tested in the 40 × 80ft wind tunnel. Design plans for the QSRA were completed during November 1974. During 1967 the DHC-8A *Buffalo* (ex-US Army 63-13687, c/n 2) arrived at Ames and was almost immediately loaned to the US National Science Foundation in Boulder, Colorado, being registered N326D. On 2 August 1974 it returned from loan, to become 'NASA 715', going to the Boeing Airplane Company at Seattle for modification to QSRA configuration. After conversion a contract flight test programme was conducted at Seattle from 6 July — when it made its maiden flight — to 2 August 1978. John A. Cochrane was the NASA QSRA programme manager and the aircraft achieved a minimum speed of 50 knots with all engines running and a maximum demonstrated lift coefficient of 9.06. Cochrane revealed that the aircraft could have flown slower than 50 knots, but for safety reasons test officials decided not to go below that speed. During the fourteen-flight contractor flight-test programme, the QSRA *Buffalo* achieved a minimum control speed of 52 knots with one outboard engine out; it was flown at a maximum speed of 190 knots at a maximum altitude of 15,000ft. Shortest take-off performed was 820ft but no STOL landings were attempted.

On 3 August 1978 the QSRA was ferried from Seattle to Ames by the NASA flight test pilots James Martin and Robert Innis, taking four hours flying time with two one-hour refuelling stops — one in Eugene, Oregon, and the other in Redding, north California. It cruised at 9,500ft at 170 knots during the ferry flights. It was envisaged that a QSRA-type aircraft the size of a Boeing 727 transport could carry the same payload at the same speeds as the 727, but could operate from small airports so quietly that it would not be heard in the surrounding airport community. On 12 February 1980, after a four-month grounding for extensive modifications, the *Buffalo* resumed its flight-test programme. The modifications included installation of underwing fairings, a speed-hold system, revised spoiler gearing, a new anti-skid brake system and alterations to the horizontal tail. A joint programme with the US Navy was to commence during April.

By 8 May 1980, the shore-based phase of the joint US Navy-NASA QSRA flight programme was completed. The first objective was to determine the best method of landing a large propulsive-lift aircraft on an aircraft carrier. Repeated landings under various conditions generated data on specific aspects of approach and landings, such as touchdown dispersion and sink rate. Between 10 and 13 July, the QSRA successfully made 37 rollers and 16 full-stop landings and take-offs from the deck of the USS *Kitty Hawk*. This was the first time that a pure-jet four-engine transport had operated aboard an aircraft carrier. The landings were made by a flight-test team consisting of one US Navy and two NASA research test pilots, and were performed without the use of arresting gear or the use of catapults. With 20 knots of headwind over the deck, the QSRA deck run for take-off was less than 375ft and the stopping distance for landing was less than 275ft. With a 30-knot wind the take-off distance was less than 300ft and the landing required less than 200ft. The exercise proved that landings and take-offs can be consistently accomplished in 700ft and 800ft respectively, with no headwind. By the end of 1980 the QSRA had completed a comprehensive flight programme in which each of twenty experimental test pilots, representing fifteen different organisations, made two flights.

The QSRA is a major milestone in STOL aircraft acoustic design. A flight demonstration test conducted at the noise-sensitive Monterey Airport, California, showed the aircraft to be undetectable either by the local residents or the monitoring microphones located around the airport.

On 12 September 1980 the Augmented Wing Jet STOL Research Aircraft 'NASA 716' completed its last research flight at Ames after over eight years of STOL flight trials. This joint programme between NASA and the Canadian Department of Industry, Trade and Commerce began in the mid-1960s with a series of model tests in the Ames 40 × 80ft wind tunnel. The aircraft was designed around a DHC-8A *Buffalo* (63-13686) of the US Army, donated to the NASA by the US Air Force; it arrived at Ames on 10 June 1967. The Boeing Airplane Company modified the aircraft while Rolls-Royce and de Havilland designed the propulsion system. A powered elevator, anti-skid brakes and a comprehensive digital avionics research system called STOLAND were installed. During 1974 the aircraft was transferred to the Avionics Research Branch, and over the next three years utilized the broad range of capabilities provided by STOLAND to investigate the operational characteristics of powered-lift transports. The first fully-automatic landing was made in 1975; later research investigated flightpath tracking and flare-control laws. On 22 September 1981 the *Buffalo* was transferred to the Canadian Government in Ottowa. On 7 August 1973 a de Havilland DHC-6 *Twin Otter* (c/n 27) arrived at Ames on loan from the Federal Aviation Administration for augmented wing modifications, as 'NASA 720', whilst a Bell UH-1H *Iroquis* helicopter (ex-US Army 69-15231) arrived at Ames on 4 May 1974 and became 'NASA 733'. It was used for variable-stability trials with a VSTOLAND system installed.

Prior to the transfer of the NASA helicopter research programme from Langley during 1976, Ames had flight-tested a variety of rotor-wing aircraft, some of which were unique. On 16 November 1961, three Hiller YROE-1 *Rotorcycles* — one-man folding helicopters — arrived at the centre. On 10 January 1957 the US Navy announced the successful first flight of the one-man collapsible *Rotorcycle* developed for the US Marine Corps by Hiller Helicopters of Palo Alto, California, but the idea was not wholly successful and it failed to achieve full production status after initial flight testing was completed in July 1957. Weighing less than 250lb, it was powered by a Nelson 4-cylinder 2-stroke, air-cooled petrol engine manufactured by Barmotive Products Inc. A single rotor — 18ft in diameter and positioned above the pilot's head — provided lift, with a small tail rotor located in a tubular boom compensating for torque. The entire helicopter was held together by quick-release pins and could be assembled quickly by just one person. In November 1958 a contract was signed for a batch of ten *Rotorcycles* to be built in the United Kingdom by Saunders-Roe at Eastleigh, near Southampton. The first Saunders-Roe XROE-1 flew at the end of October 1959, all ten examples contracted being completed by the spring of 1960. The three *Rotorcycles* received at Ames had the unique BuNos 4020, 4021 and 4024. It is reported that one NASA flight test pilot found the flight controls particularly sensitive, eventually flying round and round in ever-decreasing circles, although apparently he did not suffer the same fate as a certain mythical bird. At least six of the YROE-1 *Rotorcycles* are housed in museums, two being owned privately in California with one in San Carlos, the other at Morgan Hill.

The Vertol CH-47B *Chinook* helicopter ('NASA 544', ex-66-19138) operated by Langley arrived on the Ames inventory on 14 August 1979, to be re-registered 'NASA 737'. This heavily modified helicopter was a flying simulator with one set of conventional flight controls on the left side and a fly-by-wire, variable-stability control system on the right. A huge research console was mounted in the cargo compartment for in-flight changes in flight-control response. For the next several months the *Chinook* workhorse was to be used to study sideslip performances. After more than 18 years of faithful service at Ames, 'NASA 737' (which was the only government-operated variable-stability helicopter in the United States) was transferred back to the US Army late in 1989. Used in a joint venture with the US Army, the CH-47B had been used for the development of advanced digital flight-control systems for V/STOL aircraft. As an in-flight simulator it was used to verify results achieved by the NASA ground-based vertical motion simulator — the world's largest — located at Ames. With a safety pilot, a simulation pilot and an engineer, the *Chinook* simulated the flight characteristics of other aircraft which would fit into the CH-47's flight envelope. It was equipped with a fly-by-wire electronic control system which included three different flight computers.

Meeting the simulation pilot's need for instantaneous feedback, a rugged digital system R/1173 flight computer was able to cycle data 20 times per second, so making in-flight simulations feasible. This was a feat not possible prior to the 1980s due to the limited power of contemporary computers and high levels of vibration present in the helicopter environment.

On 14 March 1977, Hans Mark announced organizational changes to accommodate Ames' new role as lead centre in helicopter research and technology. Created within the Aeronautics & Flight Systems Directorate were the Helicopter Systems Office, responsible for integrating the various activities in helicopter systems technology; the Helicopter Technology Division, serving as the focal point for helicopter technology development within NASA; and the V/STOL Aircraft Technology Division, restructured from the Research Aircraft Projects Office. By specifying a lead centre, the Agency hoped to increase research output and to reduce costs. Both the Langley and Lewis Research Centres were to continue to be responsible for key segments of the helicopter activities. Using its unique aeronautical facilities such as the 40 × 80ft wind tunnel and flight simulators, Ames would continue to conduct its helicopter research. Whilst the activity in helicopters at Langley would be reduced, the centre expected growth in long-haul aircraft technology.

The US Army-NASA-Sikorsky Rotor Systems Research Aircraft — RSRA — made its first flight on 10 April 1978 as a compound helicopter-fixed wing aircraft, taking off from Wallops Island, Virginia, and climbing to 1500ft using both wings and rotor systems for lift. Under contract to NASA and the US Army Research & Technology Laboratories, Sikorsky built two prototypes — 'NASA 740', c/n 72-001 and 'NASA 741', c/n 72-002 — which were tested at Ames, with 'NASA 741' arriving on 12 February 1979 and 'NASA 740' on 29 September 1979. The RSRA had a 46ft wingspan and a five-blade S-61 rotor system powered by two T-58 turboshaft engines. Two auxiliary TF34 turbofan engines were mounted below the rotor system. The RSRA could also be fitted with a variety of experimental and development rotor systems for research purposes. They joined the growing fleet of short-haul research aircraft based at Ames.

The Sikorsky S.72 RSRA NASA 71 ex-72-002 seen during high-speed taxi trials at the Ames-Dryden facility at Edwards AFB on 13 November 1987. It is demonstrating an advanced rotor/fixed wing concept called X-wing, a joint NASA-DARPA programme. Pilots were G. Warren Hall from Ames and W. Richard Faull from Sikorsky. (Sikorsky CS72-125-7)

CHAPTER FOURTEEN
Earth Resources Aircraft Project

At Ames, scientists became interested in using aircraft as platforms for investigations of terrestrial as well as astronomical phenomena. Commencing in 1969, the Ames Flight Research Centre acquired a number of different research aircraft and launched several imaginative investigations that continued over the following decades. High-altitude missions relied on a pair of Lockheed U-2 aircraft, originally supplied to the Central Intelligence Agency and the US Air Force as reconnaissance platforms. For NASA they carried out Earth resources observations, compiled land-usage maps, surveyed insect-infested crops and measured damage from floods as well as forest fires. The high-flying U-2 aircraft provided information covering hundreds of square miles. For a more intensive look at details in a smaller area, Ames brought in other specialised aircraft which flew mid-altitude missions.

NACA and the U-2

Mention must be made of the elaborate cover story which was supposed to disguise the true nature of the U-2 programme, which relied on the co-operation of NACA. During May 1956 the Agency issued an official statement identifying the Lockheed U-2 as purely a weather-research aircraft, but it was not until February 1957 that the first official photo of the U-2 — suitably adorned with NACA tail markings and numbered 'NACA 320' — was released to the public. On 6 May 1960, the news that Gary Powers had been shot down over the USSR began reverberating around the world. Attention soon focused on the remote air base in Texas where the U-2 pilots were trained. Whilst the CIA operation at The Ranch and Edwards North Base remained under close wraps, people in the US Air Force 4080th Strategic Reconnaissance Wing were pointed out and questioned wherever they went. To satisfy the media, a U-2 photo-call was arranged at Edwards AFB but not at the top-secret North Base site. It was pure coincidence that the same week as Gary Powers was shot down in his U-2, the US Aviation & Space Writers Association were meeting in Los Angeles. NASA had arranged a bus to take the journalists out to their hangar at Edwards, for a first look at their aircraft and flight research facilities including the experimental rocket-powered aircraft. The tour was scheduled for Friday 6 May, just twenty-four hours after the Agency had issued the CIA-prepared cover story about one of their U-2s being missing on a weather reconnaissance flight from Incirlik, Turkey. Someone in the CIA evidently hit upon the bright idea of backing up the cover story by showing the visiting journalists a suitably-marked NASA U-2 during their visit. Early on the morning of the planned tour, an unmarked, all-black CIA Lockheed U-2A aircraft was towed the three miles along the lake-bed from Lockheed's North Base facility to the NASA hangar at the main Edwards base flight-line. Here a bemused NASA mechanic was detailed to find a stepladder, a paintbrush and a NASA decal in double-quick time, and to paint the NASA identity and a tail number on the aircraft. He chose the serial of a Radioplane OQ-19B-RP (55-741) which had been shot down in operations at Edwards some days earlier, and so when the U-2 was revealed to the press a short while later, it carried a neat yellow tail band containing the letters 'NASA' and the number 55741 below. The paint was still wet when the

journalist visitors climbed off the bus and rushed over to the unfamiliar black U-2; they almost ignored the research aircraft parked some twenty yards away with a deputation of famous flight-test pilots surrounding it. There was not a soul around the U-2 to answer questions, but the press got its pictures of a 'NASA U-2' which were duly published. The CIA's effort was in vain, however, since within a week President Eisenhower put the record straight, repudiating the cover story and taking full responsibility for the Lockheed U-2 operation known as Operation Overflight.

U-2 Research Aircraft

High-altitude photography and remote sensing for Earth-resources investigations were not, originally, research directions which Ames had clearly defined. During 1970, however, the US Air Force announced to other Federal agencies that it was ready to make available two of its high-altitude Lockheed U-2 reconnaissance aircraft for research purposes. Agencies, and even private industry, were invited to submit proposals for the use of the aircraft. At this time NASA was in the final planning stages of the Earth Resources Technology Satellite (ERTS) programme, with the first launch planned for 1972. The NASA researchers and scientists were becoming concerned that they would have trouble analyzing the data, since all available high-altitude visual-spectral-band and infra-red photography came from altitude where atmospheric distortion was minimal. The ERT satellites would be recording radiation that had passed through the entire atmosphere, and distortion would certainly be greater. The possibility that NASA might acquire the U-2s as research aircraft added inspiration to a developing idea. If photographs from above the densest part of the Earth's atmosphere could be obtained, the data should be similar to the data ERTS would produce. With U-2 photography, analysts should be able to prepare themselves to analyze the data the satellite would later produce. With this use for the high-flying U-2s in mind, the Agency requested that the US Air Force transfer the aircraft to them. Equipped with infra-red cameras, the U-2 — like the satellite — could measure the chlorophyll content of vegetation, a major feature of ERTS data. With infra-red, the reddest portions of a photograph indicate high chlorophyll levels.

Designed by an old colleague of Smith De France, Clarence L. 'Kelly' Johnson in the mid-1950s, the Lockheed U-2 has remained a thoroughbred among aircraft, capable of reaching an altitude of over 65,000ft. Sleek and fragile-looking despite its size, the single-seat aircraft had extraordinary long, narrow wings attached to an equally streamlined fuselage. Even to the untutored eye it was the epitome of refined grace, an aircraft one might imagine would require a subtle touch. Certainly the U-2 (and the later TR-1) required almost a separate education in flying and maintenance. Carelessly flown, it would crumple into a ball of aluminium foil. Carelessly maintained, it could quickly become a hazard to fly. The US Air Force, though impressed by the NASA proposal to operate the aircraft, was convinced that the Agency was somewhat unaware of the complexity and sensitivity of the U-2s, and was therefore naturally reluctant to decide in the Agency's favour until it was certain that the aircraft would be suitably provided for with both expert flight personnel and careful maintenance.

After prolonged discussion between NASA and the US Air Force, Marty Knutson, one of the first two pilots selected by the CIA for Operation Overflight, was dispatched to NASA to tutor them on the intricacies of the U-2. With his help, the Office of Space Science and Applications devised a utilization plan that was acceptable to the US Air Force, including strict requirements for flight-crew selection and maintenance to be provided by crews from the Lockheed division which had developed and manufactured the U-2 — the celebrated 'Skunk Works'.

Lockheed's role in the new NASA U-2 operation was not limited to overhauling the aircraft; it secured a contract to provide all the pilots, physiological support and maintenance crews. In the latter two categories, the Skunk Works was not short of trained personnel since it had provided similar services under contract to the CIA from the very start of Operation Overflight. Lockheed obtained the services of three experienced U-2 pilots; Jim Barnes and Bob Ericson, who had just officially retired from the US Air Force. The third pilot was British — Ivor 'Chunky' Webster, one of the RAF pilots who had trained on the U-2 in 1961, enjoying the experience so much that he resigned his Royal Air Force commission and settled down in California to continue flying the aircraft. Marty Knutson was employed by NASA as project manager for the new U-2 programme. There was, however, no vacancy for another CIA U-2 pilot — Gary Powers — who applied, but was informed that his notoriety precluded selection. The NASA headquarters wanted to downplay the connection with the famous past of the U-2 as much as they possibly could.

Whilst these negotiations were progressing, the Agency deliberated on where the aircraft should be based, with Ames making a strong case for acquiring them. Long-lived rumours have it that the Johnson Space Centre, already operating high-altitude Martin WB-57Fs, maintained that the U-2s were too dangerous for routine work and should not be accepted. The outcome was that NASA headquarters decided to place the aircraft at Ames. The two NASA U-2s were delivered to Moffett Field on 3/4 June 1971, officially on indefinite loan from the US Air Force. These former CIA U-2G aircraft had been in storage for a couple of years prior to Lockheed being asked to make them ready for service as Earth Resources Survey Aircraft. Skunk Work engineers removed the carrier-landing modifications from Articles 348 and 349 (Article is similar to the c/n) and some other items, which reduced their zero-fuel weight to 13,800lbs. They were painted in smart white and grey livery with a blue cheat line and given the NASA registrations 'NASA 708' and 'NASA 709'. This pair of high-flyers was to supplement other aircraft — two WB-57Fs, one P-3 *Orion* and a C-130 *Hercules* — which the Agency had been operating since 1965 as scientific testbeds for the sensors which it would soon send into space. The sensor system for the Lockheed U-2 was still under development at this time, but during August 1971 the two U-2 aircraft commenced flying on a regular basis over five control areas chosen for their particular ecological interest. Four of the areas were in California and therefore within easy reach of Ames, but the fifth was the Cheasapeake Bay region of the eastern US seaboard, which required a U-2 deployment at Wallops Island, Virginia. For the first time, details of the U-2 payload specifications and cameras were made freely available through NASA, which needed to acquaint the wider scientific community with the research possibilities that the high-altitude aircraft now offered.

Extra camera configurations were offered on the NASA U-2s, using Wild-Heerbrug RC-10 cameras with six- and twelve-inch focal lengths. These provided nine-inch square negatives and could be carried in pairs, or in combination with one twenty-four-inch camera. Many early NASA U-2s were flown with a multispectral camera which used 100mm lenses to record four different wavelengths simultaneously on a common film. This system emulated the main sensor on the first two ERT satellites — since renamed Landsats. By getting the U-2s into service nearly a year before the first Landsat was launched, the Agency was able to acquire preliminary imagery and prepare the processing and analysis system for the subsequent deluge of data from the satellite. After the launch, the U-2s carrying the multispectral system staged 'underflights' along the satellites' track at 65,000ft to provide a means of comparison with the imagery being relayed from space.

Rare photo taken at Edwards AFB on Friday 6 May 1960 showing the CIA Lockheed U-2 put on show for the benefit of visiting members of the Aviation & Space Writers' Association. The U-2 as a spy plane was never used by the NASA. The serial 55-741 belongs to a Radioplane OQ-19B drone shot down a few days earlier. (Gordon S. Williams)

Marty Knutson's U-2 operation initially came under the Airborne Sciences officer, which later became the Earth Resources Aircraft Project — ERAP. At that time the plan for the U-2s went no further than ERTS simulation, with the expectation that once ERTS-A was in orbit, simulation flights would no longer be required. As circumstances would have it, the future of the U-2 project at Ames became assured through pure bad luck in another NASA quarter. Although the first Landsat was operational in late 1972, there were problems with one of the sensors and Landset 2 was not launched until January 1975. The delay was catastrophic, since the satellite was to be short-lived and had been earmarked for a crop survey during the spring 1972 growing season. Goddard Space Flight Centre (managing the ERTS project) contacted Ames, who produced a plan which answered the dilemma Goddard was in. The Pre-ERTS Investigator Support programme took the planned satellite survey and restructured it for the two Lockheed U-2s, resulting in the aircraft making hundreds of flights over a three-month period. Not only did the U-2s continue to support satellite data for ERTS. Landsat and Skylab, they also provided information in other resource surveys and were used extensively in disaster assessment. Using the U-2s, the NASA — co-operating with other Federal agencies — provided, at a surprisingly low cost, information sometimes unobtainable by any other method.

Ames acquired two Lockheed U-2C aircraft, initially for flights over four ecological test areas in the USA. Depicted is NASA 708 ex-56-6091 Earth Survey 4 which served the agency from 3 June 1971 to June 1987. Today it is preserved at the Visitors Centre at Ames. (NASA 71-H-906)

With the U-2 operations extended and placed on a more permanent basis, the High Altitude Missions Branch was established within the Ames centre and the U-2 operation was given the grand-sounding title of Airborne Instrumentation Research Project — AIRP. This emphasized the fact that the aircraft were now also used for research purposes other than earth survey. In November 1973 they had been recruited into the NASA long-term Stratospheric Research Programme. The two aircraft carried a variety of sensors to 60,000ft or higher to measure various gases and particles, such as ozone, nitirc oxide and man-made pollutants. The first of what became an annual series of deployments to Eielson AFB, Alaska, took place in the summer of 1974. There was a month-long deployment to Hickam AFB in Hawaii later that year, followed by visits to Howard AFB, in Panama in the Canal Zone and Lowry AFB, Maine. By 1977 the two U-2s had sampled the stratosphere from the eastern US coast to 1,000 miles west of Hawaii, and from 10°S latitude to the North Pole region.

Surveys

During 1974 a U-2 surveyed the Dudaim melon infestation in the Imperial Valley at a fraction of what a ground survey would have cost. The melon, an alarmingly vigorous weed, was smothering the California asparagus crop; it was not only difficult to detect but ground surveying was made even worse by dreadful heat, the rough vegetation and rattlesnakes. By the time the Ames U-2 aircraft were employed, the melon had been indicted in Federal court.

In July and August 1977, a Lockheed U-2 and a Gates *Learjet* from Ames were based in the Panama Canal Zone for three weeks conducting studies of atmospheric pollution. Several governmental agencies and universities co-operated in the study, gathering information on how atmospheric pollutants such as halocarbons are carried from low altitudes into the stratosphere where they might influence the ozone balance. The study, sponsored and planned by Ames, was carried out over a 16-day period with both aircraft making daily flights. Heavily instrumented with sensing and sampling equipment, the aircraft measured atmospheric pollutants at various altitudes. The *Learjet* covered altitudes up to 14,000 metres (46,000ft), while the U-2 carried the coverage well into the stratosphere at altitudes of 70,000ft.

On 17 November 1977 the Ames Research Centre announced that measurements made by scientists utilizing a Lockheed U-2 aircraft suggested that the cosmos may have started serenely, with a powerful but tightly-controlled and completely uniform expansion. Using ultra-sensitive radio equipment, the research team measured the cosmic microwave background — the radiation left over from the Big Bang, the initial, universe-forming event — and concluded that that event was a smooth process, with matter and energy uniformly distributed and expanding at an equal rate in all directions. Scientists from the Lawrence Livermore Laboratory and the University of California at Berkeley declared, 'The big bang, the most cataclysmic event we can imagine, on closer inspection appears finely orchestrated'.

Between 22 February and 9 March 1979, one of the Ames U-2 aircraft flew several astronomy missions over Peru. The payload was an upward-looking differential microwave radio-meter for measuring the sky's background microwave radiation at extremely low temperatures. The measurements obtained were to be used to determine the movement and speed of Earth and the Milky Way with respect to far distant bodies of the universe. Similar measurements made by the same aircraft in 1976/77 supported the theory that the Milky Way and Earth were travelling through space at 1.6 billion km/hr. The mission, though accomplished successfully, was not without traumatic moments. Engine trouble with the supporting Lockheed C-130 *Hercules* aircraft, temporarily lost equipment and the hijacking of the aircraft on which the crew travelled to Peru made the mission a close-run thing. On the return flight, the U-2 was granted an emergency waiver to fly without rescue support aircraft because of more trouble with a second C-130 pressed into service.

During May and June 1980 a Lockheed U-2 took part in a NASA study to find out how volcanic eruptions affect the Earth's weather and climate by studying the plumes emitted from Mt St Helens in Washington State, which erupted on 18 May. Data collected for the Aerosol Climate Effects — ACE — programme at Ames is the most complete set of observations made of volcanic aerosols in the stratosphere. Aerosols are fine particles, either solid or liquid, suspended in gas. The ACE study began in 1979, to assess the climatic effect of aerosols in the Earth's atmosphere. Five missions were flown in the St Helens area; preliminary data analysis indicated that the volcanic plumes contained a mixture of solid ash particles and sulphuric acid, with proportions varying in different samples. The amount of sulphuric acid found in the stratosphere was several hundred times greater than that found prior to the eruptions; large increases in gaseous sulphur dioxide were also detected.

The Super Guppy, NASA 940, was purchased from Twin Fair Inc. of Buffalo, New York, on 20 February 1979, and acquired by Johnson on 13 July, 1979. It is a modified Boeing KC-97G-26-BO and used to carry shuttle hardware. It was used during the Apollo/Skylab era for hauling spacecraft etc. to the Kennedy Space Centre, Florida. Today it is based at El Paso, Texas, away from the sea corrosion of the Gulf salt water. (NASA S-79 38300)

NASA ramp scene at Edwards AFB on 5 February 1990 shortly after the arrival of Lockheed SR-71A 64-17980 with McDonnell Douglas F-18A Hornet NASA 841 ex-BuNo.161216 and TF-18 Hornet NASA 845 ex-BuNo.160781. Currently seven MDC Hornets are on the NASA inventory at Edwards. (NASA EC 90-0047-007)

Michael J. Adam seen with North American X-15 56-6670. (NASA ECN-1651)

Cecil Powell seen with the early Martin Marietta X-24A 66-13551 which first flew at Edwards 17 April 1969. (NASA ECN-2599)

Bill Dana kitted up in his pressure suit with North American X-15 56-6672 behind. (NASA EC67-1716)

John Manke and the heavily modified Martin Marietta X-24B 66-13551 which made its first flight on 1 August 1973. (NASA ECN-3778)

Gary E. Krier and the Vought F-8C Crusader Digital Fly-by-Wire aircraft NASA 802 BuNo.145546. (NASA 73-HC-107)

Don Mallick (left) and Fitz Fulton prepare for a test flight in the Mach 3 North American XB-70A Valkyrie 62-0001 which is seen in the background. (NASA ECN-2052)

Ron Gerdes in front of AV-8C NASA 719, 1986.

Don L. Mallick poses in front of the Lockheed YF-12A 60-6935 wearing a suit similar to that worn by the NASA astronauts. (NASA ECN-2978)

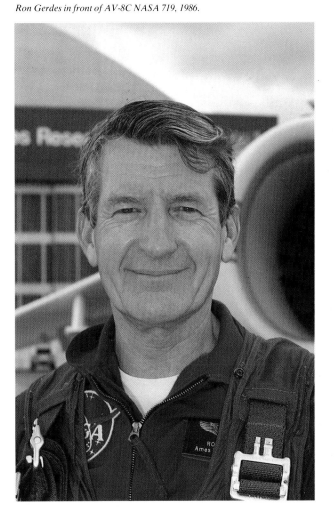

McDonnell Douglas F-18 Hornet BuNo.160780 NASA 840, in new 1989 paint scheme to aid in flow visualisation tests. It is today being flown at the Ames-Dryden facility in its high-angle of attack research programme. Flow visualisation data is gathered by generators expelling smoke from the forward portion of the aircraft and by other methods. (NASA 89-HC-292)

The second X-29A 82-0049 is today on the inventory at the Ames-Dryden research centre at Edwards AFB. Depicted over rugged terrain is the first X-29 showing the unique configuration, with forward swept wings and canards. A research programme is exploring the forward swept wing's attributes.
(Grumman History Centre)

In addition to the aerosol sampling missions, the U-2 photographed Mt St Helens on 19 June at the request of the Washington State Office of Emergency Services. The flight was the result of a month of weather map observations for predictions of skies clear enough to allow the U-2 to photograph the volcanic damage from an altitude of 65,000ft. The photography gave the state the first comprehensive coverage of the damaged area.

In both its scale and purpose, the Stratospheric Research Programme could be compared with the old High Altitude Sampling Programme (HASP) effort conducted by the US Air Force Strategic Air Command Lockheed U-2s a decade or more earlier. Once again U-2 pilots were sent up on long orbits around the sky in an attempt to measure the effect of the human race's cavalier attitude to the environment. In the late 1950s, man had caused radioactive particles to be flung into the atmosphere from the nuclear explosions. Now he is polluting it with fluorocarbons from a million aerosols.

The sampling devices on the U-2 became more sophisticated. There was an Ames-designed Q-bay sampler weighing 500lb which used chemiluminescent reactions to measure gases in situ. An alternative Q-bay payload, also designed at Ames, consisted of four cryogenically-cooled samplers plus two whole air samplers to collect gases for laboratory analysis after landing. There was also an Ames version of the traditional filter-paper type sampler for collecting aerosol and halogen particles. Fourteen years after the NASA U-2s commenced collecting this data, the world finally began to wake up to the danger, and the High Altitude Missions Branch hit the headlines as it flew the Lockheed U-2 into an ominous hole in the ozone layer over the South Pole.

Just as these flights served to warn of man's potential demise at his own hand, another series attempted to define how he had evolved in the first place. As mentioned earlier, the results suggested that the Big Bang theory of the formation of the universe needed serious modification. A team of astro-phyics scientists from Berkeley designed and placed in the U-2 an Aether Drift radiometer which looked upwards rather than downwards. A special Q-bay hatch top with two sensor ports was fabricated to accommodate the ultra-sensitive device.

The scientific data from which the discoveries were made was obtained from eleven highly demanding night flights over the western United States flown at 65,000ft. British-born 'Chunky' Webster had to fly the U-2 level to within one-half of a degree, because only then would the effects of the earth's atmospheric microwave radiation be cancelled out, thereby allowing the sensor accurately to detect the very-low-frequency light radiation coming from way beyond the quasars in outer space. The Aether Drift radiometer was, in effect, a camera which built up an image part by part spread over the eleven flights, but the frequency of the signal that the scientists were trying to detect was 20,000 times lower than the frequency of visible light. Thanks to plane and pilot, the flight criteria were met — an on-board measurement system indicated that average lateral displacement of the U-2 during Websters' flights was only plus or minus one-sixth of a degree.

Remote-sensing technology and applications expanded steadily throughout the 1970s, thus ensuring a continuing demand for the services of the NASA U-2 aircraft — 'NASA 708' and 'NASA 709'. More satellite sensors were pre-flighted on the two aircraft; these included a colour scanner for the Nimbus satellite, which would measure oceanic tides, sediments and micro-organisms, and a heat-capacity mapping radiometer which would provide continuous thermal mapping of the earth's surface.

The NASA Earth Survey Programme was conducted by the Earth Observations Division of the Manned Spacecraft Centre, Houston, Texas. Depicted are Earth Survey 1 a P-3A Orion NASA 927, and NC-130B NASA 929 Earth Survey 2. Today the Orion is based at Wallops, and the C-130 with the Science & Applications Division at Ames. (NASA 71-H-1009)

Deployments and a Replacement

During February 1977, another U-2 veteran arrived at Ames. Former CIA pilot and reconnaissance manager Jim Cherbonneaux moved from Washington DC, to become the head of the High Altitude Missions Branch. By the time he arrived, the two aircraft had acquired imagery from all fifty states, covering some 35 per cent of their total surface area. The imagery had proved useful in nearly all the earth science disciplines. Forest diseases and insect infestations had been detected, and timber harvesting practices improved. Photographic data of watersheds had indicated where pollution needed to be tackled. Large-area coverage provided the big picture on population and agricultural trends to state and local government land-use planners. The NASA U-2 photography had also proved invaluable in assessing the effects of natural disasters such as fires and floods. In rapid-response missions for the State of California, the aircraft were despatched to the scene of forest fires. The photographs taken showed the extent of the conflagration, possible access routes for fire fighters and potential firebreak locations. During the 1975/76 drought, U-2 photos helped measure water levels in rivers and reservoirs.

The NASA U-2 pilots were all veterans in their own right; Jim Barnes, Dick Davies, Jerry Hoyt and Doyle Krumrey had accumulated more than 12,000 hours in the aircraft between them. When Jim Barnes retired from Ames he had flown the U-2 for over thirty years. Not everyone understood the beating-swords-into-ploughshares situation and to some citizens of California the NASA Lockheed U-2 was still a 'spy in the sky'. During 1977, the state Coastal Conservation Commission requested a photo-survey to help its wastelands management programme. But the resulting photos also served another purpose. Unlicensed construction in the coastal zone had recently been prohibited, and licences were hard to obtain. The survey was flown with the big twenty-four-inch cameras, and therefore provided high-resolution 9" × 18" photographs from which it was possible to determine whether landholders had defied the ban on construction. Citizens of the town of Bolinas got particularly vocal about it; a resident attorney summed up his feelings, 'I do have this gut reaction to this eye in the sky able to look at all the little things in people's backyards. It is a feeling of distaste, you know, that is just one more step on the road to 1984'.

Yet all the imagery which the NASA U-2s collected was available for public inspection at Ames, via a computerized image retrieval system. Anyone could purchase full-scale reproductions of any frame from a huge data bank in South Dakota. In 1978, new ground was broken in the administrative use of the U-2 when the huge state of Alaska and ten federal agencies clubbed together to fund a photo survey of the entire state. They concluded that it was simply the most cost-effective way — in such a remote region — to gather all sorts of information about natural resources, land-settlement claims by native eskimos, wildlife-conservation, energy management, etc. The two Lockheed U-2s from Ames were joined in Alaska by a Martin WB-57F operated by the Johnson Space Centre in Texas, and in the first year of the programme they photographed 25,000 line miles of the state at a cost of $496,000. It worked out at less than $20 per data mile, which the bureaucrats reckoned was very good value compared with trekking through the tundra or hiring a light aircraft and mapping at low level. During the mapping of Alaska, the U-2s flew at altitudes between 60,000 and 65,000 ft with the dual RC-10 camera configuration. The six-inch camera was loaded with black and white film and the twelve-inch one with colour infra-red film. The flights could only be conducted in the short summer season when the snows had melted and vegetation was flourishing. The programme took three years to complete.

More technology from the secret world of military reconnaissance was gradually declassified for use on the NASA aircraft. A twenty-four-inch focal length panoramic camera which gave much greater resolution than the Itek KA-80 — albeit over a narrower strip of territory — was introduced. Two of them were mounted in the U-2 to provide convergent stereo coverage. The aircraft's utility in the firefighting disaster-assessment role was boosted by a line-scan camera which provided real-time imagery to a ground receiving station by means of a data link.

Meanwhile other NASA Earth Resources aircraft were equally active. The Lyndon B. Johnson Space Centre in Texas operated the Lockheed NC-130B ('NASA 929') marked NASA Earth Survey 2. It had arrived on 18 July 1968. On 19 October 1979, the *Hercules* returned from a four-week remote-sensing mission in the polar ocean northwest of the Spitzbergen Island Group in Norway. The aircraft took part in an ice study referred to as Norwegian Remote Sensing Experiment — NORSEX. Results of the experiment were used in developing sensors for the future space satellite systems which would monitor ice conditions in the polar regions. The experiments were international in scope, involving scientists from the USA, Norway, Canada, Switzerland and Denmark. The NASA centres at Ames, Langley and Goddard were involved. Flights also went over the Greenland icecap and the Newfoundland Grand Banks. The *Hercules* flew missions as far north as 85°N. In order to accomplish the flights and return to base at Tromso, Norway, the NC-130B refuelled on each mission at Longyearbyen in the Spitzbergen Islands — the northernmost airport in the world at 78°N. Eleven Johnson NASA personnel from the Experiments System Division flew the mission using Johnson and Langley sensors — scatterometers and radiometers. These sensors helped to define future requirements for the National Oceanographic Sea Satellite — NOSS. The Lockheed NC-130B flew twenty-five flights on the twenty-eight mission, travelling over 27,000 miles of mission and ferry flying.

Earth Survey 2 was also utilized in the NASA Airborne Instrumentation Research Programme — AIRP. The aircraft's remote-sensing capability consisted of various large-format camera combinations, multispectral scanners, passive and active microwave sensors, an inertial navigations system and a central data recording system. Payload capacity was approximately 10,000 lb and was configured to accept walk-on sensor payloads utilizing standardized racks. The NC-130B has a maximum operating altitude of approximately 30,000 ft and a flight duration of eight hours, with six hours of data acquisition. The AIRP acquired data was used for the purposes of developing techniques and applications in the disciplines of agriculture, forestry, geography and geology, hydrology and oceanography.

The aerial platform used by the NASA scientists is a heavily modified Lockheed C-130BLC — Boundary Layer Control — transport which they Agency acquired from Lockheed originally as a STOL research vehicle. It was built for the US Air Force as 58-0712 (c/n 3507) but became diverted as the BLC test aircraft fitted with blown flaps and control surfaces with compressors under the outer wings. It first flew on 8 February 1960, and prior to transfer to the NASA had standard C-130 wings fitted. Since delivery to Johnson, the aircraft has been outfitted from nose to tail with sophisticated sensor devices including large-format camera combinations, multi-spectral scanners, passive and active microwave sensors, inertial navigation and a central data recording system. Project director in the NASA AIRP project was James Lindemann. The aircraft's huge round nose came from a NASA Lockheed P-3 *Orion*, 'NASA 927', which the agency transferred to Wallops Island. The nose contained a C-band microwave antenna and a four-channel radiometer which is a key remote-sensing instrument. It also houses the standard C-130 radar navigation antenna. The transport's belly contains a camera bay and a special thermometer, eleven- and eight-channel scanners and an active microwave antenna. Another active microwave antenna is installed in the platypus extension behind the tail, while on the lower ramp two antennas can be extended out from the open ramp door — enabling a clear ground view without aircraft interferance. The *Hercules* has been heavily insulated. The spacious cargo interior includes a special laboratory featuring eight consoles, airline-type seating, lowered ceiling and carpeting. Each of four consoles contains its own built-in electrical, oxygen and intercom hook-up, enabling engineers to replace the modular units as required. The two antennas extended from the rear ramp door are passive microwave arrays which can be used in the X and K bands for measuring snow and can be switched to the L band for measuring ice.

On some missions — such as a snow research flight during 1980, conducted over the 11,000ft elevation Rabbit Ears Pass in Colorado — the rear ramp of the aircraft was opened up to enable special antennas to be extended out of the rear fuselage. Early in 1980, the NC-130B flew a series of research flights over a runaway oil spill in the Gulf of Mexico, tracing out the dimensions of the huge oil slick. 'Once we come up with a reliable signature of the spill', said James Lindemann, 'our satellite sensors of the future should be able to relay back to earth not only the area of the spills, but their depth and even the types of oil involved'. Other objectives of the NASA programme was development of sensing techniques in the disciplines of agriculture, forestry, geography and oceanography. Findings in previous research have gone into the highly successful Seasat and Landsat satellites. The 1980 research was to be used in earth/sea satellites of the future, including a new one to be called Icesat. During 1979, the AIRP *Hercules* traversed over 25,000 miles as part of the Airborne Ice Expedition out of Alaska. The data which was acquired helped pave the way for future offshore oil explorations in the Arctic regions of the United States.

During March 1980 a mission was conducted in the Great Lakes area using remote microwave sensing techniques developed by the NASA scientists. This would give navigators an hour-by-hour 'real time road-map', enabling ships and barges to ease through ice-clogged waterways and thereby adding precious weeks to the shipping season. The data will help scientists evaluate microwave identification of ice coverage, clear water passages, pressure ridges and ice thickness. This data, along with extensive 'ground truth' will be used in developing algorithms for future satellite sensors.

This Chance-Vought F8D Crusader BuNo.147085 was acquired by Ames on 18 June 1959, becoming NASA 225, serving until 15 August 1960. This single-seat carrier jet fighter was a useful test bed for the NASA. Friend Ron Gerdes flew the supercritical wing F8 on its last flight on 23 May 1973. (Peter M. Bowers)

Seen on arrival at Johnson from the Northrop factory on 27 May 1964 is 63-8181 the first T-38A Talon for the NASA which became NASA 901 and used for astronaut training. It was ferried by astronaut James McDivitt. On 28 February 1966 it was involved in a fatal accident at St Louis, Missouri, killing NASA pilots See and Bassett. (NASA 64-134)

'We think the prospects for this programme are very good', said Lindemann. 'If the sensing techniques are implemented, a ship navigator in future years facing an icy passage can call in immediately from a satellite overhead the precious information — a road map if you like — that will enable him to get his cargo through'. In conjunction with the US Coast Guard the Agency took part in Operation Icewarn, involving HC-130H *Hercules* of the Coast Guard in the Great Lakes area. Whilst based at Johnson, the NASA team and aircraft conducted far-flung research missions which ranged from the Polar regions to the fringes of a hurricane in tropical Barbados.

Update

When Lockheed won a contract in 1978 to recommence U-2 production as the TR-1, the Agency quickly became interested in the capabilities of the upgraded model. At the Johnson Space Centre at Ellington AFB, the three Martin WB-57F aircraft also in use for earth resource surveys were getting old, and had naturally been out-performed by the U-2s based at Ames. Two of the WB-57Fs were marked 'Earth Survey 3' and 'Earth Survey 6' whilst the third ('NASA 928') was unnamed and was on loan from the US Department of Energy — DoE. It nevertheless was well equipped and had been involved in Operation Airstream along with other agencies. NASA decide to retire two of the three WB-57s and acquire another Lockheed aircraft; in fact, the Agency received the very first example fro the newly-opened production line. Marty Knutson delivered 'NASA 706' (ex-80-1063) from Palmdale to Ames on 10 June 1981, and the new aircraft became designated ER-2. Knutson had now risen in the NASA ranks to become chief of the Airborne Missions and Applications Division at Ames. When a further promotion gave him the added responsibility for the NASA Dryden

Used as a small utility transport at Lewis between September 1964 and August 1968, was this Aero Commander Model 680 c/n 577-222 initially NASA 31, later NASA 631. It was transferred to Langley as NASA 503 and is seen parked at Lewis during 1965. (NASA C-65-188)

Flight Research Centre at Edwards in 1984, he finally stopped flying the U-2, having logged over 3,000 hours in the type over a 28-year period. Fellow NASA U-2 pilot Bob Ericson retired around the same time, leaving Jim Barnes as the sole remaining pilot from the early days of the U-2 to be still flying the aircraft. New aircrew for the NASA programme were recruited from the ranks of retiring US Air Force pilots. Ron Williams and Jim Barrilleaux were just two who wished to retain their association with the U-2, known as the 'Dragon Lady'.

The new Lockheed ER-2 offered significant payload and range advantages over the two U-2s. The new aircraft had large 'superpods' attached to each wing, providing extra accommodation for sensors. The other U-2 aircraft were routinely flown with wing drop-tanks which had been adapted to house various payloads, but the available volume was less than one-fifth that offered by the superpods. For the first time, a camera — typically the RC-10 — could be carried in the nose or the superpods, thus freeing the Q-bay for other experiments. The ER-2 was both more comfortable and easier to fly. Along with the new ER-2 came yet more new sensors. The Pentagon released the Itek Iris II very high-resolution optical bar panoramic camera for civilian use. This very expensive and sophisticated piece of optical technology was by now the main reconnaissance camera in use on US Air Force Strategic Air Command U-2s, and it had also been taken to the moon by the Apollo astronauts. NASA was also cleared to offer researchers a digital X-band synthetic aperture radar in the ER-2, as well as the capability to data-link to ground stations either direct of via the Tracking & Data Relay Satellite — TDRS — system. But the most significant of the new sensors was a Daedalus Thematic Mapper Simulator (TMS) a multispectral scanner radiometer which recorded radiance from the earth's surface in eleven discrete wavelength bands. The word 'simulator' appeared in the scanners' title since it was designed to replicate a similar system carried by Landsat-4. Such detailed measurement of light and heat could, with the aid of a computer, produce colour maps which vividly delineated even subtle variations in surface composition. Such manipulation of digital data provided capabilities which even the multispectral camera systems could not match.

Geologists, for instance, could map differing varieties in rock formation, picking out shales from sandstones and different types of soil. Oceanographers received maps detailing water quality and thermal pollution. Agronomists could tabulate differences in health, maturity and species of vegetation. In the TMS maps, one could distinguish redwoods from fir trees, wetlands from dry pasture and fields planted with artichokes from others planted with Brussels sprouts. Of course, colour infra-red film exposed by camera still provided greater resolution than the TMS, and the two types of sensor were regarded as complementary. They were often both used on the same surveys.

By the mid 1980s, the NASA high-altitude survey aircraft had obtained over 400,000 frames of photographic imagery covering large portions of the United States. This extensive collection of high-quality, large-scale negatives was housed in the Applications Aircraft Data Management Facility at Ames, next door to the High Altitude Missions Branch. All sorts of organizations were now anxious to review archive film. High schools and colleges, water districts, county and state planning commissions, engineering firms, utility companies, emergency sevice agencies, forestry departments, oil companies. . . the list was endless.

The demand for new surveys continued. Having been well satisfied with the results of the 1978-1980 mapping survey, the authorities in the state of Alaska contracted for further flights each summer. There were also continuing deployments to Wallops Island. In the summer of 1984, a U-2 flew from there to photograph more than 80,000sq miles in five eastern states, in order to survey the damage being caused to standing trees by the gypsy moth caterpillar and assess whether spraying programmes were having the desired effect. The following year, the ER-2 flew over Florida in a major survey of the states' citrus trees for the Department of Agriculture. In both surveys the Itek Iris camera was used, loaded with colour infra-red film. The Iris was becoming a favourite tool for regulatory agencies, thanks to its high resolution — two feet from 65,000ft using black-and-white film — and wide coverage amounting to a thirty-seven nautical mile swathe beneath the aircraft. The Environment Protection Agency used it to check up on illegal dumping of waste.

Disaster assessment flights also continued. In December 1986 a controlled fire in the San Gabriel mountains of California was started so that scientists could investigate various related phenomena. A NASA U-2 made its contribution by flying overhead to record it all on the TMS and film. Two channels of the TMS were relayed by data-link to fire-fighters on the ground. The readout of thermal data enabled them to track the active fire front, which was obscured from their vision by the dense smoke plume.

Although the NASA was not keen to publicize it, there were also a number of missions being flown by U-2 and ER-2 aircraft in support of US Department of Defense projects. The most notable of these was 'Teal Ruby', a satellite which would detect and track aircraft from space by measuring their infra-red signature. It formed part of the US Air Force Air Defense Initiative — ADI — to defend US airspace against future missile and bomber threats. While these could currently be detected by radar, alternative forms of detection might be necessary if the USSR developed stealth technology. The Teal Ruby concept was made possible by advances in charged-coupled devices and cryogenic cooling of sensors. The satellite would carry a six-foot-tall infra-red telescope which stared down at the earth's surface and registered disruptions to the normal background signal return caused by an aircraft on a mosaic of thousands of focal-plane detectors.

It was indeed complex technology, and one of the problems was establishing a data base of background measurements. The NASA ER-2 was enlisted to fly a similar multi-wavelength infra-red sensor as part of a Teal Ruby support effort codenamed 'Hi-Camp' — Highly-Calibrated Airborne Measurements Programme. This provided an atmospheric, terrestrial and oceanic background data base, making precise measurements of the clutter which Teal Ruby would have to deal with. The ability of the Hi-Camp sensor to pick out aircraft from the ER-2's 65,000ft cruising height was also tested. In a year-long series of co-ordinated flights over the western United States and Europe, a variety of US Air Force aircraft ranging in size from a Northrop T-38 *Talon* to a Lockheed C-5 *Galaxy* were flown against the Hi-Camp sensor. The Teal Ruby satellite itself was ready for launch by 1986, but the Space Shuttle *Challenger* disaster forced a three-year postponement.

Overseas TDY

A temporary duty — TDY — visit to the United Kingdom with the NASA ER-2 to RAF Mildenhall in the spring of 1985 was the first non-US territory visit by the High Altitude Missions Branch. The second came in January 1987, when 'NASA 706' was deployed to Darwin, northwest Australia. It was engaged on the Stratosphere-Troposphere Exchange Project (STEP) a continuation of earlier atmospheric studies by the Ames-based aircraft. The Mildenhall deployment was in connection with Teal Ruby. The NASA and the National Oceanic & Atmospheric Administration funded STEP to obtain yet more data on the mechanisms and rate of transfer of particles, trace gases and aerosols from the troposphere into the stratosphere. Darwin was chosen as a suitable launch point for flights into the region where the world's coldest and highest tropopause was to be found, as well as the largest and highest cumulo-nimbus clouds.

There was a more important deployment to come which involved the most hazardous flying ever carried out by the NASA team, flying long hours over inhospitable terrain from which rescue in the event of an accident might prove impossible. But the stakes were very high. During August and September 1987, the NASA ER-2 flew twelve times out of Punta Arenas, southern Chile, across deepest Antarctica. Its mission was to take detailed measurements in the recently-discovered hole in the ozone layer which was developing over the South Pole every winter. Over 170 scientists, managers and support crews descended on the desolate area at the southernmost tip of South America. In addition to the Lockheed ER-2, the Agency also deployed the Ames-based converted Douglas DC-8-72 airliner, 'NASA 717' (c/n 46082, originally delivered to Alitalia as I-DIWK on 14 May 1969). This aircraft went to Braniff as N801BN on 7 January 1979, was sold to International Air Lease Inc in 1983 and resold to NASA on 4 February 1986. The two aircraft between them carried no fewer than twenty-one separate scientific payloads to measure every conceivable variable which might be linked to the alarming phenomenon of ozone depletion. The whole

Lockheed T-33A-5-LO 58-0671 NASA 936 was acquired by the Johnson Space Centre on 31 May 1963 being originally NASA 918. On 5 March 1971 it went to Hill AFB, Utah. Johnson operated some 22 T-Birds, these gradually being replaced by the T-38A Talon. (Military Aircraft Photographs)

effort cost $10 million, but it soon became apparent that it was worth every cent. The scientists found that the hole was bigger than ever, and proved that man-made chlorofluoro-carbons — CFCs — were to blame. Since the screening effect of the delicate stratospheric layer of ozone prevents harmful amounts of the sun's ultra-violet radiation reaching earth, this was indeed significant news.

The NASA mission to Antarctica was designed to answer many questions posed over the years and provide data relevant to a number of theories. While the DC-8 would fly at the lowest extremities of the hole, with some of the sensors peering upwards, the ER-2 would fly right through the very centre of it. The latter carried some payloads from the recent STEP series of atmospheric sampling flights, but also some specially-designed experiments. A total of fourteen separate sensors were carried in the nose, Q-bay and wing pods. The most important of these experiments, if those blaming CFCs for the ozone hole were to be vindicated, was a chlorine monoxide detector designed by Harvard chemistry professor James Anderson. But, as Anderson himself pointed out, 'There isn't a single instrument on either plane that, in the long run, won't be crucial. Every chemical that can be measured must be measured, and measured precisely. If there is any ambiguity in our findings, our impact will be weakened. The burden of proof is enormous'.

Grumman OV-1B Mohawk 64-14244 was delivered to Lewis in December 1972, becoming NASA 637. Along with a C-130 '1351' from the US Coast Guard was involved in ice research programme Project Ice Warn. It is depicted flying over the snow covered shore of Lake Erie during 1973. (NASA C-73-936)

Conditions at the windswept Punta Arenas airfield were rudimentary. Prior to the deployment, the taxiways had to be resurfaced to prevent any damage to the ER-2's delicate landing gear. Even basic office accommodation had to be specially built in the draughty military hangar allocated to NASA. Under difficult conditions, the scientists laboured to perfect their experimental payloads and then crossed their fingers as they watched the ER-2 soar into the mostly grey and turbulent skies over the Magellan Strait. The missions lasted well over six hours. The weather for flying was never good and at ground level, fronts and surface winds up to sixty knots could develop rapidly. At altitude the polar vortex caused winds of up to 200 knots, and temperatures as low as −95°C were experienced. Dual VHF radios and INS were fitted to aid with communications and navigation. All the pilots took Arctic survival courses prior to leaving the United States. They realised that chances of surviving an ejection over the icecap were very slim, since there was no accompanying rescue aircraft with paramedics on board. Fortunately the Lockheed ER-2 performed flawlessly, flights were uneventful and the data 'take' was excellent. Anderson's chlorine

monoxide detector measured levels up to 500 times the normal concentration. The edge of the hole was found to have extended further north than ever, even as far as Punta Arenas.

Even as the scientists and aircrew strived to gather the vital data, a United Nations of thirty nations in Montreal agreed to a 50 per cent reduction in CFC production by 1999. This was a compromise but the United States and Scandinavian countries wanted a complete ban, and when the final results of the NASA expedition to southern Chile were published in early 1988 their case was greatly strengthened.

There was now the question — could the same phenomenon of ozone depletion be occurring over the North Pole? During late 1988 and early 1989 the Agency became involved in the Airborne Arctic Stratospheric Expedition, with Ames, Goddard and Langley centres co-operating together with many US and European agencies including the United Kingdom Meteorological Office and Cambridge University. The Ames-based ER-2 and DC-8 were deployed to Stavanger, Norway. Each aircraft flew fourteen times into the polar vortex during the two-month detachment. The ER-2 missions lasted up to eight hours, going as far as 80°N and ranging from the Barents Sea to Greenland. For flight safety the tracks were kept to within 250 nautical miles of emergency landing strips, so that the ER-2 could glide in from 65,000ft in the event of trouble. Fortunately the precaution proved unnecessary, the ER-2 performing well although the pilots endured a hard time navigating and landing due to the strong prevailing winds. With the exercise over, NASA called a press conference in Oslo to disclose their preliminary findings. The news was not good as there was every indication that the ozone layer over the Arctic was being eroded to a similar extent. The two aircraft had measured amounts of chlorine monoxide — an ozone destroyer — which were up to fifty times its normal concentration in the atmosphere. While the scientists went home to analyze their data and draw some firm conclusions, ministers and officials from 120 nations met in London to discuss the crisis. It was already clear that the Montreal agreement was inadequate. The Unites States and European countries now declared that they would phase out CFCs completely, and wanted others to agree to at least an 85 per cent reduction.

Wallops Study

Commencing on 20 April 1989, the NASA Wallops Island-based, highly instrumented remote-sensing Lockheed P-3A Orion 'NASA 428' participated in an international oceanographic experiment called the Global Ocean Flux Study — GOFS — to determine the capacity of the world's oceans to assimilate and store excess carbon dioxide CO_2 from the Earth's atmosphere. The study results were critical to predicting potential temperature increases in world climate due to the large increase in atmospheric CO_2 caused by the burning of fossil fuels. The increased atmospheric carbon dioxide may lead to a warmer Earth through the greenhouse effect. The NASA Orion is equipped to measure the concentration of phytoplankton biomass in the upper ocean layer. The primary instrument, the airborne oceanographic lidar (AOL) used a blue-green laser to stimulate fluorescence from chlorophyll contained in phytoplankton, the microscopic plants in the bottom of the marine food web.

The GOFS efforts include scientists and research vessels from West Germany, Canada, United Kingdom, the Netherlands and the United States, and the study is expected to continue until 1999. The initial experiment involved studies of the spring phytoplankton bloom in the eastern North Atlantic Ocean. The US research vessel *Atlantis II* placed instrumented moorings at two sites along the 20°W meridian. Along with research vessels from the other participating nations, it studied related phytoplankton productivity as the bloom moved northward in response to increasing solar radiation and the development of thermal stratification in the upper ocean. For the Lockheed P-3 *Orion* and its study team officials who included Bob Swift, Frank Hoge and Ed Lange, the six-week-long spring bloom study staged from Lajes in the Azores to Shannon in southern Ireland and Keflavik in Iceland.

While the Ames ER-2 was ranging far and wide on vital earth resources missions, the two NASA U-2C aircraft back at Ames were reaching the end of a thirty-year flying career. During mid-1987 'NASA 708' was retired when it reached 10,000 flying hours and today is on display outside the visitors' centre at Ames, along with a ⅓rd scale model of the Space Shuttle. The replacement 'NASA 708' was a US Air Force TR-1A (80-1069) which had served in the United

Kingdom with the 17th Reconnaissance Wing at RAF Alconbury. A second ER-2 for NASA was pending. The other U-2C ('NASA 709') made its final flight with the Agency — a record-breaking one from Dryden — during April 1989, and is now back in US Air Force livery and displayed at Robins AFB, Georgia. Only a few months earlier, NASA pilot Jim Barnes finally hung up his pressure suit after flying what he appropriately called the 'Deuce' for over thirty-one years, during which time he had amassed a remarkable 5,760 hours in the type — a record which seems very unlikely to be matched.

Excellent underside view of WB-57F 63-13503 NASA 926, which served at Johnson between 21 July 1972 to 15 September 1982 as Earth Survey 6. Photo shows the Universal Pallet System with a selection of camera ports with openings for additional sensors. (NASA S-77-28852)

NASA Earth Resource Aircraft

Title	NASA No.	Type	Service No.	Base	Dates of service
EARTH SURVEY 1	NASA 927	Lockheed P-3A	148276	Johnson	20 December 1965 to 29 September 1977
EARTH SURVEY 2	NASA 929	Lockheed NC-130B	58-0712	Johnson	18 July 1968 to 9 October 1981
EARTH SURVEY 3	NASA 925	Martin WB-57F	63-13501	Johnson	19 July 1972 to 15 September 1982
EARTH SURVEY 4	NASA 708	Lockheed U-2C	56-6681	Ames	3 June 1971 to June 1987
EARTH SURVEY 5	NASA 709	Lockheed U-2C	56-6682	Ames	4 June 1971 to April 1989
EARTH SURVEY 6	NASA 926	Martin WB-57F	63-13503	Johnson	21 July 1972 to 15 September 1982
	NASA 928	Martin WB-57F	63-13298	Johnson	24 June 1974 to 27 July 1988
	NASA 706	Lockheed ER-2	80-1063	Ames	10 June 1981 Current
	NASA 717	Douglas DC-8-62	c/n 46082	Ames	4 February 1986 Current
	NASA 708	Lockheed TR-1A	80-1069	Ames	3 June 1988 US Air Force
	NASA 709	Lockheed ER-2	80-1098	Ames	1 March 1989 Current

Notes: NASA 927 current at Wallops Island as NASA 428.
    NASA 929 current at Ames as NASA 707.
    NASA 925 preserved at Pima County Museum, Arizona.
    NASA 708 U-2C preserved at the Visitor's Centre, Ames.
    NASA 709 U-2C preserved as 56-6682 at Robbins AFB, Georgia.

CHAPTER FIFTEEN
Past, Present and Future

We are now entering the 1990s, with 75 years of aeronautical research and history with NACA and NASA behind us. For many years nostalgia has been a best-selling commodity, and fortunately the past is well documented in the NASA archives and available to any student of aeronautical research — resulting in numerous volumes being made available on a variety of subjects, as the reader will note from the bibliography which covers aeronautical history of both NACA and NASA over the years. Some volumes are sponsored by the Agency, others are published privately, but all are available to the general public. Within the Agency, most of the centres publish an internal newsletter which is a mine of varied information. As yet there is no NASA Museum, but today a wide selection of veteran research aircraft is preserved within the confines of such excellent museums as the National Air and Space Museum in Washington DC, a stone's thrown from the NASA HQ in Independence Avenue. Other aircraft are either in cold storage or being restored pending display within the United States.

The year 1989 was one of commemorating past events as the milestones passed into history, with nostalgia coming to the fore. The Ames Research Centre celebrated its Golden Anniversary, and the Agency in general commemorated the 20th anniversary of the Apollo II lunar landing. On the first day of May 1989, the Goddard Space Flight Centre — NASA's major scientific laboratory — celebrated its 30th anniversary. On 8 June 1959, the North American X-15 research aircraft made its first flight.

However, under the tranquil surface, the wars of institutional assessment raged unabated. During May 1981 Dr Alan Lovelace, the NASA acting administrator, announced that four of the centres were to be consolidated; Dryden with Ames and Wallops with Goddard. The title names would remain but would refer to 'facilities' rather than 'centres' and they would become operational elements of the parent ones. It was emphasized that the consolidations would better focus the resources of each centre to accomplish what it did best. The consolidation became effective as of the first day of October 1981. The aeronautical research activities of Ames and Dryden would be integrated and all staff functions for the two locations would be combined. It was estimated that it would take thirty months to implement fully. Selected Ames research aircraft, such as the two Bell XV-15s and the QSRA *Buffalo*, were transferred to Dryden. Only the aircraft involved in the extensive space sciences and earth resources remote-sensing programmes, such as the Convair CV990, Lockheed C-141 Kuiper Airborne Observatory, Lockheed U-2 and ER-2 aircraft etc, would be retained at Ames. These included the Douglas DC-8 and Lockheed NC-130B *Hercules*. Ames lost the useful Crows Landing field, a US Navy auxiliary strip used by Moffett Field but useful for flight-test work which required ground instrumentation and monitoring, such as with VTOL aircraft like the YAV-8 *Harrier*.

The types of aircraft employed by NASA were as diverse as the tasks for which they were procured. More and more US government agencies relied on the huge NASA organisation and its facilities for involvement in both wind tunnel and flight-test work, whilst the US armed forces became partners with the Agency in advanced flight projects which often involving a third partner, the aircraft manufacturer. This spread both workload and cost over a wide spectrum. On 14 January 1975, a Beech C-23 *Sundowner* aircraft was delivered to Langley and modified with wing thrusters during August 1978 to aid in general-aviation stall and spin tests. The aft-firing thrusters enabled the aircraft to spin into types of manoeuvres not possible without the devices and also allowed more precise measurement of spin-recovery forces. Each wingtip had two rockets with a total thrust of about 110lb and sufficient propellant for sixty seconds of firing in any desired burst sequence. It was registered 'NASA 504' (ex-N6624R, c/n M-1608) and listed in the records under GAW-2 wing research. Today (1990) the *Sundowner* is still on the Langley inventory. In the GAW-1 wing research project a Piper PA-34-200 *Senaca* (c/n 34-7250001) was used. It was delivered to Langley on 26 October 1974, remaining until 3 August 1978 and registered 'NASA 509'.

During early July 1978 it was announced that the NASA-owned Lockheed YO-3A (US Army 69-18010, registered 'NASA 718') quiet aircraft, completely modified and refurbished, would fly in formation with a US Army Bell AH-1 helicopter equipped with a new rotor system. Delivered to Ames on 27 April 1977, the YO-3A's first assignment as a noise-monitoring platform took place at Dryden during August 1978. Listed in the NASA inventory as a programme support aircraft, the YO-3A's work included noise measurements of new quiet short-haul research aircraft. The noise-monitoring role was developed by the US Army.

On 25 May 1981, Aero Spacelines Inc proposed to NASA that the Agency purchase a new *Super Guppy* aircraft for support of the Space Shuttle. The NASA had obtained a *Super Guppy* ('NASA 940') on 13 July 1979 and used the unique transport developed from a Boeing YC-97J (52-2693) to haul large pieces of cargo such as rocket assemblies. Officials of Aero Spacelines believed a second *Guppy* was needed to transport large Shuttle payloads. The company had previously held a contract with NASA since 1966 when a chartered *Pregnant Guppy* supported the Apollo programme. From this unique transporter came the current *Super Guppy*, incorporating a combination of elements of both the commercial B-377 *Stratocruiser* and its close relative, the US Air Force Boeing C-97J *Stratotanker*. Powered by more powerful Pratt & Whitney T34-P-7WA turboprop engines as used on the C-97J, the transport today carries many pieces of equipment — even a completed spacecraft. The original NASA *Super Guppy* ('NASA 940') was purchased in February 1979 and delivered to Ellington AFB, Texas, on 13 July 1979. It was previously owned by Twin Fair Inc of Buffalo, New York.

Initially based at Ellington and operated by the Johnson Space Centre, 'NASA 940' was transferred to El Paso International Airport on the Mexican border 700 miles to the north-west in February 1988. This was to escape the corrosive atmosphere of the south Houston area, with its airborne by-products of industrialism and salt air blown in from the Gulf of Mexico. El Paso has a drier desert environment, which is kinder to an ageing airframe.

The transport is flown at 190 knots indicated air-speed at all altitudes up to the limit of 25,000ft. The cabin area is unpressurised, so above 12,000ft the flight-crew wear oxygen masks. Cargoes are often unusual and sometimes exotic. The streamlined tail-cone fitted to the Space Shuttle Orbiter vehicles when they are ferried atop the NASA Boeing 747 is frequently transported. The most fabulous (and possibly the most expensive) item of cargo carried was the Hubble Space Telescope, the heart of a $5 billion programme, which was flown from Stewart Field near New York to Ames. The *Super Guppy* was accompanied by the NASA Boeing KC-135 full of scientists and technicians who made sure the precious cargo had security and protection at each stop en-route. When not required by the Agency the *Super Guppy* is often involved in carrying US armed services equipment. Notable sorties have involved conveying F-14 *Tomcats* taken out of storage for refurbishment, collecting an A-7 *Corsair* from Panama, removing three A-4 *Skyhawks* from Puerto Rico, US Navy A-6 *Intruders*, the fuselage sections of Rockwell B-1B supersonic bombers and the fuselage of a C-130 transport. A NASA T-38 was carried after its turtle-back fuel tanks were blown apart on being struck by lightning in mid-air off the coast of California with Brewster Shaw, a Shuttle astronaut, on board.

At the Dryden Flight Research Centre during 1975, two McDonnell Douglas F-15A *Eagle* fighters (71-0281 and 71-0287) were delivered for drag research. During 1980, 71-0281 was involved in flight-testing Space Shuttle thermal protective tiles. The tiles were eventually submitted to almost 1.5 times the dynamic pressure which the Shuttle attained during launch. The tiles fitted on the F-15's right wing simulated those on the leading edge of the orbiter's wing, while the tiles on the left wing represented those on the junction of the orbiter's wing and fuselage. The tests were part of the overall verification of the various Shuttle systems and involved the F-104 in addition to the F-15.

During 1979 the first of two HiMat — Highly Manoeuvrable Aircraft Technology — RPV aircraft developed by Rockwell and NASA flew at Dryden. Powered by a General Electric J-85-21 turbo-jet, it had a wingspan of 17ft and a length of 23ft. Designed as a technology demonstrator, the HiMat aircraft was a swept-wing canard configuration which should provide the technology base necessary for an advanced 1990s fighter system. It featured a composite structure of glass fibres, graphite composites and various metals. In August 1975, after the preliminary study phases, the Agency awarded Rockwell International a $11.9 million contract for two HiMat aircraft, the first of which was completed in mid-1978. After launch from the NASA Boeing B-52 mothership, the vehicle was flown through a complex series of manoeuvres at transonic speeds by a NASA pilot sited in the RPV remote-control facility. Then he guided the aircraft down for a landing on Rogers lake-bed. A chase aircraft provided an emergency backup control. By 18 December 1980, the first HiMat vehicle had completed its eighth flight; this particular flight concentrated on acquiring stability and control data as well as evaluating two major control changes. These involved adding an aileron-to-rudder interconnect and a lateral accelera-tion feedback loop to improve the lateral and directional flying qualities, especially at higher angle of attack. The twenty-eight minute flight was trouble-free and achieved Mach 0.9 at both 25,000 and 40,000ft, with a maximum 'g' level of slightly over six.

Another Remote Piloted Research Vehicle — RPRV — project 'Mini-Sniffer', commenced at Dryden in 1975. This vehicle was an attempt to develop a propeller-driven RPRV operating on hydrazine mono-propellant fuel to altitudes around 100,000ft to gather air samples from the wakes of high-flying supersonic aircraft. Three Mini-Sniffer configurations were built, and the concept led to great interest by various research facilities (including the Jet Propulsion Laboratory) in using similar vehicles for planetary sampling missions. Such an aircraft could be used on Mars as part of a planetary probe. The RPRV work at Dryden still continues to be an important aspect of NASA's research.

In January 1977 Orbiter Enterprise was moved to Dryden. This historic photo depicts the mating for the first time of the 122ft long space shuttle, to the Boeing 747 shuttle carrier NASA 905. Three attachment points only attach the oribiter to the huge carrier. Taxi trials commenced on February 1977.
(NASA 77-H-71)

Aircraft Safety

At 9.23 am on 1 December 1984, a Boeing 720-027 (c/n 18066) registered 'NASA 833' crashed at Dryden. Preparations for the crash commenced as early as 1980 in a joint Federal Aviation Administration and NASA-controlled impact demonstration. The intention was to test the post-crash fire-suppression capabilities of anti-misting kerosene and the crash worthiness of various seat and passenger-restraint systems. The Boeing 720 in question first flew on 5 May 1961; its was originally intended to become N7078 for Braniff but this registration was never taken up. It went to the FAA as N113 on 12 May 1961 and was withdrawn from use in 1971. It was re-registered N23 with the FAA fleet on 16 February 1977 and went to NASA as N2697U on 5 December 1981, becoming N833NA in December 1983. It had arrived at the Dryden Flight Research Centre on 20 July 1981 from the FAA Technical Centre. The controlled crash was planned to take place on 30 November but was delayed by 24 hours. The ensuing fire was self-extinguished in approximately ten seconds, the Boeing 720 sliding through the impact site as planned after impacting with 'wing-openers' designed to rupture the fuel tanks. Crash worthiness design experiments were also on board in the radio-controlled accident.

Today the NASA at Johnson employs four Gulfstream STA — shuttle training aircraft, NASA 944/5/6/7, flying an average of 14 landing approaches per week per astronaut. Depicted is NASA 947 c/n 147 delivered on 13 June 1974 in shuttle approach mode on a 20° glide slope, engines in reverse thrust, and main landing (NASA S85-42647)

The NASA scientists at Langley are also deliberately crashing aircraft. Working closely with the FAA, the scientists are trying to learn how to make general-aviation aircraft safer. 'General aviation' today refers to a large segment of aviation including private, corporate and executive, agricultural and air commuter aircraft. The aircraft being crash-tested at Langley are almost complete airframes of light aircraft and helicopters which were damaged in a flood several years ago. They were condemned as unairworthy because of damage by water, and instead became research tools. A giant steel girder structure — the Impact Dynamics Facility — originally constructed to suspend simulated lunar landing craft so that astronauts could practice techniques for landing on the Moon is used. Impact speeds normally reach 95 km/hr; if higher speeds are desired, small rocket engines are attached to the wings of the aircraft, thereby increasing the velocity of impact to 145 km/hr. Cameras are stationed at various locations outside and inside the aircraft and anthropometric dummies occupy the crew and passenger seats. The Langley programme has been operating for several years, and research is now expanding into the area of larger transports such as the Boeing 720 which was crashed at Dryden. Other NASA research is aimed at enhancing occupant survivability in post-crash fires.

During March 1989 a two-way digital voice terminal for land, aeronautical or maritime mobile communications was demonstrated in the field for the first time by scientists from the NASA Jet Propuslion Laboratory, Pasadena, California. During the tests the scientists used a voice terminal on an FAA Boeing 727 airliner to communicate with a ground station in Southbury, Connecticut whilst the aircraft was airborne over the US eastern seaboard. Signals were relayed between the aircraft and ground by an orbiting satellite operated by the International Maritime Satellite — Inmarsat — organization. Aircraft flying backwards and forwards across the Atlantic must relay communications at times through other aircraft on the busy ocean routes. A mobile satellite system would link each aircraft to ground stations via satellite. Dr William Rafferty, manager of the Jet Propulsion Laboratory Communications Section, headed the tests. The section also conducts the Mobile Satellite Experiment — MSAT-X — programme for the NASA.

Shuttle Trainers

Grumman *Gulfstream 2* Space-Shuttle training aircraft are today based at Ellington AFB, being assigned to the Johnson Space Centre, Texas. The NASA astronauts fly an average of fourteen landing approaches per week, these being spread between the Shuttle landing site at Edwards AFB, the return-to-launch-site abort recovery runway at the Kennedy Space Centre in Florida and Northrop Strip, White Sands Missile Range, New Mexico. The latter is the primary abort-once-around landing site and would be the primary landing site if Edwards had weather problems immediately before or during flight. This strip is also the primary Shuttle training aircraft practice area. Shuttle training aircraft (STA) — simulated Orbiter approaches commence at 35,000 ft on a 20° glide slope, with the *Gulfstream* flying in reverse engine thrust with the main landing gear down for drag. Lateral motion control surfaces on the aircraft's belly help simulate Orbiter handling qualities. The fly-by-wire *Gulfstream* is computer-driven to provide realistic Orbiter-type handling qualities under a wide variety of approach energy configurations. The left side of each cockpit duplicates primary Orbiter instrumentation, including a rotational hand controller and cathode-ray tube display. A Northrop T-38 *Talon* simulates the profile to be flown by the lead chase aircraft with the Orbiter during the Shuttle's approach to landing. The first STA *Gulfstream 2* (c/n 146, 'NASA 946') was delivered to Ellington on 20 May 1974, followed by c/n 147, 'NASA 947' on 13 June 1974. A third STA (c/n 144, 'NASA 944') was delivered on 6 May 1983, whilst a fourth ('NASA 945') was awaiting delivery from Grumman during May 1989 to be based at El Paso International Airport, Texas.

It was revealed during June 1989 that extensive tests of Space Shuttle Orbiter landing gear assemblies, from normal conditions up to and including failure modes, were to be conducted up by the Dryden Flight Research Facility using a NASA Convair CV990 *Coronada* aircraft. Planning and modification to the CV990 began in 1989, with flight tests scheduled for 1990. Data from the flight tests will give the NASA engineers information on what to expect should an Orbiter experience a flat tyre or any other anomalies on landing and will provide data to assist in developing crew procedures for various landing conditions and situations. The Ames-Dryden Project Manager is Robert Baron and the tests are part of a continuing effort by the Johnson Space Centre to upgrade and enhance the Space Shuttle landing capabilities. Officials at Ames-Dryden also hope to use the CV990 as a testbed for future landing-systems tests.

During the programme, flights tests will be conducted on both lakebed and concrete runways at Edwards, on the concrete runways at the Kennedy Space Centre and on lakebed runways at White Sands Space Harbor. The Convair will retain its normal landing gear, the Orbiters' landing gear being installed so that it can be lowered hydraulically when the aircraft first contacts the landing surface. It will be mounted on the transport's fuselage between the main tyres, and a hole will be cut in the fuselage to accommodate raising and lowering the test gear assembly. The underside of the transport's fuselage will be armour-plated to protect the aircraft from any possible damage. In addition to assessing and documenting performance of main and nose landing gear assemblies, tyres and wheels, the tests will evaluate brake and nose gear steering performance. Landing speeds of the CV990 will duplicate those of the Orbiter, approximately 225 miles per hour. Project flight test pilot is Gordon Fullerton, a NASA veteran of Space Shuttle flights. High-speed video and film cameras will record the test for thorough analysis, and the aircraft will carry other instrumentation and test gear.

Acronyms

AFTI = Avanced Fighter Technology Integration; COOP = Co-operative Control; DEEC = Digital Electronic Engine Control; HARV = High-Alpha Research Vehicle; HIDEC = Highly Integrated Digital Electronic Control; INTERACT = Integrated Research Aircraft Control Technology, and PROFIT = Propulsion/Flight Control Technology are just a few of the acronyms used by the NASA over the years, many current today at the NASA Ames-Dryden establishment.

The aircraft that is now known as the HIDEC McDonnell Douglas F-15 *Eagle* ('NASA 835', ex-71-0287) first arrived at Edwards back in 1976. It is one of two *Eagles* used by NASA, the other being YF-15A 71-0281 which was employed for the aerodynamic testing of the Space Shuttles' thermal protection tiles as mentioned earlier. This F-15 is today preserved at Langley. Project manager for HIDEC is Jim Stewart and project pilot is Jim Smoker. The *Eagle* was chosen since it represents a typical modern fighter aircraft, was available in the NASA fleet and had already been converted to digital — as opposed to hydromechanical — engine control under an earlier programme. The HIDEC F-15 is powered by advanced versions of the Pratt & Whitney F100 turbofan with reheat, and being unique evaluation-demonstration power plants they carry the special designation F100 EMD — Engine Model Derivative. Jim Smoker has also flown such aircraft as the Mission Adaptive Wing AFTI F-111 and the AFTI F-16

Fighting Falcon. Jim is one of three NASA pilots assigned to the HIDEC F-15. The other two are Tom McMurtry and Bill Dana. Einar Enevoldson (now retired) and Lieutenant Colonel Dave Spencer (US Air Force) also flew this unique aircraft. It is more than coincidental that two good friends of the author — Squadron Leader Rick Pope and Flight Lieutenant Gordon McClymont of the Royal Air Force on secondment to the 6515th Test Squadron at Edwards — have flown the aircraft. Both are ex-test pilots with the Royal Aerospace Establishment, Bedford. Rick Pope was replaced by Gordon on 23 February 1990.

Prior to entering the NASA fleet inventory at Dryden, the HARV McDonnell Douglas F-18A *Hornet* ('NASA 840', ex-BuNo.160780) had served with the US Naval Air Test Centre at Patuxent River, Maryland. It was the sixth F-18 built and being a non-standard pre-production aircraft was put in storage pending further use. NASA wished to obtain a test-bed for its projected high angle-of-work or 'high alpha' research programme, the ultimate aim of which was to flight-test on onboard thrust-vectoring control system to investigate its influence on manoeuvrability. The *Hornet* arrived at Dryden in a number of large and small pieces on a truck during September 1985; it required eighteen months' work to be refurbished, joining the HARV flight test programme in April 1987. Einar Enevoldson made the first flight and Ed Schneider the second. Schneider and Bill Dana have flown virtually all the HARV test flights since. The aircraft is equipped with camera pods on the wingtips in lieu of the AIM-9L Sidewinder missiles. These and other cameras view streams of white smoke emitted from the forward fuselage, which reveal much about the behaviour of the airflow as it passes across the aircraft at different angles of attack. To make the white smoke trails stand out better, the aircraft's upper surfaces are painted matt black.

Smoke streams reveal off-surface airflow behaviour well but on-surface flow visualization can also be accomplished, by analysis of the flow patterns created by a special red liquid emitted from dozens of tiny holes in the aircraft's nose cone.

Two-seat Lockheed TF-104G NASA 824 an ex-West German Air Force aircraft is currently held in storage at Edwards AFB. It was ex-USAF 61-3065 and acquired in 1975 going into storage on 31 August 1988 and is depicted with wing-tanks fitted. (Military Aircraft Photographs)

The liquid is poly-glycol methyl ether — PGME — mixed with red dye. NASA has a contract with McDonnell Douglas covering design and installation of a thrust-vectoring control system, which will take the HARV programme — and fighter aircraft technology — into new realms. Externally, the most obvious modification to the HARV will be the addition of three curved steel plates called 'turning vanes' arranged at 120° intervals around the circular endpipe of each engine in place of the existing nozzles. Powerful hydraulic rams — borrowed F-18 aileron actuators — will depress the turning vanes into the exhaust plume, diverting the flow of thrust to improve manoeuvrability greatly in the pitch and yaw axes, or in both simultaneously. Flight testing was due to commence in January 1990, at which point TVCS — Thrust-Vectoring Control System — will be added to the aircraft's title. With the HARV-TVCS F-18 *Hornet*, the Agency is poised to undertake supermanoeuvrable-type flying for real.

One of the most significant items about the General Dynamics/Boeing F-111 operated in conjunction with the Mission Adaptive Wing AFTI programme is not so much the changing wing shape but that it changes shape without resorting to the need for conventional hinged flaps, spoilers and fairings. Instead it presents a smooth, unbroken surface to the oncoming air, thanks to an advanced wing-covering material which overlays a multiplicity of internal hydraulic devices linked to a sophisticated computer system within the fuselage. A Boeing KC-135 tanker of the US Air Force from Edwards is airborne twice a day, and by prior arrangement the AFTI F-111 collects 20,000lbs of fuel as many as three times during a single sortie. The flight-test programme demands frequent use of the afterburner, which consumes fuel at a prodigious rate. Gross take-off weights are up over 80,000lb, and a Northrop T-38 *Talon* chase aircraft takes off prior to the F-111 to avoid jet-wash and wingtip vortices. An abort on take-off at 165 knots would use up the entire 15,000ft of the long Edwards AFB runway. NASA is interested in the pure aerodynamic aspects of the AFTI F-111, whilst the US Air Force tend to focus on the systems aspects which implies discovering what can be done using complex interplay between computers and control systems in flight. It is in the combination of these two technological disciplines that the strength of the AFTI programme lies. Project pilot was initially Rogers Smith, who has now moved over to the F-16 AFTI programme; as of 1990 it was Roger Fullerton, assisted by Jim Smoke from NASA and Tim Seeley and Ron Johnston from the US Air Force. The AFTI F-16 Joint Test Force is directed by Lieutenant Colonel Lawrence 'Spike' Davis and the aircraft is 75-750. This is a modified pre-production model F-16A which has accumulated more than 1,000 flight hours and flown 475 technology integration sorties since the AFTI programme testing began in June 1982. As we write, the AFTI F-16A testbed is at General Dynamics' Forth Worth, Texas, plant where it is undergoing extensive updates and modifications.

Veterans

The veteran Boeing NB-52 *Stratofortress* (52-008) was originally acquired by NASA at Edwards AFB on 8 June 1959, and is still in service with Dryden today employed in a wide variety of tasks. It was the eighth 'B' model built and is the oldest B-52 flying; ironically it also has the lowest number of flying hours on the airframe, a mere 2,200 hours. It was never used in anger as a bomber, arriving direct from the Boeing factory to be modified by North American Aviation into a mothership for the legendary X-15. Currently *Pegasus*, the revolutionary winged satellite-launcher being fielded by the Orbital Sciences-Hercules Aerospace industrial team with support from the US Air Force, NASA and DARPA, is fast approaching its first space mission as we write, having been put through its paces at Edwards on several taxi-tests beneath the wing of its NASA Boeing NB-52 mothership and having made two captive flights up to February 1990. Normal pilots for the NB-52 is the NASA flight test pilot Gordon Fullerton, a former Shuttle astronaut who was pilot on the STS-3 *Columbia* mission in March 1982 and Commander on the 51-F *Challenger* Spacelab 2 in July/August 1985. Other NASA NB-52 jockeys have included Ed Schneider, Fitz Fulton and Don Mallick. During captive-carrying test flights with the Pegasus which is 50ft long and weighs 42,000lb on board, the *Stratofortress* has a voice and telemetry link-up with the US Air Force Western Test Range headquarters at Vandenburg AFB and with the NASA control room at Dryden. The flight path takes the aircraft over the Pacific Ocean abeam Big Sur to the prescribed launch point.

Early in April 1990, the *Pegasus* air-launched booster was dropped by its Boeing NB-52 carrier aircraft off the coast of California, making its first operational flight. After take-off from Edwards the NASA B-52 headed to the drop point about 60 miles south-west of Monterey, California, at an altitude of 43,000ft. The booster dropped cleanly away from its pylon under the right wing of the B-52. The first-stage motor was ignited about five seconds later when the booster was about 350ft beneath the aircraft. An earlier launch had been scheduled for 4 April but was postponed prior to take-off because of weather conditions in the Edwards AFB area and along the flight path to the drop point off the Californian coast. The 40,000-lb *Pegasus* vehicle on this first mission carried a multi-mission payload developed by the Defence Advanced Research Projects Agency and the NASA Goddard Space Flight Centre and a US Navy communications satellite.

This was the first time a space-bound rocket had been launched from an aircraft; it was proof of the viability of a commercial venture to develop a relatively inexpensive way of sending a spacecraft into orbit.

The Mission Adaptive Wing AFTI F-111 developed by General Dynamics and Boeing completed its last flight at the end of 1988, the flight-testing embracing 59 flights and 143 hours and proving the validity of the concept. The aircraft has now been declared redundant by the US Air Force Wright Research and Development Centre and is to be transferred to the Flight Test Centre Historical Foundation Museum at Edwards AFB. The F-111 first flew from Edwards on 18 October 1985, after Boeing had modified it with a variable-camber wing. Project pilot was intially Rogers Smith, followed by Roger Fullerton when Smith transferred to the ATFI F-16 programme. Other pilots were Jim Smoker from NASA and Tim Seeley and Ron Johnson from the US Air Force. One of the most significant items about the AFTI F-111 was not so much the changing wing shape, but that these changed shape without resorting to the need for conventional hinged flaps, spoilers and fairings. Instead, it presented a smooth, unbroken surface to the oncoming air, thanks to an advanced wing-covering material which overlays a multiplicity of internal hydraulic devices linked to a sophisticated computer system within the fuselage. A Boeing KC-135 tanker of the US Air Force from Edwards was airborne twice a day, and by prior arrangement the AFTI F-111 collected 20,000lbs of fuel as many as three times during a single sortie. The flight test programme demanded use of the after-burner, which consumed fuel at a prodigious rate. Gross take-off weight was up over 80,000lbs and a Northrop T-38 *Talon* chase aircraft which took off prior to the F-111 so as to avoid jet-wash and wingtip vortices. An abort on take-off at 165 knots would have used the entire 15,000ft of the long runway at Edwards. Yes, the F-111 had brakes which were powerful. The NASA were interested in the pure aerodynamic aspects of the AFTI F-111, whilst the US Air Force tended to focus on the systems aspect. A combination of two technological disciplines using complex interplay between computers and control systems in flight, provided strength for the AFTI programme.

The prototype Boeing 747 flew on 9 February 1969, and is still in production; NASA today has two 'jumbos' on its inventory. It is not generally known that there were other candidates in the running prior to the Boeing 747 being selected as the Shuttle Carrier Aircraft — SCA. These included the Lockheed C-5A Galaxy and even a totally new aircraft configured so as to carry the Space Shuttle Orbiter beneath a stub-wing linking two fuselages positioned side by side. At one point it was even considered feasible to fit turbofans to the Orbiter to enable it to fly under its own power. The Boeing 747-100 (c/n 20107) started its airline career in October 1970 with American Airlines as N9668. After accumulating 8,899 flying hours and 2,985 landings, it was purchased by NASA on 22 June 1974 for $15 million. It was returned to Boeing during April 1976 for modification, its main structure being reinforced to bear the Orbiter's 150,000 lb weight. Two additional vertical fins were fitted to the ends of the horizontal tail surfaces and braced by streamlined struts; these enhanced directional stability since the presence of the Orbiter atop the SCA partially masks the 747's central fin/rudder assembly from the airflow. Many other modifications were made and a slide-escape system was provided so that the flght crew could abandon the Boeing 747 in an emergency. A mass inertia-damper was installed in the forward fuselage, consisting of a 1,000 lb mass which moves its position laterally by means of rollers. This damps out oscillations caused by air flowing over the Orbiter and inducing turbulence across the airliner's tail. The Pratt & Whitney JT9D-3A turbofan engines were changed for JT9D-7AH units, giving a useful increase in take-off thrust. During ferry flights two pilots and two engineers occupy the flight deck, and when flying coast-to-coast with the Orbiter they stage through Kelly AFB, San Antonio, Texas to refuel. Contingency stopping points are available should weather conditions dictate, these being primarily Sheppard AFB, Texas, and McDill AFB, Florida. In addition there are a number of emergency landing sites en-route. Bergstrom AFB, is about 30 miles from Kelly AFB, and Orlando International Airport would be used if for some reason the SCA was unable to land at the Kennedy Space Centre. Although based at Edwards AFB, 'NASA 905' is carried on the Johnson Space Centre inventory and most of the pilots are based at Ellington AFB, Texas. In 1990 they included project pilot Ace Beale, David Finney, Joe Algranti, Kenneth Haugen and A. R. Roy, whilst Gordon Fullerton and Tom McMurtry are based at Edwards.

So vital to the NASA programme is the SCA that the Presidential Commission which investigated the Challenger Orbiter accident made recommendations pertaining to the future conduct of the programme, advising that a second Boeing 747 be converted as soon as possible. On Monday 17 April 1989, the NASA's second SCA (registered 'NASA 911') was delivered to the Boeing plant at Wichita, Kansas, for conversion. It was expected to enter service later in 1990. The original SCA, 'NASA 905', has transported Orbiters since 1977, when Enterprise was released from its back over Edwards AFB for glide-tests. Since then all the current Orbiter vehicles — Enterprise, Columbia, Challenger, Discovery and Atlantis — have been ferried atop N905NA, to quote its US civil registration. During early June 1983 'NASA 905' and Enterprise ventured across the Atlantic to be displayed in the United Kingdom at Stansted International Airport in Essex. On 5 June 1983, the strange pair made a slow and gentle flypast over London (Heathrow) Airport.

NASA Workhorse

Any agency as diverse as NASA, renown since its beginings as NACA for its exotic flying machines, is bound to have its 'workhorse' amongst the many types of aircraft operated. The ubiquitous Douglas C-47 has been universal over the years, and indeed served the NACA and the NASA well over many years at most of the establishments. However, the current workhorse (restricted today to two on the Edwards flight-line) is the Lockheed F-104 Starfighter. Over the years, since 1956, eleven F-104s have served with NASA, with 'NACA 818' (c/n 1007, ex-55-2961) a pre-production YF-104A on loan from the US Air Force being the first; it arrived during August 1956. It was employed on basic research, primarily stability and control, propulsion and handling qualities. During October 1976 this aircraft was donated to the National Air & Space Museum in Washington DC. During 1963 three purpose-built F-104N's were delivered by Lockheed to Edwards for the Agency. They were initially registered 'NASA 011', 'NASA 012' and 'NASA 013', later becoming 'NASA 811/812/813'. The latter was lost with Joe Walker in the horrific collision with the XB-70 on 8 June 1966. It was replaced by 'NASA 820' (ex-56-0790) which was withdrawn from use on 30 October 1983 but is still at Dryden to be preserved for inclusion in the Flight Test Historical Foundation Museum. Remembered by some pilots as an exotic hangar queen for various reasons which kept it on the ground, this F-104 claimed the distinction of being only the second vehicle to bridge the gap between atmospheric and space flight, the first being the X-15. The unofficial record altitude achieved by an NF-104 was 120,800 ft set by Major Robert W. Smith of the US Air Force on 6 December 1963.

This Lockheed YF-104A 55-2961 c/n L183-1007 was acquired at Dryden during August 1956, and is depicted prior to going into the paint shop, having just minimal markings. Today this NASA workhorse is in the National Air & Space Museum, Washington DC. It was donated in October 1976.
(Military Aircraft Photographs)

The original concept of providing a flying machine which could facilitate spaceflight training at the Aerospace Research Pilots School — ARPS — was proposed in 1961 by a small group of staff members who included the author's ex-astronaut friend Frank Borman. By modifying production Starfighters, the considerable costs of development for a purpose-built aircraft were saved. Apart from the addition of the LR-121 rocket engine, the NF-104 incorporated other modifications. The J79-3A turbojet was replaced by an uprated J79-3B variant and the wingtips were extended by two feet on either side to house X-15 hydrogen-peroxide reaction control-system thrusters. There were eight nozzles in the nose-cone for pitch and yaw control and two nozzles in each wingtip for roll control, for use at altitudes where full aerosurface authority could no longer be assured. A large 'G' model vertical tail replaced the standard 'A' model unit for better directional stability at high Mach numbers; the M-61 gun and other items of military hardware, including the drag-chute were removed and a Mercury capsule-specification Collins 718B-1 radio system replaced the standard VOR radio. With flight operations being limited to the Edwards control zone, other items of equipment including the ILS could be dispensed with. In addition to Frank Borman, the present day 'Shuttle-naut' John Blaha also flew the types.

X-15 Simulator

The NASA veteran flight-test Bill Dana made 15 flights in the North American X-15, including the 199th and last flight of that research rocket aircraft. The Lockheed F-104 was used in preparing pilots for X-15 flights, becoming an excellent simulator for practice landings not only on the dry lakebed at Edwards but other dry lakebeds specified as emergency landing sites for the X-15. The furthest away was called Smith's Ranch, some 75 miles northwest of Tonopah, Nevada and 270 nautical miles away from Edwards. Three other lakebeds were 200 miles away and included Mud Lake, Tonopah, Nevada; Railroad Valley Dry Lake, 75 miles east of Tonopah and Delamar Lake, near Alamo, Nevada, some 75 miles north of Nellis AFB, Las Vegas. Closer strips were at Hidden Hills, California, on the state border with Nevada between Las Vegas and Edwards, and an even closer one called Silver Lake. The F-104's fuel capacity enabled five practice landings at the furthest strips, and up to ten in the Edwards area. Milton Thompson, NASA flight test pilot was involved with the F-104 on research projects.

Possibly the most bizarre use of the F-104 in NASA service was air-launching a rocket housing a high-altitiude research balloon to a target height of 1,000,000ft! The intention was to measure the rate of descent of the balloon and thereby calculate air density at varying altitudes to calibrate data gathered during high-altitude flights of the X-15 which were soon to commence. The F-104 used was 56-749, specially modified. The aircraft had been used by the US Air Force for flight-testing so was already modified with a launcher-rack already fitted. The rocket was launched from a loop in the vertical at approximately 60,000ft at Mach 1.4. In 1962 Milt Thompson ejected from 56-749 during a flight to check uprange weather prior to a X-15 launch whilst he was practising X-15 type approaches at Edwards. The F-104 went into an asymmetric roll configuration which Thompson intially could handle, but the aircraft subsequently went completely out of control and he ejected at 20,000ft.

Space Shuttle orbiter thermal protection tile test articles were included in these flight tests. Up to 1980 NASA F-15 10281 was included in the programme. On behalf of the Shuttle programme office at NASA's Johnson Space Centre there have been five different thermal protection-system test programmes conducted with the Flight Test Fixture to date. When the fixture was first introduced into service in 1964, it was specifically to support 'flutter panel research'. Thin sheets of different types of materials were bolted onto the fixture so that a given sample's flutter characteristics could be analyzed at high Mach numbers and high dynamic pressures.

Prior to the testing of Space Shuttle TPS materials, the fixture was employed for skin-friction measurements. Before the Shuttle flew there were five critical areas of Orbiter-vehicle TPS coverage which NASA and the prime contractor, Rockwell International, wanted to test. Two of these areas were tested using an F-104 equipped with the Flight Test Fixture as a flying testbed, whilst three were trialled on a NASA F-15 *Eagle* as mentioned earlier, testing tile and blanketing materials. Flight test pilots participating in this programme have included Bill Dana, Ed Schneider. Rogers Smith and Jim Smoker. The NASA F-104 manager was Roy Bryant.

Another important test programme involving the F-104 'NASA 826' was an analysis of moisture impact damage on Shuttle TPS materials, essentially the combination of water and aerodynamic forces. These specimens were mounted on the Flight Test Fixture on the F-104 and flown in a series of test conditions in which rain was simulated by a mist of water droplets streamed from a Boeing KC-135 tankers' refuelling boom. Moisture impact test flights began at a relatively low speed and increased in increments until the specimens began to sustain physical damage. All the data was recorded for subsequent laboratory analysis. The Flight Test Fixture was also used to test the installation of air-pressure sensors with Shuttle TPS materials. These sensors form part of the SEADS experimental pressure-measurement system currently fitted to the Orbiter *Columbia*, and the research could also support future sensor installations in Shuttle Orbiters.

Historical Background

As many as eleven Lockheed F-104s served with NASA at Edwards, currently spanning some 34 years of research and test flying. It was August 1956, in the days of the NACA, when Lockheed YF-104A 65-2961 (c/n 1007, an early production *Starfighter*) was loaned to the agency by the US Air Force. It later became 'NASA 818' and was employed on basic flight research; primarily stability and control, propulsion, and general handling qualities. The NACA flight test pilots and engineers were equally enthusiastic about discovering the intimacies of one of the world's most capable flying machine. A bizarre quirk of circumstances led to the YF-104A being entrusted with several flight programmes of tremendous and vital importance to Lockheed when they and the US Air Force lost all their instrumented F-104s in accidents. The NACA aircraft was the only one available with a full test instrumentation suite, so it was decided to save time by employing it to investigate a particular thorny handling problem, ironically the same handling problem which had affiliated the Douglas X-3 — inertia and 'roll coupling' phenomena. A major 'roll coupling' test programme was conducted, essentially a series of test conditions in which various roll-rates were performed at incrementally increasing speeds all the way from Mach 0.8 out to Mach 2.

It was NASA which inspired a subsequent flight-test programme concerned with the spate of flame-outs which were dogging the *Starfighter's* initial entry into operational service. The problem was eventually traced to unsymmetrical airflow into the engine compartment from two engine inlets located either side of the forward fuselage. One such flame-out incident was experienced by Joe Walker with Neil Armstrong flying chase, but Joe managed to get it restarted. The NACA flight-test programme resulted in the incorporation of a vertical 'duct-splitter' in production F-104s. Situated just in front of the compressor blades to balance-out the intake airflow from each side, it alleviated the type's asymmetric flow problem and greatly enhanced its reliability.

At a later date 'NASA 818' was fitted with a ballistic control system, partly to facilitate research into control capabilities in the essentially airless regions at high altitude where conventional aerosurfaces become useless, and partly to give pilots experience in operating such systems in preparation for the North American X-15 programme. The early Bell X-1B research rocket aircraft had had a ballistic control system, but the turbojet-engined F-104 would allow pilots greatly extended flight durations by allowing them to make repeated climbs to high altitudes on each given flight. The Bell X-1B, on the other hand glided back to earth soon after its 'one-time' propellant load was expended. Reaction-control thruster units were situated on a modified F-104 nose-cone unit for pitch and yaw control, and similar devices were installed in small pods on either wing-tip for roll control. Only one F-104 was modified, and it successfully performed zoom-climbs to altitudes of 85,000 and 90,000ft.

Later 'NASA 818' was employed on smaller-scale experiment primarily concerned with base drag, skin friction, drag due to surface friction, and dynamic-pressure studies, using an ingenious add-on experiment-carrying device, the Flight Test Fixture described earlier. 'NASA 818' was gracefully retired during 1975, being transferred for display to the public in a gallery of the National Air & Space Museum in Washington DC. The Flight Test Fixture was transferred to F-104 'NASA 826' (c/n 8213) a Fokker-built F-104G which went initially to the West German Air Force and then to Luke AFB, Arizona, going to NASA during 1975.

The three F-104N aircraft specially built for NASA by Lockheed were eventually replaced by three aircraft purchased for the West German Government, which were first used in the USA for pilot training for West German Air force pilots at the 4510th Combat Crew Training Wing at Luke AFB. These were two TF-104G aircraft, 'NASA 824' (c/n 5735, ex-US Air Force 61-3065 and 27 + 37, currently in storage) and 'NASA 825' (c/n 5939 ex-66-13628 and 28 + 09, still in service) whilst 'NASA 826' already mentioned is also still in service at Dryden. These three were all used as safety-chase and proficiency-training aircraft and as useful airborne simulators for pilots to practice X-15 and Manned Lifting Body approaches and landings.

Of the original three purpose-built NASA aircraft 'NASA 813' was lost in the mid-air collision with the XB-70 Valkyrie Mach 3 research bomber described earlier. A US Air Force F-104B (56-0790, c/n 1078, later 'NASA 820') was seconded to replace the lost F-104 and become involved in the pilot agility programme, which analyzed factors that could degrade a pilot's ability to control his aircraft. This involved a tail-chase sequence and performance of a series of controlled wind-up turns ranging from 2g to 4g. Test subjects employed a gunsight operated by a control-column button linked to a camera-gun to verify results. 'NASA 820' was withdrawn from use on 30 October 1986 and is currently on display at Edwards pending the creation of the US Air Force Flight Test Centre's projected museum.

During 1990 'NASA 825', the two-seat TF-104, was grounded. It was expected to re-emerge at Dryden later in 1990, modified to support the new hypersonic X-30 National Aero Space Plane (NASP) programme. This is the next-generation reusable shuttlecraft, a single-stage-to-orbit vehicle which can double as a suborbital military reconnaissance platform or a transcontinental commercial passenger carrier. It is already advertised as a hypersonic 'Orient Express' flying from Los Angeles to Tokyo in two hours. In the interest of reducing 'parasite drag' — aerodynamic drag attributable to such external protuberances as cockpit canopies — the new X-30/NASP will likely feature a windowless cockpit design, in which the flight crew's view of the outside world is assured via a periscope device. Currently 'NASA 825' is earmaked to serve a NASP-related visibility-study programme, performing a series of flights in which the pilot's ability to perform landing approaches with limited visibility is varied. The test subject will occupy the rear cockpit, with the pilot-in-command handling the remaining flight duties from the front cockpit.

Various ideas are under consideration to simulate the restricted flying conditions involved. One is to fit a prototype X-30 style periscope device. Another is to cover the rear cockpit canopy and helmet visor with materials of different colours, these combining to create varying visibilty levels when the visor is raised and lowered. If NASA does elect to have the TF-104 modifed with the addition of a periscope for the rear cockpit canopy, it will not be the first time such a configuration has been tested. Back in the 1960s a similar modification was completed as part of a programme devoted to pilot 'depth-perception' experiments. This was one of a series of biomedical studies conducted at Edwards involving the *Starfighter*. A NASA TF-104 was fitted with what was called a 'binocular system' in which the back-seat pilot peered through an apparatus not unlike a set of field-glasses, with his outside vision assured through a couple of lenses mounted up on top of the canopy area between the forward and aft cockpits and aft cockpits. In serving the new X-30 programme, the Lockheed F-104 *Starfighter* will enter a new decade at the forefront of aerospace technology — just where it was 36 years ago when the prototype of the aircraft known as 'missile with a man in it' first took to the air.

Tilt Fan V/STOL

During the mid 1970s the Grumman Aerospace Corporation conceived a 'Design 698' V/STOL aircraft to meet the US Army requirement for the Special Electronic Mission Aircraft — SEMA-X — for reconnaissance, electronic warfare and surveillance and communications relay, together with target acquisition and designation. Design 698 was also offered to the US Navy to undertake a variety of missions from proposed Sea Control Ships. Of twin-tilt-fan configuration, the two-seat aircraft was to have been controlled through horizontal and vertical vanes located in the exhaust flow of pod-mounted 9,065-lb thrust General Electric TF34-GE-100 turbofans. Aerodynamic forces generated by the high-speed flow over these surfaces were to provide the control in pitch, yaw, and roll required for hover and transition. Spoilers, an all-movable stabiliser and a rudder were provided for control in conventional flight.

Grumman commenced work on Design 698 during 1976, and after 5,000 hours of testing were ready to embark on the detailed design and fabrication of a manned flight demonstrator. The aircraft would use an existing fuel-efficient, high-bypass turbofan engine, the TF34 as used in the US Air Force Fairchild A-10. This offered the required high static thrust, and fuel-efficiency computer simulations, radio-controlled model tests and small-scale wind tunnel trials were followed by full-scale experiments. Sponsored jointly by NASA, the US Naval Air Systems Command and Grumman, the full-scale tests were conducted in the NASA 40 × 80 ft wind tunnel at Ames, these confirming the feasibility of the tilt-fan concept. Nine different kinds of models were used to study STOL and cruise aerodynamics.

However, lack of funds resulted in the US Army cancelling its SEMA-X requirement, and the US Navy lost interest in the project when Congress terminated funding for the Sea Control Ship. Principal characteristics and performance were: span 36 ft 8 in, length 38 ft 5½ in. height 14 ft 11 in, empty weight 11,723 lb, maximum VTO weight 15,430 lb, VTO power loading 0.85 lb/lb st, maximum speed 500 mph, cruising speed 409 mph and ferry range 1,150 miles. The two GE-TF34-GE-100 turbofans provided 18,130 lb of total thrust.

This Convair F-106B Delta Dart 57-2516, NASA 816, arrived at Langley from Lewis on 29 January 1979 and is now retired with a replacement being sought. Flight tests began on 2 August 1988 with modified leading edge vortex flaps, developed by Langley. Flaps could be deflected as much as 56 degrees. Photo taken on 30 July 1981 with NASA pilot Perry Deal and Major Jerry Keyser, USAF, flying the fast, rugged aircraft. (NASA 81-H-518)

Investigation of rotary-wing aircraft continued, even as expansion of the experimental XV-15 tilt-rotor craft evolved into the larger V-22 *Osprey* built by Boeing-Vertol and Bell Helicopter Textron for the US armed services. A joint programme linked technology in the United Kingdom, the Agency and the US Department of Defense — DoD — for investigation of advanced short-takeoff and vertical landing aircraft. Based on the successful concepts used in the British *Harrier* and US AV-8A/B in use with the US Marine Corps, designers commenced wind-tunnel tests of aircraft which could fly at supersonic speed while retaining the *Harrier's* renowned agility.

Kestrel and Harrier

Interest in the British-designed and built Hawker P.1127 V/STOL aircraft was initially shown by the Douglas Aircraft Company during the late 1960s; the aircraft had made its first flight on 21 October 1960. The *Kestrel* V/STOL tactical strike aircraft followed during December 1964. During April 1962, when V/STOL techniques had been explored with the six prototypes, the Governments of the United Kingdom, the Federal German Republic and the United States agreed to share the cost of procuring nine developed P.1127s for use by a unique three-nation evaluation squadron. In the meantime development continued and the sixth prototype embodying various modifications flew in February 1964, powered by a 15,200lb-thrust Rolls-Royce Pegasus 5. During 1964 and 1965 the nine tri-partite aircraft XS688–695 were built, the first flying on 7 March 1964. Named *Kestrel* FGA.Mk 1 in October 1964, they were certified for service flying during December of the same year, commencing operations at RAF West Raynham, Norfolk, on the first day of April 1965. Some 600 hours were flown from a variety of grass strips, roads and small airfields. Early in 1966 six *Kestrels* were shipped to the USA to equip the XV-6A Tri-Service team, flying with the US Army, Navy and Air Force until the end of July. Four aircraft then went to Edwards AFB, the remaining two to NASA at Langley.

Hawker Siddeley XV-6A *Kestrel* 'NASA 520' arrived in the USA during February 1966 and became 64-18267 at Patuxent River prior to going to Edwards AFB with the US Air Force. It subsequently went to NASA at Langley as 'NASA 52' on 23 July 1966, being written-off on 7 August 1967. It apparently was rebuilt using XV-6A 64-18266. XV-6A 'NASA 521' also arrived in the USA during February 1966 as 64-18263, and after service at Edwards went to Langley on 23 July 1966; it was withdrawn from use on 20 June 1974. It was donated to the National Air & Space Museum in Washington DC. 'NASA 520' was withdrawn from use on 13 May 1970 and donated to the Hampton Aerospace Park, Virginia, where it remains on display today.

During the pre-merger period of the Douglas Aircraft Company and McDonnell of St Louis, discussions between Douglas and Hawker Siddeley towards the possible manu-factures of the new P.1154 *Harrier* by the United States reached an advanced stage. These discussions later led to McDonnell Douglas obtaining the licence rights for the *Harrier* if the US Government decided to have the aircraft ordered for the US Marine Corps built in St Louis. However all 112 single-seat AV-8A's and eight two-seat TAV-8A's for the US Marine Corps, first ordered in 1969 and entering service in April 1971, were built in the United Kingdom with McDonnell Douglas providing support through its St Louis facilities. The two companies continued to co-operate on the development of advanced versions of the *Harrier*, including the projected AV-8C and AV-16 — respectively, a minimum development of the AV-8A and a more extensive redesign of the *Harrier*.

The anticipated cost of the AV-16 programme was found excessive and the US Department of Defense chose to sponsor the development of a less complex version. Designated AV-8B, this version was first proposed in 1975 and approval of its development was announced on 27 July 1976. Using the AV-8A's Rolls-Royce Pegasus 11 vectored-thrust turbofan and the basic AV-8A fuselage but with improvements which increase maximum thrust to 21,500lb, the AV-8B incorporated major changes primarily designed to increase range and load-carrying ability and to improve reliability and maintainability. These changes include the use of a McDonnell Douglas-designed superficial wing made largely of composite materials and incorporating a new high-lift system with large slotted flaps and drooped ailerons, and the installation of lift improvement devices consisting of fixed strakes on the twin under-fuselage 30mm Aden cannon pods and a moveable flap-like panel at the forward end of the pod. The engine air inlets and rotating exhaust nozzles have also been redesigned to increase their efficiency whilst internal fuel capacity has been increased 45 per cent. The net result of these changes is a 6,000lb increase in short-take-off weight-lifting ability, based on using a 1,000ft run.

An AV-8A was modified to represent an AV-8B for testing during 1976 and 1977 in the 40 × 80ft wind tunnel at Ames. Two YAV-8B prototypes were converted from AV-8A airframes and the first flight took place at St Louis on 9 November 1978. One YAV-8B operated by NASA at Ames was NASA 704, which went into temporary storage during October 1986. An AV-8C currently at Ames is 'NASA 719' (ex-BuNo.158387) which has a sticker on the port Pegasus engine nacelle declaring 'Thank God it's British'.

Something Old, Something New

During April 1990 it was announced that the Boeing Model 367-80 — the historic 'Dash 80' prototype of the Boeing 707 family — was being restored to flight condition by a Boeing team at the huge desert storage facility located at Davis-Monthan AFB, Arizona. Impetus for the restoration came when it was discovered the aircraft was deteriorating rapidly after 16 years of storage in the desert. The transport was due to be flown on 19 May to Boeing Field, Seattle, where the exterior will be restored, and then flown to the Paine Field, Everett site of the Boeing 747 and 767 final assembly facility. There it will be displayed in the Museum of Flight annex, along with the recently retired prototype of the Boeing 747. The Model 367-80 belongs to the National Air & Space Museum and will be eventually displayed at the NASM annex when it is established at Dulles International Airport.

During 1964 the Boeing 707 'Dash 80' (N70700, c/n 17158) was used for tests by NASA at Langley with blown flaps, which were conventional trailing-edge types but which could be extended to a full 90-degree position with high-velocity air bled from the jet engines being blown over them. It was also used for SST variables-stability simulation tests, also at Langley. During 1967 it was operated under a NASA contract at Ames and involved in a programme lasting two months. This was in connection with noise reduction of commercial airliners during airport approaches. It was heavily modified and had a 15ft spike or nose probe which contained aircraft noise-response sensing equipment. Its operations with the NASA qualified the aircraft to carry the NASA 'meat ball' logo and the words 'NASA TEST' on the fuselage. Its first flight was on 15 July 1954.

On 18 January 1990 the author's good friend Major Tom McCleary from Detachment 4, 9th Strategic Reconnaissance Wing at RAF Mildenhall, Suffolk, departed back home to Beale AFB, California, as pilot of Lockheed SR-71A 64-17964. The *Blackbird* was about to go into retirement. Speculation and strong rumours that NASA would acquire up to three Lockheed SR-71A aircraft on loan from the US Air Force were confirmed on 14 February 1990 with the publication of NASA News Release 90-24. The first SR-71 was scheduled for arrival at the NASA Ames-Dryden Flight Research Facility at Edwards AFB on 15 February, with the second SR-71A arriving on 20 February. Arrangements for the third *Blackbird* was to be determined later. The aircraft were placed in flyable storage at the NASA facility, with Agency officials currently assessing the research opportunities and experimentation which can benefit the administration as a result of operating these high-speed flying test beds. A loan agreement between NASA and the US Air Force has been prepared.

The Dryden facility operated the YF-12, similar to the SR-71, from 1969 to 1979, gaining much useful research data on structures and stability and control of air-breathing aircraft at high speeds and altitudes. The Lockheed SR-71 is capable of flying in excess of three times the speed of sound. The aircraft's 101-ft long titanium structure is coated with a special black paint which helps dissipate heat caused by high speeds.

During the first eight weeks of 1990, the NASA Wallops Flight Facility was the operations centre for Project COWAX — Convective Waves Experiment — conceived by a team of scientists from the Goddard Space Flight Centre, the University of Colorado and the US National Centre of Atmospheric Research. It conducted a study of the vertical propagation of atmospheric gravity waves with convection in the atmospheric boundary layer. Convection is initiated when cold air strikes a warm ocean surface and an updraft of moisture-laden air come in contact with elevated layers in the atmosphere. The initial energy of the gravity waves originates from these updrafts. Gravity waves are responsible for momentum exchange between the atmosphere close to the Earth's surface and are thought to slow down the horizontal propagation of the large-scale high and low pressure systems that affect our day-to-day weather.

The Wallops Flight Facility Lockheed L-188C *Electra* aircraft 'NASA 429' (c/n 1103, acquired on 27 September 1978 from the Federal Aviation Administration) was instrumented with Goddard's downward-looking lidar system and Langley's gust probe system in the nose boom. The lidar, operated at about 10,000ft provided a high-resolution picture of aerosol layers and convection in the boundary layer below the aircraft while the gust-probe system sensed the presence of the gravity waves at flight altitude. A total of six successful flights were carried out during which the *Electra* flew co-ordinated flights with two other instrumented aircraft operated by the National Centre of Atmospheric Research. A Beech *King Air* flew in the boundary layer to sample the convection near the ocean surface, whilst a *Sabreliner* flew at upper levels in the tropopause to probe the vertical propagation of the gravity waves. During the flights, the tracking of the radiosonde system provided high-resolution soundings of temperature, moisture and wind speed. A navigation system supported the lidar system in the *Electra*.

X-Planes

On 23 March 1990, the X-31 enhanced fighter manoeuvrability test aircraft made its first taxi run with Rockwell chief test pilot Ken Dyson at the controls. It reached a speed of 30 knots and the nosewheel steering and brakes were successfully tested. First flight of the X-31 was expected during May.

Any research student in aircraft will realise that not all the family series of X-planes were operated by NACA and NASA. Today, seventy-five years after the inception of NACA, the series has reached X-31 and no doubt will continue. Jay Miller has catalogued them all in great detail in his excellent volume which we highly recommend.

The Grumman Aerospace Corporation X-29A was designed to explore the advantages and disadvantages of advanced composites, variable camber, relaxed static stability, a close-coupled canard configuration, and a thin supercritical airfoil section forward-swept wing. The integration of these state-of-the art technologies into a single airframe represents a major step forward in aircraft structural and aerodynamic engineering.

Two aircraft have been built, US Air Force 82-0003 and 82-0049, utilizing parts from a US Air Force Northrop F-5A (63-8372) retrieved from the storage unit at Davis-Monthan AFB, Arizona, and a Northrop F-5A (65-10573) of the Royal Norwegian Air Force. The US Defense Advanced Research Project Agency, the US Air Force and Grumman are listed as the sponsors and provided the funding. Competition came from Rockwell International, who produced the *Saberbat* full-scale mock-up study during the mid-1970s. General Dynamics also submitted a forward-swept wing design. The concept of forward-swept wing technology is not new, having been explored as part of the vast German aerodynamic research effort during World War II. Grumman initiated the

During June 1990 Lewis concluded the latest series of tests in a long-term programme that is examining hot gas ingestion in advanced ASTOVL aircraft. The tests, a joint effort with McDonnell Douglas were conducted with a 1/10th scale model in the NASA Lewis 9 × 15 ft low-speed tunnel as depicted. (NASA C-87-7312)

full resurgence of interest in forward-swept wing technology in 1975, following a contract loss to Rockwell International during the HiMat — Highly Manoeuvrable Advanced Technology — remote piloted research vehicle competition. During World War II the Germans had produced the Junkers Ju.287 swept-forward wing bomber and the invading Soviets captured several incomplete airframes. Post-war the Germans built the Hansa HFB-320 Hansajet corporate aircraft, having some sales success.

During 1977, a request for proposals was issued by DARPA and the US Air Force Flight Dynamic Laboratory. Grumman Aerospace, Rockwell International and General Dynamics responded. Three years of design studies followed the preliminary DARPA request, culminating in the Grumman Aerospace Corporation being selected as prime contractor on 22 December 1981 based on their Design 712 submission. A $71.3 million letter contract was signed several months later. Officially designated X-29A by the US Air Force in mid-1981, it was to have performance characteristics comparable to those of existing lightweight fighters. Several thousand hours of wind tunnel tests were eventually logged, and metal was cut on the first of two prototypes during January 1982 following a final design review by the US Air Force during December 1981. Due to expense and a relaxed programme priority, the USAF moved the aircraft to Edwards AFB by sea. Following weather delays of over a week, the first flight with Charles A. 'Chuck' Sewell in the cockpit was finally successfully completed on 14 December 1984 in 82-0003 with a flight time of sixty-six minutes.

No further flights were completed in 1984 due to unusually poor weather at Edwards. The second and third flights took place on 4 and 22 February 1985. Grumman pilot Kurt Schroeder completed the fourth flight on the first day of March, and on 12 March the aircraft was officially turned over to the US Air Force to begin the second phase of the flight-test programme. More pilots were assigned to fly the X-29A including two from NASA — Stephen Ishmael and Rogers Smith, who flew the fifth and sixth sorties respectively. On 13 December 1985, on its twenty-sixth flight, the X-29A became the first forward-swept wing aircraft in the world to fly supersonically in level flight, when Mach 1.03 was reached with Ishmael in the cockpit.

In December 1986 the US Air Force, NASA and DARPA jointly funded a $30.2 million X-29A follow-on flight research programme covering high-angle-of-attack studies to 90°, utilising both X-29A aircraft. Initial flight trials had only involved 82-0003 the second (82-0049) remaining in storage with Grumman. A new plan commencing in October 1987 called for the use of both aircraft to achieve the test programme objectives more rapidly, while concurrently expanding the angle-of-attack envelope. The high-angle-of-attack research project was to last for at least a year as part of a drive to improve the tactical agility of future fighters. A $4.65 million contract to design a system which gave the X-29A a 70° angle-of-attack capability was awarded to Grumman by the US Air Force Systems Division. Flight tests of 82-0003 continued during 1986, and between 23 December 1986 and 19 June 1987 the X-29A was grounded to permit installation of a specially calibrated General Electric F404 engine. By the end of June 1987, a total of 110 flights and been completed; these resulted in a maximum speed of Mach 1.87 and a maximum altitude of 50,200ft. Many major and secondary test objectives were accomplished.

The first NASA flight took place on 2 April 1985, a pilot familiarization flight by Ishmael. In addition to utilizing the forward fuselage of a Northrop F-5A aircraft, the Menasco-manufactured main landing gear, the emergency power unit and the servo actuators are from a General Dynamics F-16A. The nose landing gear is stock F-5A. A Martin-Baker IRQ7A ejection seat is fitted. Length is 48.1ft, wingspan 27.2ft and wing area 188.84sqft. The NASA aircraft inventory dated 1 March 1990, indicates that the second X-29A aircraft (82-0049) is now at Dryden for research and development.

Hypersonic Trans-atmospheric Vehicle

Research into hypersonic trans-atmospheric vehicle design has been on-going in the United States for several decades, with finance provided by many aircraft manufacturers such as Boeing, McDonnell Douglas and Rockwell International. Many models and designs on paper appeared, and even full-scale mock-ups, but it was not until the early 1980s that serious interest in the concept was expressed publicly by the Department of Defense. The Agency, during this same time period, had lobbied on several occasions for a hypersonic research aircraft programme, but for some reason had stopped short of declaring that it should be capable of trans-atmospheric performance. Abortive projects such as the X-24C and the earlier Boeing X-20 *DynaSoar* had generated a more conservative atmosphere within NASA and it was finally realised that any TAV or hypersonic research aircraft programme would require the financial support of the DoD before it could reach fruition.

The birth of President Ronald Reagan's highly controversial 'Strategic Defense Initiative' indirectly played a key role in giving a rationale to the NASP. The ability to deliver heavy equipment, parts and other hardware into space on a near-routine basis, as would be required by the SDI project, dictated that a trans-atmospheric vehicle be seriously considered for development. Coupled with its peripheral capabilities — reconnaissance, weapons delivery and commercial transportation — the trans-atmospheric vehicle idea gained considerable momentum during 1984 and 1985. Thousands of miscellaneous studies were conducted under DoD, DARPA and NASA auspices during this period. The result was an announcement by President Reagan during his State of the Union Address on 4 February 1986 that the DoD had agreed to fund what was referred to as the National Aerospace Plane — NASP. In a civil guise it became nicknamed *The Orient Express*. Ronald Reagan's announcement placed heavy emphasis on the peaceful uses of the NASP, heralding the fact that the configuration would play a key role in the design of hypersonic commercial airliners of the future.

A total of seven contracts for work leading to the development of the NASP, worth about $450 million over a 3½-year period, were awarded during the first week of April 1986 by NASA and the DoD. Five of the contracts went to airframe manufacturers including Boeing, General Dynamics, Lockheed. McDonnell Douglas and Rockwell International. Two went to powerplant manufacturers which included General Electric (in partnership with Aerojet Tech Systems) and Pratt & Whitney. A third engine contractor, Rocketdyne, has entered the competition using its own funding. The US government contracts cover the key technology development phase and are managed by DARPA. A jointly-operated programme office with the US Air Force, the US Navy and NASA has been established at Wright-Patterson AFB, Ohio.

The number of X-30As to be manufactured has been stated by the DoD to be two. The airframe study contracts, valued at approximately $32 million each, will serve to permit the five manufacturers to consolidate their respective trans-atmospheric vehicle research efforts and to generate viable designs in response to the US Air Force request for proposals. The powerplant contracts, valued at approximately $175 million each, permitted the respective contractors to design powerplant modules appropriate for use on the X-30A and later the NASP. Heavy emphasis is placed on hydrogen-fuelled supersonic combustion ramjet (scramjet) technology. Contract awards were made on 7 October 1987 to three of the five initial contenders, these being General Dynamics, McDonnell Douglas and Rockwell International. Each was awarded $25.5 million to move into the next NASP development phase.

First flight of the X-30A is tentatively scheduled for 1993 and it represents the most ambitious X-aircraft programme in history. Because of its basic performance requirement and mission objectives the demands placed on its design, airframe and propulsion system are the most stringent ever. As a research vehicle, it almost certainly will become a major contributor to the entire aerospace industry data base at virtually every level. The X-30A test programme will remain important to the aerospace industry for many decades to come. It will be the first aerospace aircraft, and thus the progenitor of all manned space transportation systems to follow.

The NASP plans call for a hydrogen-fuelled aircraft which could reach Mach 25 and be capable of operating in a low Earth orbit much like the Shuttle, or cruise within the Earth's atmosphere at hypersonic speeds of Mach 12. It would have the ability to cruise from the United States to Asia in about three hours. Propulsion-system testing at Langley and Lewis has resulted in successful low-speed work utilizing a Mach 3.5 ramjet configuration. By 1993 there is no doubt that the $3.3 billion NASP concept will have increased in cost.

Fuel Efficiency

Flight research into fuel efficiency continues. Despite jet fuel prices being reduced in the mid-1980s, the cost was still five times that prevailing in 1972, representing a significant percentage of operating costs for the world's airlines. For example, the early Boeing 727s were well known as 'fuel guzzlers' but then fuel was only six cents a gallon (US). Propfan engines have been looked at initially with intense interest, instigated by NASA's earlier Aircraft Energy Efficiency Progamme. Using a gas turbine, the new engine featured large external fan blades which were swept and shaped so that their tips could achieve supersonic velocity. Propfan-powered airliners would retain the speed, but achieve fuel savings of up to 30%. Different trial versions of multi-bladed propfan systems were in flight-test configuration by the beginning of 1986, with operational use projected by the early 1990s. Lockheed at Marietta, Georgia flight-tested a propfan-equipped Grumman Gulfstream II ('NASA 650', registered N650PF) to gather data on the fuel-saving propulsion system. Following flight-envelope clearance flights, the propfan testbed aircraft was flown on a 150-hour test assessment research flight programme to evaluate the structural integrity of the propfan blades as well as noise generated by the propfan inside and outside the aircraft cabin.

Investigation of rotary wing aircraft continues. The experimental Bell XV-15 tilt-rotor craft evolved into the larger V-22 *Osprey*, built by Boeing Vertol and Bell Helicopter for the US armed services. However, US Armed Forces budget constraints have taken their toll of the mass production envisaged. The XV-15 'NASA 703' is currently at Ames, and during April 1990 was supporting the development of a fibreglass blade made by Boeing. The first XV-15, 'NASA 702', is on loan to Bell and during April 1990 was operating in the Washington DC area for a series of flight demonstrations. On 19 March 1990, this XV-15 established five unofficial world class records at Fort Worth, Texas. These included time to climb to 3,000 metres in four minutes 24.36 seconds; time to climb to 6,000 metres in eight minutes 28.96 seconds; maximum altitude achieved at 22,660 ft, horizontal flight at altitude at 22,600 ft and horizontal flight at altitude with payload at 22,570 ft. On its way to St Petersburg, Florida, to an air show, 'NASA 702' established an international speed record between Arlington, Texas, and Baton Rouge, Louisiana. It flew 380 statute miles averaging 247.5 knots — 285 mph — in one hour, nineteen minutes and 59 seconds. Flight hours on both XV-15s total a little over 600 out of a permissible airframe life of 1,100 airframe hours.

A joint programme links the United Kingdom, NASA and the Department of Defense for investigation of advanced short take off and vertical landing aircraft. Using the successful British *Harrier* concept, designers have begun wind-tunnel tests of V/STOL aircraft which can fly at supersonic speed whilst retaining the renowned agility of the *Harrier*.

Several new NASA facilities promise to make significant contributions to these and other futuristic NASA research programmes. The NASA Numerical Aerodynamic Simulator Facility located at Ames was declared operational in 1987, relying on a scheme of building-block supercomputers capable of one billion calculations per second. For the first time, designers can routinely simulate the three-dimensional airflow patterns around an aircraft and its propulsion system. The facility gives greater accuracy and reliability in aircraft design, so reducing the high costs of extensive wind tunnel testing. At Langley, a new National Transonic Facility permits the NASA engineers to test models in a pressurized tunnel in which air is replaced by the flow of super-cooled nitrogen. As the nitrogen vaporises into gas it provides a medium more dense and viscous than air, offsetting scaling inaccuracies of smaller models — usually with wing spans of three to five feet — tested in the tunnel.

For many years, the world's largest wind tunnel was the 40 × 80 ft closed-circuit tunnel located at Ames. It is a low-speed tunnel — 230 mph — but its size permits tests of comparatively large scale models of aircraft. As Ames became more involved in tests of helicopters and new-generation V/STOL aircraft, the need for a full-size, low speed tunnel became more apparent. This has resulted in a new tunnel section, built at an angle to the existing 40 × 80 ft structure. Completed in 1987, the addition boasts truly monumental dimensions, with a test section 80 ft high and 120 ft wide — three times as large in cross-section as the parent tunnel. Overall, the new structure is 600 ft wide and 130 ft high. The original tunnel fans have been replaced with six units which have increased the available power by four times and raised the speed of the original tunnel from 230 to 345 mph. Although the tunnels can not be run simultaneously, the facility allows the NASA technicians to set up one test section whilst the other is in operation.

Flight research into fuel efficiency produced different trial versions of multi-bladed propfan systems. During February 1987, Lockheed at Marietta, Georgia, were contracted to flight test this Grumman Gulfstream II NASA 650 c/n 188 to gather data on fuel saving propulsion systems. (Lockheed RP 3451-1)

NASA has evolved into an agency carrying out myriad of activities, many representing a significant contribution to economic and commercial development. The 'commercialization' of space', a theme of President Ronald Reagan's space policy in the late 1980s, promised many more benefits stemming from renewed Shuttle missions and an operational space station named *Freedom*. In 1980 the NASA budget stood a $5 billion, rising to $10.7 billion for the 1989 fiscal year. Manned space flights account for over half of the budget, with space science accounting for another $1.9 million or about 18%, reflecting a consistent pattern over the years and averaging about 20 cents of each NASA dollar.

As 1990 passes us by — the monumental 75th Anniversary Year of its founding as the National Advisory Committee for Aeronautics — the National Aeronautics and Space Administration is a robust and very diverse agency, with approximately 22,000 employees and possibly ten times that number employed on NASA contracts. It still has its critics and its challenges, and is still breeding its aerodynamic explorers and continuing in a diversified environment of air and space it has helped to create.

Lockheed F-104 Starfighters used by the NASA

NASA No.	Model	c/n	USAF	Acquired	Remarks
NASA 811	F-104N	L683C-4045		August 1963	Ex NASA 011
NASA 812	F-104N	L683C-4053		September 1963	Ex NASA 012
NASA 813	F-104N	L683C-4058		September 1963	Ex NASA 013 destroyed 8 June 1966
NASA 818	YF-104A	L183-1007	65-2961	August 1956	To National Air & Space Museum October 1976
NASA 819	F-104B		57-1303	December 1959	
NASA 820	F-104A/G	1078	56-0790	1966	WFU 30 October 1983. For AFFTC Museum
NASA 824	TF-104G	5735	61-3065	1975	WGAF 27+37. In storage
NASA 825	TF-104G	5939	66-13628	1975	WGAF 28+09. Currently in use
NASA 826	F-104G	8213		Fokker built 1975	WGAF 24+64. Currently in use
	F-104A		56-1734		
	F-104A		56-0749		Modified for high-altitude rocket launch
	F-104A		56-745	17 July 1958 to 12 August 1958	40 × 80 ft wind tunnel tests
	F-104B	283-5015	57-1303	3 October 1958 to 16 December 1959	Variable stability
	JF-104A		56-745	19 March 1959 to 6 May 1960	

Notes: NASA 813 was lost in collision with the XB-70 Valkyrie killing pilot Joe Walker.
Last three aircraft were used by Ames, the rest used at Edwards.

APPENDIX ONE
The NACA-NASA Aircraft Marking and Numbering System

The huge NACA-NASA organisation is not only unique for the wide variety of aircraft and flight-test vehicles it has flown over the years, but to the student of aircraft livery and identity, the variety of markings introduced over the years is more than unique. Fortunately the archives record the NACA aircraft back to the beginning of flight testing in 1919, and as will be seen from the following Appendices, the system continues to be used today by NASA.

It was a letter dated 29 October 1928, addressed to the Secretary of Commerce from John F. Victory, Secretary of NACA, requested that an exclusive numbering system be approved for use on the NACA aircraft, commencing at NACA 1. The same letter requested approval of a design symbol for use on the tail surfaces of the aircraft. This was approved in a reply dated 2 November 1928 from the Office of the Director of Aeronautics — Department of Commerce and signed by Clarence M. Young. So was born the forerunner of the unique marking and numbering system.

With reference to the aircraft appearing in the following Appendices, some participated long and extensive research probes, others served as flying test beds for the study of specific innovations, some were given merely cursory flight evaluations and still others were only visitors at the laboratory in question. The early NACA Langley aircraft numbering system can be very confusing because it does not conform sequentially to the order of receipt. The type of arrangement by which the aircraft was assigned — purchase, temporary or permanent loan — dictated the markings. A good example is the Lockheed P-80A-5-LO *Shooting Star* (44-85352, acquired

The fin and rudder on this NACA aircraft reveals all. It is a Fairchild XR2K-1 BuNo.9998 NACA 82 all inscribed on the aircraft and the NACA emblem. (NACA)

on 5 November 1946) which was listed in early records as 'NACA 281', but still continued to carry the US Army Air Force 'buzz number' PN-352 until it was transferred to the ownership of NACA on 1 May 1950, when it became 'NACA 112'.

Over the years, in fact, more than one aircraft could have the same NACA-NASA number — not at the same time of course. The current NASA numbering system begins at NASA 1 — which is one of a small but efficient fleet of executive-type aircraft. The number carried by the aircraft has no relationship to the NASA facility from which it operates, and all of them can be observed at any aviation airport or facility one cares to mention in the USA. Most are the property of the Agency whilst occasionally the odd one is leased pending purchase, and all are used on a wide variety of administrative duties. The current NASA 1 (ex-N18LB) is a Gulfstream III delivered in April 1989. The old NASA 1 — a Gulfstream I, c/n 98, delivered in March 1963 — became NASA 2 in April 1990 and the old aircraft was sold as N29AY. The current executive and administrative NASA fleet is as follows:

N1NA	NASA 1	Gulfstream III	c/n 309	Langley
N2NA	NASA 2	Gulfstream I	96	Johnson
N3NA	NASA 3	Gulfstream I	92	Marshall
N4NA	NASA 4	Gulfstream I	151	Kennedy
N5NA	NASA 5	Gulfstream I	125	Lewis
N6NA	NASA 6	not currently used		
N7NA	NASA 7	Beech King Air 200	c/n BB997	Pasadena
N8NA	NASA 8	Beech King Air 200	BB950	Wallops
N9NA	NASA 9	Beech King Air 200	BB1091	Johnson

PROPOSED STANDARDIZED MARKINGS
FOR NACA AIRPLANES

October 29, 1928

The Honorable,
    The Secretary of Commerce,
        Washington, D. C.

Sir:

    Pursuant to the provisions of existing law I have the honor to submit herewith for your approval the design of a symbol for use on the tail surfaces of this Committee's airplanes.

    It is respectfully requested that authority be granted for the exclusive use of this symbol by the Committee.

    It is intended that our airplanes will, in addition, be numbered serially as follows: NACA 1, NACA 2, etc., similar to the manner in which Army airplanes are numbered.

Respectfully,

NATIONAL ADVISORY COMMITTEE,

FOR AERONAUTICS,

J. F. Victory
Secretary

THE NACA-NASA AIRCRAFT MARKING AND NUMBERING SYSTEM

Above: *The proposed symbol markings for NACA airplanes is seen applied to this Fairchild XR2K-1 BuNo.9998 NACA 82, delivered to Langley on 16 September 1935. The shield was red and white, the NACA 82 in silver on a blue background, with the wing and tail surfaces finished with aluminium pigmented dope with the exception of a yellow centre panel extending from wingtip to wingtip. The fuselage was blue, as were the struts and wheel covers.* (NACA 22160)

Opposite: *The NACA symbol seen here applied to a Douglas C-47 Skytrain transport, the winged emblem being ochre or yellow with black lettering. The background panel was apparently also yellow.* (NACA A 16544)

There was a period when two-digit numbers were used by the NACA and NASA but today the system appears to be more stable and standardized, using a three-digit system for research and test aircraft. The first digit indicates which of the NASA centres is the aircraft's home base. Langley, for instance, today operates aircraft commencing with '5' such as 'NASA 515' — a Boeing 737-130 (c/n 19437, ex N73700) which is a heavily modified research aircraft.

A wide variety of military aircraft have been used over the years by the NACA and NASA on temporary loan for specific test programmes requested by the US armed forces. These never appear in the Agency's numbering system but often had the NASA logo on the fin together with the last three digits of the military service serial. Tri-service and manufacturers' test aircraft were often well decorated with a variety of inscriptions and logos, as is well illustrated by the photos in this volume. Early post-war markings, especially on aircraft used by Ames, included the word TEST in large letters on the aircraft's fuselage and wings, an airborne warning to any inquisitive aircraft to keep well clear. Other aircraft which creep into the NASA inventory but which are rarely seen are those delivered from the US armed forces for use as spares. These include helicopters as well as fixed-wing aircraft. A good example is the North American T-39 (61-670) of the US Air Force with fin serial 10670, delivered to Wallops early in 1990.

From 1969 onwards, the majority of the NASA aircraft fleet have appeared on the US civil register and carry a civil-type registration on the fuselage, normally with an NA suffix. A good example is the newly-acquired Boeing 747-100 N911NA (c/n 20781), sold to NASA on 27 October 1988, which is the ex-Japan Air Lines JA 8117. It appears that using a civil registration allows NASA certain Federal concessions and is also essential in obtaining diplomatic clearances when the aircraft is operated overseas on international research missions. The highest-sequenced NASA number recorded is 'NASA 963', a Northrop T-38 *Talon* acquired by the Johnson Space Centre during May 1989. The highest unofficial NASA number recorded to date is 'NASA 991', a Grumman F-14A *Tomcat* (BuNo.157991) in use by the Dryden Flight Research Centre at Edwards AFB using the last three digits of the US Navy serial. There are oddball aircraft operating within the huge NASA organisation, an example being 'NASA 933' (N933NA), a Learjet 23 operated by the John C. Stennis Space Centre located at Bay St Louis, Mississippi.

The NACA-NASA numbering system has undergone a few changes over the years. The '100' series was once used by both Lewis and Ames, and the '200' series was once in use by Ames. The '300' series was used by the Marshall Space Flight Centre, and the '400' series by Patrick AFB-based aircraft and for use by aircraft using the Kennedy Space Centre. Today, three helicopters are based at Kennedy for programme support but they appear numerically in the Wallops Flight Facility system. These are 'NASA 416' (N416NA), a Bell UH-1B (ex-62-2064); 'NASA 417' (N417NA), a Bell UH-1M and 'NASA 418' (N418NA), another Bell UH-1M.

The following system is in use today:

'400' Wallops Flight Facility, Wallops Island, Virginia 23337.
'500' Langley Research Centre, Hampton, Virginia 23665.
'600' Lewis Research Centre, Cleveland, Ohio 44135.
'700' Ames Research Centre, Moffett Field, California 94035.
'800' Dryden Flight Research Facility, Edwards AFB, California.
'900' Johnson Space Centre, Houston, Texas 77058.

The NACA yellow-backed winged symbol was replaced in 1958 by the well-known NASA 'meatball' which in turn was replaced in the 1970s by what has been described in some circles as that 'gawdawful' red inch worm officially described as a 'graphics improvement' and no doubt designed by a Madison Avenue consultant at vast expense.

DEPARTMENT OF COMMERCE
OFFICE OF THE
DIRECTOR OF AERONAUTICS
WASHINGTON

November 2, 1928

National Advisory Committee for Aeronautics,
Mr. J. F. Victory, Secretary,
3841 Navy Building,
17th & B Streets, N. W.,
Washington, D. C.

Gentlemen:

In acknowledgement of your communication of October 29, regarding the markings for the airplanes of the National Advisory Committee, this method of numbering meets with no objection on the part of the Department of Commerce. It should be the responsibility of the Committee to maintain its own records, and in the event it becomes necessary to identify a given plane, the Department shall merely call on your office to supply the necessary information.

Regarding the proposed insignia, it is not believed we are in a position to exercise any authority as to the exclusive use of any symbol. In this matter, your situation is not unlike that of the Army Air Corps or other governmental agencies. However, as far as the Department of Commerce is concerned, there again is no objection to the use of the insignia.

It is believed, however, that the displaying of your identification markings — NACA followed by a numeral — should be accomplished in the manner required by the Air Commerce Regulations, particularly when commercial type aircraft is used. This manner of marking civilian aircraft has become quite well established and to do otherwise would undoubtedly give rise to requests for exceptions on the part of a number of civilian operators.

Yours very truly,

Clarence M. Young,
Director of Aeronautics.

The Doyle O-2 manufacturer's emblem appears on the forward fuselage, with a blank NACA emblem on the rudder. The aircraft was NACA 34 which appeared on the top of the wing. It arrived at Langley during 1939. (NACA)

The NACA yellow winged emblem is clearly defined in this photo depicting a Douglas R4D-6 Skytrain BuNo.50831 which served with the NACA at Langley between November 1948 and July 1952. (NASA 61386)

The Vought F8C Crusader BuNo.145546 NASA 802 was converted as the digital fly-by-wire — DFBW — aircraft, making its first flight on 25 May 1972. It is seen in the hangar at Edwards AFB having the full title of the NASA centre inscribed on the tail. (Military Aircraft Photographs)

Douglas C-118A Liftmaster NASA 428 ex-NASA 28, c/n 43575 ex-51-3828 has only tail identify and NASA 'meatball' aft of the cockpit. It was acquired at Goddards during September 1964 and transferred to Wallops on 28 July 1975 going into storage at Davis-Monthan on 12 September 1976. (Military Aircraft Photographs)

Records indicate this photo is dated 30 July 1971 and depicts Sikorsky SH-3A Sea King still in US Navy livery, but with NASA 933 added. It is BuNo.149723, delivered to Johnson on 26 August 1965 serving until 20 July 20 1970 going to Langley as NASA 538, and on 1 November 1977, going to Ames as NASA 735 being current. Used on helicopter noise measurement tests. (NASA HQ RA72-15052-3)

Minus any national markings, but retaining the BuNo.152399, with NASA 'meatball' added, the Sikorsky CH-53A Sea Stallion is seen at Langley on 14 February 1977. It became NASA 543 and on 5 May 1978 went to the Federal Aviation Administration, Atlantic City, New Jersey. (NASA HQ RA77-1794(3))

This Douglas DC-4 N88937 c/n 18337 retained its civil registration with 'NATIONAL AERONAUTICS AND SPACE ADMINISTRATION' on the fuselage, small NASA 'meatball' logo behind the cockpit. It was used at Goddard for programme support from February 1960 to February 1965, being replaced by DC-4 N67566 c/n 10344. (Peter M. Bowers)

Amongst the many logos on this Canberra PR.9 WH793 from RAE Bedford, includes one from 1967 when it was based at Ames and involved in HICAT — a research programme into clear air turbulence at all altitudes. Close cooperation exists today between NASA and the Royal Aerospace Establishment in the United Kingdom. (British Aerospace AW 33652 via Bob Jackson.)

APPENDIX TWO

Langley Memorial Aeronautical Laboratory Aircraft

Order of receipt	Type and designation	Serial	Date Arrived	Date Departed	NACA No.	Name and remarks
1	Curtiss JN-4H	N 6249	1919	8 Nov 1923		*Jenny.*
2	Curtiss JN-4H	AS 44946	1919	Nov 1923		*Jenny.*
3	Curtiss JN-4H	AS 38131	1919	10 Apr 1923		*Jenny.*
4	Vought VE-7	A 5669	Apr 1921	29 Jul 1929	4	
5	British Se.5A	AS 58049	5 Sep 1922	10 Sep 1926		
6	de Havilland DH.4	AS 22830	7 Sep 1922	28 Jul 1927		
7	Fokker D.VII	6328		1923	7	Fatal mid-air collision with MB-2.
8	Martin MO-1	AS 63335	7 Sep 1922	1 Nov 1923		
9	de Havilland DH.9A	AS 31839	27 Sep 1922	6 Dec 1927		
10	SPAD VII	AS 7142	5 Sep 1922	24 Feb 1925		
	Nieuport 23		1922			
	Thomas-Morse MB-3		Jan 1923	1924		
11	Douglas DT-2	A 6125	11 May 1923	15 Jun 1925		
12	Curtiss JN-6H	AS 44946	4 Sep 1923	20 Aug 1924		
13	Consolidated CBS					Weight and balance sheet dated 11 April 1929.
14	Vought VE-7	A5950			14	Ice formation research. 22 June 1928.
15	Curtiss TS-1	A 6249	9 Nov 1923	4 Dec 1928		Twin-float.
16	Sperry M-1	AS 68473	19 Jan 1924	1927		*Messenger.*
17	Supermarine Vickers	A 6073	1923	1924		*Viking IV.*
18	Curtiss F4C-1	6689				
19	Curtiss JNS-1	AS 24232	1 Nov 1924	3 Jun 1926		
	Martin MB-2		1924			Fatal mid-air collision with NACA-7.
20	Sperry M-1	AS 64226	9 Jan 1925			*Messenger.*
21	Vought UO-1					
22	Boeing PW-9		5 Jan 1927	15 Oct 1928		
23	Curtiss F4C-1	6690				
24	Douglas M-3		1927	1931		*Mailplane.*
25	de Havilland DH.4B	AS 31839	7 Dec 1927	1 Jan 1930	25	
26	Fairchild FC-2W2		1928		1	First aircraft purchased by NACA.
27	Curtiss XF7C-1	A 7653				*Seahawk.*
28	Atlantic Fokker C-2A	AC 28-123				
29	Douglas O-2H	AC 29-168	28 Jan 1929	1935		
30	Curtiss H-16		Feb 1929	Jul 1929		Flying-boat.
31	Stearman C3B	NS 7550				
32	Curtiss P-1A	AC 25-411		Jul 1929		*Hawk.*
33	Consolidated NY-1					
34	Doyle O-2				34	*Oriole.*
35	Vought O2U-1					*Corsair.*
36	Consolidated/Fleet XN2Y-1	A 8019	1929			
37	Boeing F3B-1					
38	Boeing F2B-1					
39	Boeing XF4B-1	A 8128				
40	Consolidated NY-2					
41	Verville Sportsman					Single-engined biplane. Radial engine.
42	McDonnell Doodlebug					*Doodlebug.* Low-wing monoplane.
43	Vought O3U-1					*Corsair.* 1st aircraft tested in Full Scale tunnel.
44	Pitcairn PCA-2 Autogiro		21 Jul 1931	13 Sep 1933	44	
45	Curtiss O2C-1					*Helldiver.*
46	Loening XSL-1	BuNo.8696				
47	Fairchild C-7A				47	
48	Martin XBM-1	BuNo.9212				
49	Douglas YO-31A					
50	Fleet N2Y-1					
51	Detroit/Lockheed XRO-1	BuNo.9054				*Altair* c/n 179.
52	Curtiss F6C-4					*Hawk.*
53	Curtiss O2C-1	BuNo.8455	23 Aug 1932	23 Feb 1933		*Helldiver.*
54	Boeing F4B-2	BuNo.8628	24 Jul 1932	20 Jun 1934		
55	Boeing P-12					
56	Curtiss P-6E					*Hawk.*
57	Fleet N2Y-1		13 Sep 1933			
58	Vought O2U-1					*Corsair.*

Order of receipt	Type and designation	Serial	Date Arrived	Date Departed	NACA No.	Name and remarks
59	Pitcairn PAA-1 Autogiro					
60	Fairchild 22		15 May 1933		60	
61	Fairchild C-7A		10 June 1936		47	Modified into low-wing — designated J-2.
62	Boeing F3B-1					
63	Consolidated NY-2					
64	Vought					
65	Vought XO4U-2	BuNo.8641	10 Apr 1933	22 May 1933		
66	Boeing F4B-1					
67	Boeing P-12C					
68	Curtiss P-6E					*Hawk.*
69	Vought XF3U-1	BuNo.9222				
70	Weick W-1A	NS-67	12 Feb 1934	27 Apr 1935		Pobjoy engine.
71	Boeing P-26A	AC 33-56	26 Jun 1934			
72	Grumman JF-1		1 Jun 1934	2 Aug 1934		*Duck.*
73	Northrop XFT-1	BuNo.9400	8 Jun 1934	9 Jul 1934		
74	Martin T4M-1					
75	Vought SU-2					
76	Boeing F4B-4	BuNo.8912				
77	Boeing XFB-1	BuNo.8975	4 Aug 1934			
78	Curtiss N2C-2		22 Nov 1934			
79	Vought O2U-2 Consolidated XB2Y-1	BuNo.9221				NACA photo dated 9 August 1934.
80	Boeing YP-29A	AC 34-24	2 Mar 1935	5 Jul 1940		
81	Kellett KD-1		2 Apr 1935	19 Apr 1935		
82	Bellanca Pacemaker		25 Apr 1935	29 Apr 1935		
83	Fairchild XR2K-1	BuNo.9998	16 Sep 1935	1946	82	
84	Curtiss BF2C-1	BuNo.9586	28 Oct 1935	11 Dec 1935		
85	Great Lakes XTBG-1	BuNo.9723	13 Nov 1935	26 Nov 1935		
86	Grumman XSF-2	BuNo.9493	2 Dec 1935	10 Jan 1936		*Scout.*
87	Taylor E-2	NC13Y	5 Dec 1935			*Cub* c/n 121.
88	Taylor E-2	NC13117	5 Dec 1935			*Cub* c/n 47.
89	Kellett YG-2 Autogiro	AC 35-279	17 Dec 1935	30 Mar 1936		X14776. Crashed 30 March 1936.
90	Kellett YG-1 Autogiro	AC 35-278	3 Jan 1936	26 May 1936		
91	Curtiss XBFC-1	BuNo.9219	13 Jan 1936			c/n 13644.
92	Aeronca E113A	NR13089	30 Jan 1936			c/n 801.
93	Franklin PS-2 glider	BuNo.9615	27 Apr 1936			
94	Franklin PS-2 glider	BuNo.9614	27 Apr 1936			
95	Stinson SR-8E		31 Jul 1936	Surveyed 1951	94	*Reliant.*
96	Aeronca C-2N				95	
97	Ryan ST		6 Aug 1938	Surveyed 1947	96	
98	Curtiss XSBC-3	BuNo.9225				
99	Cunningham-Hall	X14324				NACA photo dated 23 October 1936.
100	Grumman XSBF-1	BuNo.9998				
101	Pitcairn XOP-2					
102	Vought O2U-4	BuNo.8104				NACA photo dated 16 March 1937.
	Consolidated PB2A					NACA photo dated 7 May 1937.
103	Curtiss SOC-1					
104	Boeing P-26A					
105	Curtiss XF13C-3	BuNo.9343				
106	North American BT-9A					
107	Martin B-10B					
108	Vought SB2U-1					
109	Douglas DC-3	NC16070	Sep 1937			c/n 1910. United Air Lines. Stall and icing tests.
110	Douglas XB-7	AC 30-228				ex XO-36.
111	Seversky P-35	X1254/1390				
112	Grumman F3F-2					
113	Fairchild F-46					
114	Douglas B-18					*Bolo.*
115	Kellett YG-1B Autogiro					
	Brewster XF2A-1					
116	Stearman-Hammond Y	NS73				
117	Wilford XOZ-1 Autogiro		1939	Surveyed 1941		US Navy. 1939 and 1940 flight measurement tests.
118	Grumman XF4F-2	BuNo.0383				*Wildcat.*
119	Northrop A-17A	AC 85				
120	Grumman F3F-2	BuNo.0967				
121	Fairchild XC-31	AC 34-26				
122	Vought SB2U-1	BuNo.0726				
123	Douglas OA-4A	AC 32-404	30 Jun 1938			*Dolphin.*
124	Curtiss AT-5	AC 28-66				NACA cowl investigation.
125	North American BT-9B	AC 37-227	3 Aug 1938	10 Feb 1941		
126	North American BC-1		30 Aug 1938	10 Sep 1938		
127	Vought XSB3U-1	BuNo.9834	31 Aug 1938	3 Jan 1939		

Order of receipt	Type and designation	Serial	Date Arrived	Date Departed	NACA No.	Name and remarks
128	Vought SB2U-2	BuNo.1326	28 Nov 1938	24 Mar 1939		
129	Vought SB2U-1	BuNo.0770	8 Nov 1938	28 Dec 1938		
	Sikorsky XPBS-1	BuNo.9995	1938			
130	Douglas XBT-2					
131	Lockheed 12 c/n 1268	NC17396	20 Jan 1939	4 Oct 1940	97	Transferred to Ames, California.
132	Northrop A-17A	AC 36-184	13 Feb 1939	11 Jun 1939		
133	Brewster XSBA-1	BuNo.9726	16 Feb 1939			
134	Lockheed XR-40 Model 14	BuNo.1441	20 Feb 1939			c/n 1482.
135	Curtiss P-36A		22 Mar 1939	3 Nov 1939		*Hawk.*
136	Curtiss XP-40	AC 38-10	29 Mar 1939	29 Mar 1944		*Warhawk.*
137	Seversky P-35		7 Apr 1939			
138	Seversky XP-41	AC 36-430	10 Mar 1939	3 Apr 1942		
139	Fairchild	NX18689	15 May 1939	3 Apr 1942		
140	Bell XP-39	AC 38-326	6 Jun 1939			*Airacobra.*
141	Grumman XF4F-3	BuNo.0383		10 Aug 1939		*Wildcat.*
142	Aeronca 65-C		23 Aug 1939	20 Sep 1939		
143	Douglas B-18		1 Sep 1939	18 Nov 1941		*Bolo.*
144	Kellett YG-1B Autogiro	AC 37-635	11 Sep 1939	3 Jan 1940		
145	Piper Cub	NC26899	19 Mar 1940	21 Oct 1947	98	Surveyed.
146	Piper Cub 2		23 Sep 1939	9 Oct 1939		
147	Taylorcraft		9 Oct 1939			
148	Douglas R2D		5 Dec 1939	Surveyed 1941		*DC-2.*
149	Stinson 105		26 Jan 1940	24 Mar 1940		Flying quality tests.
150	Beechcraft	NC20780	6 Jun 1940	12 Jun 1940		c/n 339.
151	Beechcraft	NC19494	17 Jun 1940	20 Jun 1940		
152	Curtiss P-40		18 Jun 1940	15 Jul 1940		Ground loop tests.
153	Bellancair	NC15690	19 Jun 1940	1 Jul 1940		*Cruisair.*
154	Bellanca	NC25303	1 Jul 1940	27 Jul 1940		*Cruisair.*
155	Douglas C-39		11 Jul 1940	22 Jul 1940		
156	Curtiss P-40		18 Jun 1940	15 Jul 1940		*Warhawk.*
157	Curtiss XP-42	AC 38-4	4 Sep 1940	9 Nov 1942		
158	Curtiss XSO3C-1	BuNo.1385	18 Sep 1940	16 Nov 1940		*Seamew.*
159	St Louis PT-LM-4	NX25500	23 Sep 1940	3 Oct 1940		
160	Lockheed XC-35 c/n 1060	AC 36-353	2 Oct 1940	5 Feb 1943		First pressurised aircraft.
161	Brewster XF2A-2	BuNo.0451	25 Oct 1940	9 Jun 1941		*Buffalo.*
162	Grumman F4F-3	BuNo.1845	8 Nov 1940	28 May 1942		*Wildcat.*
163	Curtiss P-36C		10 Dec 1940	5 Jun 1941		*Hawk.*
164	Seversky XP-41	AC 36-430	11 Dec 1940	3 Apr 1942		
165	Douglas A-20A		9 Jan 1941	4 Aug 1941		*Havoc.*
166	Brewster XSBA-1	BuNo.9726	13 Jan 1941	21 Sep 1945		
167	Republic YP-43		4 Feb 1941	7 Feb 1941		*Lancer.*
168	Bell YP-39		6 Feb 1941	27 Jul 1944		*Airacobra.*
169	Curtiss YP-37	AC 38-474	10 Feb 1941	12 Jan 1942		
170	Grumman F4F-3	BuNo.2538	15 Apr 1941	28 May 1942		*Wildcat.*
171	Lockheed 12A		30 May 1941	13 Mar 1960	99	c/n 1292.
172	Vought XF4U-1	BuNo.1443	14 Jun 1941	1 Mar 1943		*Corsair.*
173	Fleetwing		20 Oct 1941	12 Dec 1941		
174	Martin B-26		4 Nov 1941	18 Dec 1941		*Marauder.*
175	Hawker Hurricane		24 Nov 1941	28 Dec 1941		25017.
176	Supermarine Spitfire	R7347	24 Nov 1941	13 Jan 1943		Mk.V. 35497
177	Lockheed YP-38	AC 39-690	27 Nov 1941	4 Feb 1942		*Lightning.*
178	Supermarine Spitfire	W3119	24 Dec 1941	13 Feb 1942		Mk.V 37147.
179	North American XP-51	AC 41-38	27 Dec 1941	14 Dec 1942		*Mustang.*
180	Grumman F4F-3	BuNo.3990	23 Feb 1942	28 May 1943		*Wildcat.*
181	Curtiss P-40E	AC 41-5534	13 Mar 1942	20 Jul 1942		*Warhawk.*
182	Republic P-47B	AC 41-5897	18 Mar 1942	20 Oct 1942		*Thunderbolt.*
183	Curtiss SNC-1	BuNo.6295	22 Apr 1942	1 May 1942		*Falcon.*
184	Grumman XTBF-1	BuNo.2540	18 May 1942	11 Jun 1943		*Avenger.*
185	Curtiss P-40K-1-CU		6 Jul 1942	16 Oct 1944		*Warhawk.*
186	Republic P-47B	AC 41-5901	13 Jul 1942	20 Oct 1942		
187	Curtiss P-40E	AC 42-45801	23 Jul 1942	1 Sep 1942		
188	Brewster F2A-2	BuNo.1426	3 Jul 1942	17 Jun 1944		
189	Curtiss P-40F	AC 41-13600	23 Jul 1942	1 Sep 1942		
190	Fairchild 24		22 Sep 1942	29 Oct 1942	100	Transferred to Ames.
191	Republic P-47C		7 Oct 1942	1 Dec 1942		
192	Republic P-47C	AC 41-6102	7 Oct 1942	1 Dec 1942		
193	Vought F4U-1		19 Oct 1942	1 Mar 1943		
194	Republic P-47C-1-RE	AC 41-6130	24 Oct 1942	25 Aug 1944		
195	Curtiss SB2C-1	BuNo.00014	5 Dec 1942	27 May 1943		
196	North American SNJ-3C	BuNo.01847	11 Dec 1942	15 Jul 1945		
197	Bell P-39D-1-BE	AC 41-28378	5 Jan 1943	23 Apr 1943		
198	Grumman F6F-3	BuNo.04776	3 Feb 1943	24 Jun 1946		Surveyed.
199	Curtiss SB2C-1	BuNo.00056	10 Feb 1943	22 Jul 1943		
200	Republic XP-47F	AC 41-5938	22 Feb 1943	15 Oct 1943		
201	Japanese Zero		5 Mar 1943	11 Mar 1943		4593.

Order of receipt	Type and designation	Serial	Date Arrived	Date Departed	NACA No.	Name and remarks
202	North American XP-51	AC 41-39	9 Mar 1943	15 Jan 1944		ALD Tests. Surveyed.
203	Curtiss SB2C-1	BuNo.00140	19 May 1943	28 Nov 1943		
204	Martin RA-30	AC 41-27687	10 Jun 1943	10 Mar 1944		
205	Consolidated B-24D	AC 42-40223	26 Jun 1943	3 Apr 1944		*Liberator.*
206	Vought F4U-1	BuNo.02161	22 Jul 1943	27 Sep 1943		
207	Bell P-63	AC 42-68861	26 Jul 1943	25 Nov 1943		
208	Curtiss P-40F	AC 41-14119	31 Jul 1943	9 Feb 1945		
209	General Motors TBM-1	BuNo.24820	10 Aug 1943	20 Nov 1943		
210	North American P-51B	AC 43-12105	16 Aug 1943	4 Jan 1951		Surveyed. *Mustang.*
211	North American P-51B-1-NA		11 Sep 1943	17 Oct 1943	104	*Mustang.*
212	Republic P-47D-3-RE	AC 42-8207	14 Oct 1943	13 Mar 1944		*Thunderbolt.*
213	North American XP-51	AC 41-38	11 Jan 1944	25 Jul 1945		N51NA c/n 73-3101. EAA Museum, Oshkosh.
214	Sikorsky YR-4B		23 Jan 1944	20 Oct 1948		*Hoverfly.* First military helicopter.
215	Bell P-63A-1-BE	AC 42-68889	4 Feb 1944	27 Aug 1945		*Kingcobra.*
216	Douglas A-26B-2-DL	AC 41-39120	13 Mar 1944	16 Jun 1944		*Invader.*
217	Grumman XF6F-4	BuNo.02981	1 Apr 1944	1 Mar 1945		
218	Sikorsky HNS-1	BuNo.39034	17 Apr 1944			US Navy version of R-4B *Hoverfly.*
219	Curtiss SB2C-1C	BuNo.18294	17 Apr 1944	21 Jun 1945		
220	Bell P-63A-1-BE	AC-42-68881	10 May 1944	1946		
221	Cessna UC-78-18-1-CE	AC 43-31957	19 Jun 1944	10 Oct 1945		*Bobcat.* Assigned to USAAF Liaison Office.
222	Republic P-47D-15-RE		1 Jul 1944	8 Jul 1945		
223	Republic P-47D-28-RE	AC 42-28541	17 Jul 1944	23 Apr 1948		
224	Grumman TBM-3	BuNo.22857	11 Aug 1944	17 Aug 1944		
225	Brewster F3A-1	BuNo.11213	19 Aug 1944	12 Sep 1944		*Corsair.*
226	Supermarine Spitfire	EN474	19 Aug 1944	17 Nov 1944		Mk.IX.
227	de Havilland F-8	AC 43-34928	Aug 1944	19 Jan 1945		*Mosquito.*
228	de Havilland F-8	AC 43-34960	Aug 1944	19 Jan 1945		*Mosquito.*
229	Douglas A-26B-10-DL	AC 43-22280	7 Sep 1944	16 Oct 1945		*Invader.*
230	Douglas SBD-5	BuNo.28373	22 Sep 1944	8 Oct 1945	101	
231	Curtiss SC-1	BuNo.35324	21 Oct 1944	20 Feb 1945		
232	Grumman XF8F-1	BuNo.90460	13 Dec 1944	5 Feb 1945		*Bearcat.*
233	North American P-51D-5-NA	AF 44-13257	22 Dec 1944	12 May 1957	108	Salvaged.
234	Curtiss SB2C-3	BuNo.19332	22 Jan 1945	19 Jun 1945		
235	General Motors FM-2	BuNo.74507	17 Jan 1945	18 Apr 1945		*Wildcat.*
236	Culver PQ-14B	AF 44-21896	18 Jan 1945	11 Apr 1949		
237	North American P-51D-5-NA	AF 44-14017	18 Jan 1945	5 Jun 1952	102	
238	Republic P-47N-1-RE	AF 44-87790	23 Jan 1945	9 Jul 1945		Crashed.
239	Republic P-47D-30-RE	AF 44-33441	25 Jan 1945	2 Oct 1948		*Thunderbolt.*
240	Vought F4U-1D	BuNo.82716	9 Feb 1945	20 Mar 1945		*Corsair.*
241	Beechcraft UC-45F	AF 44-47264	9 Mar 1945	1951		*Expeditor.* Assigned to USAAF Liaison Office.
242	Sikorsky JRS-1	BuNo.1063	26 Apr 1945	20 Nov 1946		Twin engined flying-boat.
243	Boeing B-29B	AF 44-83927	9 May 1945	10 Sep 1946		*Superfortress.*
244	Grumman XF8F-1	BuNo.90461	10 May 1945	Apr 1946		
245	North American P-51B	AF 43-12114	11 Jun 1945	24 Jul 1945		Crashed.
246	North American P-51B	AF 43-12491	11 Jun 1945	30 Oct 1945		
247	North American P-51H	AF 44-64164	20 Jun 1945	20 Sep 1946		Acquired from Ames.
248	Grumman JRF-5	BuNo.34094	10 Jul 1945	1 Apr 1946	103	*Goose.* Used on shuttle to Wallops Island.
249	North American SNJ-3	BuNo.05475	19 Jul 1945	1 Oct 1948		*Texan.*
250	Boeing TB-29	AF 44-69700	31 Jul 1945	6 Nov 1950		*Superfortress.*
251	Vought F4U-1D	BuNo.50378	2 Aug 1945	14 Aug 1945		
252	Consolidated PBY-5A	BuNo.2473	4 Aug 1945	19 Sep 1945		*Catalina.*
253	Republic P-47N-25-RE	AF 44-89303	9 Aug 1945	27 Dec 1950		
254	North American TP-51D	AF 44-63826	14 Aug 1945	24 Apr 1946		*Mustang.*
255	Douglas C-47-DL	AF 41-18392	23 Aug 1945	20 Oct 1945		*Skytrain* c/n 4430.
256	North American P-51D-25-NT	AF 44-84864	27 Aug 1945	12 Jul 1957	126	*Mustang.* Salvaged.
257	Douglas BTD-1	BuNo.09060	27 Aug 1945	28 Nov 1946		*Destroyer.* Laboratory tests.
258	Curtiss SC-1	BuNo.93334	28 Aug 1945	18 Apr 1946		
259	North American P-51D-25-NT	AF 44-84900	4 Sep 1945	5 Jun 1952	127	
260	North American P-51D-25-NT	AF 44-84944	20 Sep 1945	5 Jun 1952	128	
261	North American P-51D-25-NT	AF 44-84953	21 Sep 1945	5 Jun 1952	129	
262	North American P-51D-25-NT	AF 44-84958	28 Sep 1945	25 Aug 1950	148	Transferred to Muroc High Speed Flight Test Unit.
263	Douglas C-47A-25-DK	AF 42-93791	4 Oct 1945	Apr 1946		*Skytrain* c/n 13742.
264	Douglas BTD-1	BuNo.09058	8 Oct 1945	28 Nov 1945		
265	Curtiss SB2C-4E	BuNo.82877	5 Nov 1945	25 Oct 1946		
266	Boeing B-29-96-BW	AF 45-21801	16 Nov 1945	15 Dec 1955	124	Transferred to Aberdeen Proving Ground.
267	Grumman F8F-1	BuNo.90448	4 Jan 1946	18 Feb 1946		*Bearcat.*
268	Beechcraft UC-45	AF 44-47110	11 Jan 1946	May 1947	105	Transferred to Muroc High Speed Flight Test Unit.
269	Douglas C-47B-51-DK	AF 43-49526	18 Jan 1946	28 Oct 1971		Transferred to Lewis Research Centre.
270	Northrop P-61C	AF 43-8327	5 Feb 1946	19 Feb 1946		*Black Widow.*
271	Grumman JRF-5	BuNo.34088	1 Apr 1946	11 Jun 1948	103	*Goose.*
272	Grumman F8F-1	BuNo.94812	21 May 1946	17 Apr 1947		*Bearcat.*

Order of receipt	Type and designation	Serial	Date Arrived	Date Departed	NACA No.	Name and remarks
273	Fairchild PT-19A	AF 42-83595	22 Feb 1946	1950		*Cornell.* Surveyed.
274	Grumman F8F-1	BuNo.94873	6 Jun 1946	1951		
275	North American AT-6	AF 44-81682	8 Jul 1946	8 Jan 1959	117	*Texan.* Surveyed.
276	Douglas R4D-6	BuNo.50795	27 Aug 1946	4 Oct 1946		Transferred to Lewis Research Unit.
277	Douglas R4D-6	BuNo.50826	27 Aug 1946	9 Dec 1948		c/n 26924
278	Douglas R4D-6	BuNo.50812	27 Aug 1946	7 Oct 1946		c/n 26708. Transferred to Ames.
279	Bell L-39-1	BuNo.90060	22 Aug 1946	12 Dec 1949		Swept-wing mod. Transferred to Lewis.
280	Bell L-39-2	BuNo.90061	11 Dec 1946	12 Dec 1949		Swept-wing mod. Transferred to Lewis.
281	Douglas C-54D	AF 42-72713	28 Sep 1946	4 Aug 1947		*Skymaster.*
282	Lockheed F-80A	AF 44-85352	5 Nov 1946	8 Jan 1959		*Shooting Star.* Salvaged NACA-112.
283	Vultee L-5E-1-VW	AF 44-17984	28 Mar 1947	8 Oct 1947		*Sentinel.*
284	Vultee L-5E-1-VW	AF 44-17939	28 Mar 1947	14 Nov 1947		*Sentinel.*
285	Grumman J4F-2	BuNo.32972	28 Aug 1947	27 Jun 1951		*Widgeon.*
286	Sikorsky HO3S-1	BuNo.122520	24 Feb 1948	6 Sep 1961	201	*Dragonfly* helicopter.
287	Grumman J4F-2	BuNo.32976	20 Mar 1948	1951		*Widgeon.*
288	North American XP-82	AF 44-83886	6 Jun 1948	5 Oct 1955	114	*Twin Mustang.* Salvaged.
289	Grumman JRF-5	BuNo.37778	11 Jun 1948	7 Jul 1958	103	*Goose.* Salvaged.
290	Cessna 190	N3477V	6 Aug 1948	25 Jun 1958	104	c/n 7177. Transferred to C.A.A.
291	North American B-45	AF 47-21	1 Oct 1948	15 Aug 1952	121	*Tornado.* Fatal crash Herbert H. Hoover.
292	North American SNJ-5	BuNo.84839	12 Oct 1948	3 Jun 1957		Transferred to VR-31 Sq. Norfolk NAS.
293	Douglas R4D-6	BuNo.50831	25 Nov 1948	6 Jul 1952		c/n 26978
294	Bell H-13B	AF 48-839	26 May 1949	18 Apr 1950		*Sioux.*
295	Lockheed TO-1	BuNo.33870	27 May 1949	28 Oct 1949		*Shooting Star.* AF 48-381.
296	Beechcraft C-45F	AF 43-35906	22 Aug 1949	18 Mar 1959	125	*Expeditor.* Gust alleviation research.
297	Republic YF-84	AF 45-59490	22 Aug 1949	18 Nov 1949	134	Transferred to Muroc High-speed Flight.
298	Vought F4U-4B	BuNo.97392	21 Mar 1950	1951		*Corsair.*
299	Vultee L-5	AF 45-34927	2 May 1950	29 Nov 1950		*Sentinel.*
300	Beechcraft 35	N5094C	24 Aug 1950	28 Nov 1950		*Bonanza.*
301	McDonnell F2H-	BuNo.122540	10 Oct 1950	20 Jun 1951		*Banshee.*
302	Grumman F9F-2	BuNo.122560	5 Jan 1951	7 Jul 1960	215	*Panther.*
303	Lockheed TO-2/TV-1	BuNo.124933	15 Jan 1951	15 Feb 1960		*Shooting Star.* Transferred NAS Norfolk.
304	Piasecki HRP-1	BuNo.111813	13 Feb 1951	1951		*Rescuer.*
305	North American F-86A-1-NA	AF 47-620	23 Feb 1951	11 Jul 1958	136	*Sabre.* Salvaged.
306	McDonnell F2H-1	BuNo.122530	8 Aug 1951	22 Sep 1959	214	*Banshee.*
307	Boeing B-47A	AF 49-1900	11 Jul 1952	17 Mar 1953		*Stratojet.* Transferred to Ames.
308	Sikorsky HRS-1	BuNo.127783	11 Mar 1953	Aug 1964		Towed to woods — disposal?
309	McDonnell XF-88B	AF 46-525	13 Jul 1953	16 Sep 1958		Salvaged.
310	Hiller HTE-1	BuNo.128646	16 Jul 1953	7 Feb 1956		Transferred to NAS Norfolk.
311	Bell H-13G	AF 52-7834	17 Sep 1953	25 Jul 1967		Transferred to Fort Walter, Texas.
312	Lockheed F-80B	AF 45-8683A	10 Dec 1953	31 Oct 1958	152	Salvaged.
313	Grumman F9F-7	BuNo.130864	20 Jan 1954	25 Nov 1959		*Panther.* Transferred to NAS Norfolk.
314	Vertol H-25A/HUP-1	AF 51-16574G BuNo.147631	24 Feb 1954	26 Jan 1960		*Retriever.* Transferred to NAS Norfolk.
315	McDonnell F2H-3	BuNo.126300	31 Jul 1954	29 Sep 1959	210	*Banshee.*
316	Grumman JRF-5	BuNo.37816	26 Aug 1954	3 Nov 1954	103	*Goose.* Crashed at Wallops Island.
317	Vertol H-25	AF 51-16637	14 Sep 1954	1 Feb 1955		Transferred to Fort Eustis, Virginia.
318	Grumman JRF-5	BuNo.87748	24 Feb 1955	23 Mar 1960	202	Replaced NACA-103. To US Dept of Interior.
319	McDonnell XF-88A	AF 46-526	25 Feb 1955	7 Jul 1958		Salvaged.
320	Beechcraft C-45F	AF 44-47106	24 Mar 1955	17 Mar 1959		*Expeditor.* Salvaged.
321	North American EJF-86D	AF 50-459	24 Jun 1954 21 Feb 1956	16 May 1955 25 May 1960	204	Returned to USAF for 'Project Pullout' on 16 May, 1955.
322	North American F-86D-31-NA	AF 51-5959A	14 Mar 1956	19 Sep 1958		*Sabre.*
323	North American JF-86D-5-NA	AF 50-509	6 Apr 1956	12 Jul 1960	205	*Sabre.* Salvaged.
324	Sikorsky HSS-1	BuNo.137855	29 Jun 1956	18 Feb 1957		*Seabat.*
325	McDonnell F-101A	AF 53-2434	22 Aug 1956	1 Mar 1960	219	Ex Ames. Transferred to Davis-Monthan.
326	Hiller YH-32	AF 55-4968	30 Nov 1956	6 Nov 1958		*Hornet.* To Fort Eustis, Virginia.
327	Hiller YH-32	AF 55-4970	30 Nov 1956	6 Nov 1958		*Hornet.* To Fort Eustis, Virginia.
328	Vought F8U-1	BuNo.141354d	10 Dec 1956	5 Feb 1959		*Crusader.* Returned to US Navy.
329	Grumman F11F-1	BuNo.138623	4 Jan 1957	17 Aug 1961		Returned to US Navy.
330	Grumman SA-16A	AF 49-088A	21 May 1957	15 Mar 1958		*Albatross.* To California ANG.
331	Noth America T-28A	AF 50-279A	26 Jun 1957	25 Mar 1959		*Trojan.* Transferred to Davis-Monthan.
332	North American F-100C-25-NA	AF 54-2024A	13 Sep 1957	12 Oct 1959		*Super Sabre.* To Davis-Monthan.
333	Lockheed T-33A	AF 49-939A	26 Nov 1957	3 Feb 1958		Transferred to Ames.
334	McDonnell F-101A	AF 54-1442A	18 Apr 1958	22 Dec 1958	220	Transferred to SperryGyroscope.
335	Lockheed T-33A-1-LO	AF 49-945A	23 Nov 1958	27 Jun 1959		Returned to US Air Force.
336	Convair T-29B	AF 51-7909A	12 Jan 1959	5 Apr 1963	250	Transferred to Lewis as 'NASA-5'.
337	North American T-28A	AF 51-3725A	23 Mar 1959	30 Nov 1979	223 502	*Trojan.* To Maryland Aviation History Society.
338	Lockheed TV-2/T-33A	BuNo.131878	1 May 1959	1 Oct 1962	224	AF 51-9157. to NASA Houston, Texas.
339	Vought F8U-3	BuNo.146340	26 May 1959	2 Aug 1960	226	To NAS Norfolk, Virginia.
340	Vought F8U-3	BuNo.146341	26 Jun 1959	2 Aug 1960	227	To NAS Norfolk.
341	Cessna L-19A	AF 51-12770	17 Jun 1959	18 Aug 1959		*Bird Dog.* To Fort Eustis, Virginia.
342	Vertol VZ-2	AF 56-6943	23 Sep 1959	22 Mar 1961		Model 76.
343	Sikorsky H-34A	AF 53-4475	21 Apr 1960	30 Oct 1961		*Chocktaw.* To Fort Eustis, Virginia.

Order of receipt	Type and designation	Serial	Date Arrived	Date Departed	NACA No.	Name and remarks
344	Douglas C-54G	AF 45-529A	1 May 1960	10 Jun 1960	231	*Skymaster*. To NASA Goddard, Maryland.
345	Douglas C-54G	AF 45-637A	5 May 1960	15 Jun 1960	232 432	*Skymaster*. To NASA Goddard, Maryland.
346	Doak VZ-4DA	AF 56-6942	5 May 1960	21 Aug 1972		VTOL Flying platform. To Fort Eustis.
347	Curtiss-Wright X-100	AF 62-12197	11 Oct 1960	24 Oct 1961		Later X-19A-CU. N853 *c/n* 1.
348	North American F-100C	AF 53-1709A	Mar 1961	2 Nov 1960	25	Returned to Ames.
349	Douglas C-47D-10-DK	AF 43-49106A	5 Jun 1961	30 Jun 1967		*Skytrain*. c/n 26367.
350	Sikorsky JH-19D	AF 56-1552	7 Oct 1961	28 Jan 1970	535	*Chikasaw*. Salvaged on base.
351	Lockheed T-33A-5-LO	AF 51-9086A	10 Nov 1961	11 Jan 1963		To City of Hampton for display.
352	Sikorsky CH-37C/HR2S-1	BuNo.140316	18 Jan 1962	30 Oct 1963		Ex Federal Aviation Administration — NAFEC.
353	Vertol CH-46C	AF 58-5514	9 Feb 1962	24 Mar 1975	533	YHC-1 prototype. To Lewis crash tests.
354	Lockheed T-33A-5-LO	AF 53-5172	26 Mar 1962	22 Apr 1971	510	Salvaged on base.
355	Vertol VZ-2	AF 56-6943	18 Sep 1962	4 Jan 1965		Model 76. To Smithsonian Air Museum.
356	Grumman G-159		28 Mar 1963	Current	1	*Gulfstream* c/n 96.
357	Boeing Model 367-80	N70700 c/n 17158	1964			Blown-flaps. SST. Variable-stability tests.
358	Bell 204B/UH-1B		20 Sep 1964	Current	29 530	c/n 2017.
359	Lockheed XH-51N	c/n 103/151263	10 Dec 1964	14 Mar 1973	30 531	Rigid rotor research. RAE Bedford 1970.
360	Bell OH-4A	AF 62-4204	1 Apr 1965	23 Jan 1973	532	Ex N73917. to Miss. State Highway Patrol.
361	Mooney M-20E	c/n 582	7 May 1965	7 Jul 1965		
362	Hiller OH-5A	AF 62-4207	11 Oct 1965	17 Dec 1968		To US Army Aberdeen Proving Ground. MD.
363	Cessna T-37B	AF 54-2737A	20 Oct 1965	13 Jan 1967		Ex Ames. To Wright-Patterson AFB, Ohio.
364	Northrop T-38A-60-NO	AF 65-10329	15 Dec 1965	Current	511	From NASA Houston, Texas. Talon.
365	Boeing 720-023B	N7545A	1965	1965		c/n 18031. Hydroplaning tests.
366	Hawker Siddeley XV-6A	AF 64-18267	23 Jul 1966	13 May 1966	520	P.1127 *Kestrel*. 64-18263 also used.
367	Hawker Siddeley XV-6A	AF 64-18263	23 Jul 1966	20 Jun 1974	521	P.1127 *Kestrel*. To NASM Washington DC.
368	Bell OH-13A	AF 57-6207	16 Jun 1968	5 Apr 1972	931 537	From NASA Houston, Texas as NASA 931.
369	Aero Commander 680	c/n 577-222	1 Aug 1968		631 503	From Lewis as NASA 631.
370	Sikorsky UH-19D	AF 56-1522	10 Oct 1968	6 Nov 1969	535	To City of Hampton, VA for display.
371	LTV-Ryan-Hiller XC-142	AF 62-5924	13 Oct 1968	5 May 1970	522	Tilt-wing. To USAF Museum, Dayton, Ohio.
372	Lockheed T-33A	AF 55-3025	17 Dec 1969	27 Mar 1973	512	Donated to USMC Museum, Quantico, VA.
373	Helio Courier	N4153D or N4153F	22 Jan 1970	1 Mar 1972		Leased.
374	Sikorsky SH-3A	BuNo.149723	21 Jul 1970	1 Nov 1977	933 538	From Houston, Texas. To Ames.
375	Lockheed T-33A	AF 56-3689	28 Sep 1970	30 Mar 1971	939	Ex Houston, Texas NASA 939. To Nebraska ANG.
376	Hawker Siddeley XV-6A	AF 64-18266	30 Jul 1970	19 Jul 1974	520	For spare parts. To City of Hampton, VA.
377	Cessna U-3A	AF 57-5921A	7 Feb 1971	Current	505	*Blue Canoe*. From Ames.
378	Piper Pa-28-180	N7852W	13 Sep 1971 17 Oct 1972	17 Apr 1972 14 Feb 1975	506	*Cherokee*.
379	Sikorsky CH-54B	AF 69-18467	18 Mar 1972	15 Jan 1974	539	Boron-eopoxy mods. To Fort Eustis, VA.
380	Sikorsky CH-54B	AF 69-18490	18 Mar 1972	15 Jan 1974		Boron-epoxy mods. to Fort Eustis, VA.
381	Cessna 172K	172-58729	8 May 1972	Current	507	*Skyhawk*. Purchased as N7029G.
382	de Havilland DHC-6	c/n 4 N508NA	24 Jul 1972	25 Oct 1983	508	Cross-wind landing gear research.
383	Bell OH-58A	AF 71-20702	1 Aug 1972	Current	540	*Kiowa*. c/n 41564.
384	Ayers S2R-800	c/n 206R	9 Aug 1972			*Thrush*.
385	Northrop T-38A	AF 65-10328	22 Sep 1972	30 Jan 1980	908	*Talon*. From Houston as NASA 908.
386	Bell AH-1G	AF 66-15248	18 Dec 1972	1 Mar 1978	541	To Ames as NASA 736. *Cobra*.
387	Grumman American AA-1	AA1-0001	22 Feb 1973	Current	501	*Yankee*. Spin research.
388	Martin B-57B	AF 52-1576	19 Mar 1973	8 Feb 1975	237	Ex Lewis as NASA 637. To Ames as NASA 809.
389	Bell UH-1H	AF 64-13628	4 Apr 1973	1 Mar 1978	542	*Huey*. To Ames as NASA 734.
390	Piper PA-24-260	24-4569	14 Nov 1973	16 Aug 1974	540	N9094P.
391	Boeing 737-130		17 May 1974	Current	515	c/n 19437.
392	Vertol CH-47B	AF 66-19138	1 Jul 1974	14 Aug 1979	543 544	*Chinook*. To Ames as NASA 737. Ident changed on 9 April 1974.
393	Vertol CH-47B	AF 66-19121	9 Oct 1974	2 Jun 1975		*Chinook*. To Fort Eustis, VA.
394	Piper PA-34-200	34-7250001	26 Oct 1974	3 Aug 1978	509	*Senaca*. GAW-1 wing research.
395	Beech C-23	N6624R M-1608	14 Jan 1975	Current	504	*Sundowner*. GAW-2 wing research.
396	Sikorsky CH-53A	BuNo.152399	25 Jun 1975	5 May 1978	543	To Federal Av. Administration, Atlantic City, N.J.
397	Martin B-57B	AF 52-1576	26 Jun 1975	18 Mar 1975	516	*Canberra*. To Ames.
398	Beech 65-B80	c/n LD-507	1 Jun 1977	Current	506	Ex N1530L. *Queen Air*.
399	Sikorsky RSRA 1	AF 72-001	21 Jul 1977	Mar 1979	545	*Heathcliff*. To Ames as NASA 740.
400	Ayers S2R-800	c/n 2069R	9 Aug 1977 26 Jul 1978	14 Dec 1982 14 Dec 1982	517	Full-Scale tunnel tests. Flight tests. Ex N4308X.
401	Sikorsky RSRA 2	AF 72-002	22 Dec 1977	Sep 1978	546	*Gertrude*. To Ames as NASA 741.
402	Grumman OV-1B	AF 64-14244	22 Apr 1978	24 Mar 1980	637 518	From Lewis as NASA 637. *Mohawk*.

Langley Memorial Aeronautical Laboratory Aircraft — continued

Order of receipt	Type and designation	Serial	Date Arrived	Date Departed	NACA No.	Name and remarks
403	Beech T-34C	GL-108	1 Jun 1978	Current	510	*Mentor*.
404	Piper PA-28R-200	28R-7635243	2 Aug 1978	Current	519	*Arrow*. Modified T-tail.
405	Grumman OV-1B	AF 62-5880	15 Nov 1978		518	*Mohawk*.
406	Convair NF-106B	AF 57-2516	29 Jan 1979	Current	816	Ex Lewis as NASA 616. On loan from Dryden.
407	North American T-39A	AF 61-0649	19 Jun 1979	23 Sep 1982		*Sabreliner*. To McClellan AFB, California.
408	Grumman OV-1B	AF 62-5880	5 Apr 1980	24 Apr 1989	512	*Mohawk*. To ACRAD.
409	Convair NF-106B	AF 57-2507	12 May 1981	24 Feb 1984		*Delta Dart*. Cut in half. Full-Scale tunnel tests.
410	Convair F-106A	AF 59-123	19 Mar 1982	Current		*Delta Dart*. Spares for NASA 816.
411	Cessna 402B	402B-0313	8 May 1982	Current	503	
412	Bellanca	001	24 Mar 1981	28 Mar 1982		*Skyrocket II*. N14666.
	Cessna 402B	402B-0313	8 May 1982	Current	503	
	Cessna 206	U20602927	7 Jul 1983	7 Sep 1983		Anti Icing Research (Air) Group.
	General Dynamics F-106B	AF 57-2545	30 Jan 1985	Current		Non-flyable.
	Schweizer SGS-1-36	001	13 Feb 1986	Current	502	N3616X.
	Gates Lear 28	25-064	12 Mar 1988	Current	566	N266GL.
	Northrop F-5F	AF 73-0889	21 Aug 1989	Current	550	Ex 425th TFTS. Williams AFB, Arizona.

Langley Research Center, Hampton, Virginia

The following list of aircraft have been related to research either financed or performed by the Langley Research Center. Unfortunately, detailed data or records have not been located that would include these aircraft in my regular listing of Langley aircraft.

Type Aircraft	Serial No.	Remarks
Fairchild M-224/ZV-4/VZ-5	56-6940	V/STOL, R&D.
Ryan XV-8A 'Fleep' Model 164	63-13003	Flex-wing. Tunnel tests, R&D.
Convair F-106B Delta Dart	59-158	Astro-proficiency aircraft, photo w/7 Mercury astros 4-61.
Convair F-106A Delta Dart	58-782	Astro-proficiency aircraft. Shown in hangar at Langley on cover of *Aviation Week* 5-21-62 (Air Force loan). 21 May 1962.
Lockheed YO-3A		R&D, Loaned to University of Illinois (circa 1972–73).
Lockheed C-130H Hercules	73-1592	R&D, modified with light weight boron epoxy wing box. In use with 314th TAC Airlift Wing, Little Rock, Arkansas (circa 1975–76).
Lockheed C-130H Hercules	73-1594	
Rockwell Aero Commander		R&D, Full Scale Tunnel Tests. Modified with JT15D P&W jets mounted atop wing (Conda effect).
Boeing B-727-25	c/n 18272	R&D, at Wallops Island, Va., in fall of 1967. Steep approach, landing and take-off tests.
Eastern Airlines	N8121N	
Boeing B-727 FAA		R&D, at Wallops Island, Va.
Rutan, Vari Ezi (Homebuilt)		R&D, Evaluation of flying characteristics 20 January 1979.
Piper PA-23 Apache		R&D, Loaned to University of Cincinnati.
Piper PA-30B Twin Comanche		Static tests. Probably Flight Research Center's NASA 808.
Bell X-22A	151521	R&D. V/STOL
Cessna 177B Cardinal	177-00002 N9110F	R&D, Project REDHAWK, at University of Kansas.
North American Navion A	NAV-4-345 N91566	R&D, horizontal tail-planes on wings. Princeton University, New Jersey.
Cessna 337 Skymaster	377-663 N3769C	R&D, Wankel engine tests with shroud on rear engine.
Dornier Do.31	c/n E-1	R&D. V/STOL tests concluded in Germany on 17 April 1970. (*Av. Week*).
Beech T-34B Mentor		R&D, US Navy loan at Mississippi State University. Engine cooling tests.
Piper J-3 Cub		R&D. Stall-proof aircraft tests at Texas A&M University.
Hawker-Siddeley Buccaneer S (Mk.1, Blackburn B.103)		Fatal crash in U.K. with NASA Langley Research Pilot Bill Alford on 12 October 1959.
Vertol CH-47C Chinook		Crash tests, several of this type used.
Piper PA-13 Navaho's	N7471L	Many used for crash tests. Acquired from Piper due to damage in floods at factory in Pennsylvania.

APPENDIX THREE
Ames Research Center Aircraft

Type and designation	Serial	Date arrived	Date departed	No.	Remarks
North American 0-47-1	AF 37-323	5 Sep 1940	13 Mar 1946		Ice research.
Lockheed 12A *Electra*	NC 17396	20 Jan 1941			c/n 1268 Salvaged.
Douglas SBD-1 *Dauntless*		24 Oct 1941	21 Feb 1942		
Lockheed 12A *Electra*	NC 17397	16 Jan 1942			
Vought-Sikorsky OS2U-2 *Kingfisher*	BuNo.2189	9 Mar 1942			Full span ZAP flap tests.
Consolidated XB-24F-CO *Liberator*	AF 41-11678	13 May 1942			
Brewster F2A-3 *Buffalo*	BuNo.01516	21 May 1942			
Boeing B-17F *Flying Fortress*	AF 42-5474	28 Aug 1942			Photo shows 41-24613.
North American O-47A-1	AF 37-279	6 Nov 1942	21 Mar 1946		
Fairchild Model 24		6 Nov 1942	8 Oct 1953	100	Ex Langley.
Lockheed P-38F *Lightning*	AF 41-7632	30 Dec 1942			
Bell P-39D-1 *Airacobra*		31 Dec 1942			
Bell P-39D-1 *Airacobra*	AF 41-28268	4 Jan 1943	4 Jan 1943		
Bell P-39D-1 *Airacobra*	AF 41-28328	5 Jan 1943	6 Jan 1943		
Vought-Sikorsky OS2U-2 *Kingfisher*	BuNo.3075	6 Feb 1943	24 May 1946		ZAP flap tests.
Douglas A-20A *Havoc*	AF 39-726	10 Mar 1943	31 May 1943		
Curtiss C-46A-5-CU *Commando*	AF 41-12293	10 Mar 1943	24 Mar 1949		Ice research.
North American B-25D *Mitchell*	AF 41-29983	26 Mar 1943			
Bell P-39N-1 *Airacobra*	AF 42-18476	10 May 1943	11 May 1943		
Bell P-39N *Airacobra*	AF 42-8849	11 May 1943	17 May 1943		
Vultee A-35A *Vengeance*	AF 41-31174	16 Jun 1943			
Douglas A-24 *Dauntless*		29 Jul 1943	30 Jul 1943		
North American P-51B *Mustang*	AF 43-12111	11 Aug 1943	7 Sep 1947		Handling quality tests.
Martin B-26 *Marauder*	AF 41-31702	27 Sep 1943	27 Oct 1943		Handling quality tests.
Bell P-39Q-10 *Airacobra*	AF 42-20790	7 Jan 1944	30 Jan 1944		
Lockheed PV-1 *Ventura*	BuNo.48871	6 Jan 1944			
Bell P-63A *Kingcobra*	AF 42-68892	17 Feb 1944			
North American P-51D *Mustang*	AF 44-13257	28 Mar 1944	2 Jun 1944	108	
Northrop P-61A-5 *Black Widow*	AF 42-5572	20 Apr 1944	16 Nov 1944		
Vultee BT-13B *Vibrator*	AF 42-90461	22 Apr 1944	22 Oct 1945		
Vultee BT-13B *Vibrator*	AF 42-89854	25 Apr 1944	17 May 1945		
Douglas SB2D-1 *Destroyer*	BuNo.03552	12 Jun 1944	10 Jan 1946		Crashed.
Douglas BTD-1 *Destroyer*	BuNo.04968	28 Jul 1944	30 Jun 1947		Salvaged.
Douglas XP-70	AF 39-735	19 Aug 1944	21 Aug 1944		
Lockheed P-38J-15 *Lightning*	AF 43-28519	30 Aug 1944	15 Mar 1946		
Grumman XF7F-1 *Tigercat*	BuNo.03550	2 Sep 1944	19 Jun 1948		
Lockheed XP-80A *Shooting Star*	AF 44-83023	19 Sep 1944	27 Jan 1947		
Douglas XSB2D-1 *Destroyer*	BuNo.03551	9 Oct 1944	24 May 1946		
North American P-51B *Mustang*	AF 43-12094	16 Nov 1944	9 Sep 1947		
General Motors P-75A-1 *Eagle*	AF 44-44550	22 Nov 1944	7 Feb 1946		Surveyed by US Army Air Force.
Republic P-47N-1 *Thunderbolt*	AF 44-87806	12 Dec 1944			
Howard GH-3(DGA) *Nightingale*		12 Jun 1944			Transport on temporary loan.
Howard GH-1(DGA) *Nightingale*		11 Jan 1945			Transport on temporary loan.
Bell P-63A-6 *Kingcobra*	AF 42-68941	27 Jan 1945	18 Jun 1946		
Republic XP-47M-1 *Thunderbolt*	AF 42-27385	2 Feb 1945	8 Aug 1945		
Douglas SBD-5 *Dauntless*	BuNo.28279	6 Feb 1945			
Ryan FR-1 *Fireball*	BuNo.39650	17 Feb 1945	10 Mar 1947		
Lockheed PV-2 *Harpoon*		27 Feb 1945			
Grumman FM-2 *Wildcat*	BuNo.73700	13 Mar 1945	25 Feb 1946		
Douglas A-26B *Invader*	AF 44-34307	12 Apr 1945	6 Aug 1945		
Republic P-47D-25 *Thunderbolt*	AF 42-26408	27 Apr 1945	7 Sep 1945		
Ryan FR-1 *Fireball*	BuNo.39659	28 Apr 1945	2 Jun 1945		
North American XP-51F *Mustang*	AF 43-43332	30 Apr 1945	6 Nov 1947		
Ryan FR-1 *Fireball*	BuNo.39657	5 May 1945	1 Jun 1947		
Ryan FR-1 *Fireball*	BuNo.39656	11 May 1945	14 Jul 1945		
Douglas BTD-1 *Destroyer*	BuNo.04971	18 May 1945	31 Oct 1947		Modified to two place for ice research.
Douglas XB2D-1 *Skyraider*	BuNo.09085	21 May 1945	28 May 1945		
Ryan FR-1 *Fireball*	BuNo.39660	31 May 1945	3 Jul 1945		
North American P-51H-1-NA *Mustang*	AF 44-64164	6 Jun 1945	16 Jun 1945		To Langley.
Grumman F6F-3 *Hellcat*	BuNo.42874	22 Jun 1945	9 Sep 1960	158	Variable stability trials. Salvaged.
Douglas A-26B *Invader*		4 Jul 1945	Jan 1951		
Douglas BTD-1 *Destroyer*	BuNo.09059	20 Jul 1945	5 Nov 1947		

Type and designation	Serial	Date arrived	Date departed	No.	Remarks
Douglas SBD-5 *Dauntless*	BuNo.54669	30 Jul 1945	5 Aug 1945		
Ryan FR-1 *Fireball*	BuNo.39665	11 Aug 1945	21 Sep 1945		
Vought F4U-4 *Corsair*	BuNo.97028	21 Aug 1945	30 Apr 1947		
North American AT-6 *Texan*	AF 41-32079	26 Sep 1945	22 Jul 1947		Fitted with armament.
Lockheed P-80A *Shooting Star*	AF 44-85099	30 Jan 1946	10 Apr 1950		Crashed.
Lockheed P-80A *Shooting Star*	AF 44-85169	7 Feb 1946	23 Sep 1947		
Grumman F7F-3 *Tigercat*	BuNo.80521	8 Feb 1946	29 Apr 1947		
Douglas XBT2D-1 *Skyraider*	BuNo.09086	11 Mar 1946	4 Sep 1947		
McDonnell XP-85 *Goblin*	AF 48-3886	26 Mar 1946			40 × 80 wind tunnel tests.
Grumman F8F-1 *Bearcat*	BuNo. 94819	2 Apr 1946	1 Jun 1953		Crashed. Rudolph Van Dyke killed.
Grumman F7F-3 *Tigercat*	BuNo.80372	2 May 1946	29 Apr 1947		
Grumman F7F-3 *Tigercat*	BuNo.80546	31 May 1946	1 Feb 1949		
North American P-51H *Mustang*	AF 44-64415	18 Dec 1946	Apr 1961	130	To US Navy, sold 23 June 1961 for $268.90.
Lockheed P-80A-1 *Shooting Star*	AF 44-85299	18 Dec 1946	6 Jun 1955	131	Salvaged by US Navy.
Douglas R4D-6 *Skytrain*	BuNo.50812	9 Oct 1946	13 Dec 1948		From Langley.
Douglas B-26B *Invader*	AF 41-39553	25 Oct 1946	29 Jan 1951		
North American P-51H *Mustang*	AF 44-64691	25 Jan 1947	17 May 1948		Crashed. Ryland D. Carter killed.
Beechcraft AT-11 *Kansan*	AF 42-37209	7 Feb 1947	2 Sep 1947		
North American P-51D *Mustang*	AF 44-74944	15 Apr 1947	23 Nov 1949		
Ryan XFR-4 *Fireball*	BuNo.39665	11 Jun 1947			40 × 80 wind tunnel tests.
North American B-25J *Mitchell*	AF 44-86891	29 Aug 1947	11 Aug 1949		
Fairchild C-82A *Packet*	AF 44-23056	31 Aug 1947	7 Feb 1961		Utility transport — Gust research.
North American P-51H *Mustang*	AF 44-64703	6 Nov 1947	17 May 1956	110	Salvaged by US Navy.
Convair XP-92A *Delta*	AF 46-682	Nov 1947			40 × 80 wind tunnel tests.
Republic YP-84A-5 *Thunderjet*	AF 45-59488	2 Dec 1947	5 Oct 1948		
McDonnell XP-85 *Goblin*		8 Jan 1948			40 × 80 wind tunnel tests.
Northrop F-15A-1-NO *Reporter*	AF 45-59300	6 Feb 1948	Oct 1954	111	Salvaged by US Navy.
Taylorcraft L-4 *Cub*	NC 254	5 Oct 1948	25 Oct 1948		Cross wind landing gear tests.
Republic F-84C *Thunderjet*	AF 47-1530	19 Oct 1948	29 Oct 1948		
Douglas R4D-6 *Skytrain*	BuNo.99827	14 Dec 1948	9 Sep 1965	701	Utility transport and gust research.
Douglas C-47J *Skytrain*		17 Dec 1948	5 Mar 1963		Stricken off records.
Curtiss SB2C-5 *Helldiver*	BuNo.83135	18 Dec 1948	Jun 1955	147	Salvaged by US Navy.
Curtiss SB2C-5 *Helldiver*	BuNo.83292	22 Dec 1948	14 Jan 1949		
Grumman F7F-3 *Tigercat*	BuNo.80526	28 Jan 1949	9 Dec 1949		
Republic YF-84A *Thunderjet*		14 Feb 1949	20 Dec 1950		
Republic F-84C *Thunderjet*	AF 47-1530	4 Aug 1949	11 Aug 1949		
North American F-86A *Sabre*	AF 48-291	29 Aug 1949	11 Jan 1960	116	Spin tests, tracking, A-1 sight etc.
Lockheed XR60-1 *Constitution*	BuNo.85163	21 Nov 1949	18 May 1950		
North American F-86A *Sabre*	AF 47-609	10 Apr 1950	15 Mar 1956	135	Salvaged by US Navy.
Grumman F6F-5 *Hellcat*	BuNo.79669	19 Jun 1950	9 Sep 1960	208	Salvaged by US Navy.
Vought F6U-1 *Pirate*	BuNo.122483	23 Jul 1950	5 Aug 1953		Salvaged by US Navy.
Vought F6U-1 *Pirate*	BuNo.122491	24 Aug 1950	5 Oct 1953	138	Salvaged by US Navy.
Northrop ERF-61C-1-NO *Black Widow*	AF 42-8330	5 Feb 1951	10 Aug 1954	111	To Smithsonian Museum.
North American YF-93A	AF 48-317	5 Feb 1951		139	Salvaged.
North American YF-93A	AF 48-318	5 Jun 1951		151	Salvaged.
Northrop EF-61C-1-NO *Black Widow*	AF 43-8357	13 Jul 1951	8 Aug 1954	130 146	Salvaged by US Navy.
Republic YF-84F *Thunderstreak*	AF 51-1345	12 Feb 1952	3 May 1952		40 × 80 wind tunnel tests.
Lockheed F-94C *Starfire*	AF 50-956	25 Feb 1952	20 Mar 1952	156	40 × 80 wind tunnel tests.
North American F-86A *Sabre*	AF 50-580	8 Apr 1952	18 Apr 1952		
McDonnell XV-1 *Convertiplane*	AF 53-4016	1952			40 × 80 wind tunnel tests.
North American YF-86D *Sabre Dog*	AF 50-577	26 Jun 1952	15 Feb 1960	149	Control sensitivity research.
North American YF-86D *Sabre Dog*	AF 50-578	16 Sep 1952			
North American F-86D *Sabre Dog*	AF 51-5986	12 Jun 1953	7 Nov 1957		
North American F-86F *Sabre*	AF 52-435	10 Oct 1953	13 Sep 1965	228	Intercept fire control target aircraft.
Lockheed TV-1 *Shooting Star* P-80C	BuNo.33868 AF 48-379	12 Oct 1953	Feb 1960	206	Bullpup missile simulator. Salvaged.
Republic F-84F-5-RE *Thunderstreak*	AF 51-1364	31 Oct 1953	7 Mar 1957	155	Crashed.
Republic F-84F *Thunderstreak*	AF 51-1346	1 Mar 1954	11 Mar 1954		
Lockheed F-94C-1 *Starfire*	AF 50-956	29 Jul 1954	18 Nov 1958	156	In flight thrust reverse tests. Crashed.
North American FJ-2 *Fury*	BuNo.132015	26 Aug 1954	14 Sep 1954		
North American FJ-3 *Fury*	BuNo.135800	3 Sep 1954	30 Apr 1956		
Lockheed T 33A	AF 53-4919	15 Oct 1954	18 May 1956		40 × 80 wind tunnel tests.
Grumman F9F-4 *Panther*	BuNo.125156	21 Oct 1954	10 Aug 1955		
Grumman F9F-8 *Cougar*	BuNo.131086	6 Jan 1955	7 Feb 1955		
North American F-86D-5 *Sabre Dog*	AF 50-509A	6 Jan 1955	3 Apr 1956		
North American F-86D-60 *Sabre Dog*	AF 53-787	17 Mar 1955	1 Feb 1960	216	E,I,T-CSTI presentation.
Grumman F9F-6 *Cougar*	BuNo.128138	9 May 1955	3 Aug 1955		
Chance-Vought F7U-3 *Cutlass*	BuNo.129656	10 Jun 1955	4 Oct 1955		Burnt after transfer.
North American F-86E *Sabre*	AF 50-606A	30 Jun 1955	Nov 1959	157	Variable stability tests.
Douglas XA3D-1 *Skywarrior*	BuNo.125413	6 Jul 1955	2 Aug 1956		40 × 80 wind tunnel tests.
Douglas F4D-1 *Skyray*	BuNo.134759	4 Apr 1956	16 Oct 1959		Miscellaneous research and development.
North American F-100C *Super Sabre*	AF 53-1709A	4 Sep 1956	2 Nov 1960	703	Variable stability flying qualities.
North American F-100A *Super Sabre*	AF 53-1585A	2 Oct 1956	15 Feb 1960	200	Boundary layer control tests.

Type and designation	Serial	Date arrived	Date departed	No.	Remarks
McDonnell F3H-1 *Demon*	BuNo.135502	19 Jun 1956	14 Mar 1958		40 × 80 wind tunnel tests.
North American F-100C *Super Sabre*	AF 54-1964	22 Mar 1957	15 Feb 1960		T-ALCS.
Cessna T-37A/B	AF 54-2737	2 Aug 1957	19 Oct 1965		Thrust attenuation tests.
Bell XV-3	AF 54-148	4 Aug 1957	28 Oct 1957		40 × 80 wind tunnel tests. V/STOL.
Douglas F5D-1 *Skylancer*	BuNo.139208a	20 Aug 1957	16 Jan 1961	212	Inlet research — OGEE wing.
Douglas F5D-1 *Skylancer*	BuNo.142350b	20 Aug 1957	15 Jun 1961	213	Target for auto manoeuvre interceptions.
Lockheed T-33A-5	AF 49-920A	27 Nov 1957	15 Sep 1965	720	Check ground simulator for realism.
Convair F-102A *Delta Dagger*	AF 56-1358	23 Dec 1957	21 Mar 1960		Fire control auto manoeuvres.
Convair F-102A *Delta Dagger*	AF 56-1304	10 Apr 1957			Auto attack evalution.
Ryan VZ-3RY	AF 56-6941	20 May 1958	24 Feb 1959	235 705	V/STOL.
Douglas F5D-1 *Skylancer*	BuNo.139209	16 Jun 1958	Feb 1959		Used for spare parts.
Douglas F5D-1 *Skylancer*	BuNo.142349b	18 Jun 1958			Used for spare parts.
Lockheed F-104A *Starfighter*	AF 56-745A	17 Jul 1958	12 Aug 1958		40 × 80 wind tunnel tests.
Convair F-106A *Delta Dart*	AF 57-235	4 Sep 1958	14 Dec 1959		MA-1 Fire control and auto manoeuvre intercept.
Bell XV-3	AF 54-148	16 Sep 1958	29 Oct 1958		40 × 80 wind tunnel tests.
Lockheed F-104B *Starfighter*	AF 57-1303A	3 Oct 1958	16 Dec 1959		Variable stability. High altitude reac controls.
Hiller H-23C *Raven*	Army 56-2288	3 Nov 1958	28 Apr 1959		VTOL — Photo and utility helicopter.
North American F-100F *Super Sabre*	AF 56-3725A	30 Jan 1959	5 Mar 1959		40 × 8 wind tunnel tests.
Stroukoff YC-134C	AF 54-556	6 Mar 1959	31 May 1961	222	Boundary layer control research.
Lockheed JF-104A *Starfighter*	AF 56-745A	19 Mar 1959	6 May 1960		
Collins Aerodyne		21 Apr 1959			40 × 80 wind tunnel tests — VTOL.
Stroukoff YC-134A	AF 56-1672A	30 Apr 1959	2 Nov 1960		Salvaged for spare parts.
Chance-Vought F8U-3 *Crusader* III	BuNo.147085	18 Jun 1959	15 Aug 1960	225	
Hiller H-23C *Raven*	Army 56-2288	5 Aug 1959	27 Jun 1961		
Bell XV-3	AF 54-148	12 Aug 1959	9 Jun 1965		Flight tests. V/STOL.
Ryan VZ-3RY	AF 56-6941	24 Aug 1959	20 Jun 1966		V/STOL — To Smithsonian Museum.
Bell X-14A/B	AF 56-4022	2 Oct 1959	29 May 1981	234 704	Accident — Ron Gerdes, no injuries.
Vanguard Omniplane		15 Dec 1959	19 Aug 1960		40 × 80 wind tunnel tests.
Avro Car I	AF 53-7496	22 Dec 1959			40 × 80 wind tunnel tests. VTOL research.
Kaman K-16B	US Navy	28 Sep 1960	6 Nov 1962		40 × 80 wind tunnel tests.
Bell HU-1 *Iroquois*	Army 57-6098	4 Jan 1961	23 Oct 1961		40 × 80 wind tunnel tests.
Grumman YAO-1 *Mohawk*	Army 57-6463	15 May 1961	21 Jun 1961		40 × 80 wind tunnel tests.
Hiller H-23D *Raven*	Army 59-2758	27 Jun 1961	11 Apr 1962		
Vanguard Omniplane		26 Sep 1961	15 Oct 1962		40 × 80 wind tunnel tests.
Lockheed NC-130B *Hercules*	AF 58-712	30 Jun 1961	20 Dec 1961		Boundary layer control research.
Hiller YROE-1 *Rotorcycle*	US Marines 4020	16 Nov 1961			One-man folding helicopter.
Hiller YROE-1 *Rotorcycle*	US Marines 4021	16 Nov 1961			One-man folding helicopter.
Hiller YROE-1 *Rotorcycle*	US Marines 4024	16 Nov 1961			One-man folding helicopter.
Hiller UH-12E *Raven*	Army 56-2265	1 Mar 1963	17 Dec 1976	706	
Curtiss Wright X-100	N853	14 Mar 1963	8 Jul 1963		40 × 80 wind tunnel tests.
Lockheed NC-130B *Hercules*	AF 58-712	27 Feb 1963	23 May 1963		Flight tests.
		16 Oct 1963	5 Jan 1967		To Johnson Space Centre, Texas.
Convair Model 340	N73103	21 May 1963	3 Sep 1976	19 707	STOL research, c/n 003. To NASA Lewis, Ohio.
Douglas F5D-1 *Skylancer*	BuNo.139208a	4 Mar 1963	9 Apr 1968	212 708	
North American F-100C *Super Sabre*	AF 53-1709A	11 Mar 1964	21 May 1972	25 703	Hybrid, A-fuselage, C-wing, D-tail.
Ryan XV-5A	Army 62-4505	27 Mar 1964	30 Jan 1974	705	40 × 80 wind tunnel tests — flight tests.
Lockheed XV-4A *Hummingbird*	Army 62-4500	15 Jul 1964	16 Sep 1964		40 × 80 wind tunnel tests.
Convair CV.990 *Coronado* 'Galileo'	c/n 001	2 Apr 1965	12 Apr 1973	711	Mid-air crash with Lockheed P-3, crew 16 lost.
Learjet Model 23	c/n 23-049	17 Sep 1965	11 Jan 1980	701	To Johnson Space Centre, Texas. NASA 960.
de Havilland C-8A *Buffalo*	Army 63-13687	1967	1967		Loaned to National Science Foundation, Boulder, Colorado.
de Havilland C-8A *Buffalo*	Army 63-13689	1967			Landing accident on arrival at Ames. $500K in damages — salvaged.
de Havilland C-8A *Buffalo*	Army 63-13688	10 Jun 1967			Transferred to NoAA (ESSA 88).
de Havilland C-8A *Buffalo*	Army 63-13686	10 Jun 1967	22 Sep 1981	716	Transferred to Canadian Government, Otawa. Augmented wing — JSRA.
North American OV-10A *Bronco*	BuNo.152881	8 Apr 1968	7 Oct 1976	718	Cylinder flaps research.
Grumman F-111B	BuNo.151974	30 Oct 1968			40 × 80 wind tunnel tests — salvaged.
Bell UH-1B *Iroquois*	Army 62-1908	14 Oct 1970	10 Feb 1980	732	NASA 414 Wallops Island.
Lockheed T-33A	AF 53-5400A	12 Jan 1971	19 Sep 1973	715	To College of Alameda. California.
Lockheed U-2	AF 56-6681	3 Jun 1971	1988	708	Earth Resources Survey Programme.
Lockheed U-2	AF 56-6682	4 Jun 1971	28 Apr 1989	709	Earth Resources Survey Programme.
Lockheed Model L 300-50A C-141A	c/n 6110	3 Feb 1972	Current	714	Gerard P. Kuiper Airborne Observatory.
Northrop T-38A *Talon*	AF 65-10357	20 Dec 1972		717	From Johnson Space Centre. NASA 915.
de Havilland DHC-6 *Twin Otter*	c/n 27	7 Aug 1973		720	FAA- on loan for augmented wing mods.
Convair CV.990 *Coronado* 'Galileo II'	c/n 37	10 Dec 1973		712	Replacement for 'NASA 711'.
Bell UH-1H *Iroquois*	Army 69-15231	4 May 1974		733	V/STOLand system. Variable stability.
de Havilland C-8A *Buffalo*	Army 63-13687	2 Aug 1974		715	Return from loan — N326D. QSRA aircraft.
		3 Aug 1978			Back from Boeing after modification.

Ames Research Center Aircraft — continued

Type and designation	Serial	Date arrived	Date departed	No.	Remarks
Learjet Model 24	c/n 24-102 N365EJ	28 Mar 1973			Leased — ASSESS, airborne telescope.
		10 Jun 1974	Current	705	Purchased.
Convair Cv.990 *Coronado*	c/n 29	8 May 1975	Nov 1983	713	To Davis Monthan AFB, Arizona.
		Dec 1978		710	Re-registered.
Cessna Model 402B *Businessliner*	c/n 402-0313	5 Jun 1975		719	To Langley 'NASA 503'.
Hughes OH-6A *Cayuse*	Army 67-16219	2 Apr 1976	11 Sep 1981	731	Crashed — Dave Barth killed.
Lockheed YO-3A	Army 69-18010	27 Apr 1977	Current	718	Helio noise measurements – programme supp.
Sikorsky SH-3G *Sea King*	BuNo.149723	9 Nov 1977	Current	735	Helio noise measurments — ex Langley.
Bell TH-1S *Huey Cobra*	Army 66-15248	1 Mar 1978	Current	736	Rotor blade experiments — ex Langley.
Bell UH-1H *Iroquois*	Army 64-13628	1 Mar 1978	Current	734	Chase helicopter — proficiency etc.
Bell XV-15 Model 301	c/n 001	23 Mar 1978	Current	702	V/STOL research and development.
Sikorsky RSRA	c/n 72-002	12 Feb 1979	Current	741	Rotor Systems Research Aircraft ex Langley.
Vertol CH-47B *Chinook*	Army 66-19138	14 Aug 1979	Dec 1989	737	Flying simulator ex Langley — to US Army.
Sikorsky RSRA	c/n 72-001	20 Sep 1979		740	Rotor Sytems Research Aircraft ex Langley.
Bell XV-15 Model 301	c/n 002	30 Oct 1980	Current	703	V/STOL research and development.
Lockheed ER-2	AF 80-1063	10 Jun 1981	Current	706	Earth Resources Survey.
Lockheed NC-130B *Hercules*	AF 58-712	9 Oct 1981	Current	707	Earth Resources Survey — Ex Johnson.
Beech Model 200 *Super King Air*	c/n BB-1164	5 Aug 1983	Current	701	Leased with option to purchase.
Lockheed ER-2	AF 80-1098		Current	709	Earth Resources Survey.
Hawker Siddeley AV-8C *Harrier*	BuNo. 158387		Current	719	Proficiency.
Hawker Siddeley YAV-8B *Harrier*	BuNo. 158394		Current	704	Ex-Patuxent River. Research.

Unusual non-standard NASA markings on Douglas R4D-Skytrain BuNo.99827 NASA 18 c/n 33110 which served at Ames between 14 December 1948 and 9 September 1965. After service with the US Army it became N48066 and is still flying today. (AP Photo Library)

APPENDIX FOUR
Lewis Research Center

Manufacturer and Type	Serial No.	NACA No.	Arrived	Departed	Remarks
Martin B-26B *Marauder*	AF 41-17604		1-43	10-43	Radial air-cooled engine performance tests.
Republic P-47G *Thunderbolt*	AF 42-24929		1-43	6-45	Radial air-cooled engine performance tests. Turbo-supercharging.
Cessna UC-78 'Bamboo Bomber'	AF 43-7283		3-43	10-45	Utility, low speed airspeed calibration and pace aircraft.
Bell P-63A *King Cobra*	AF 42-68889		10-43	6-45	Knock limited performance for liquid-cooled engines.
Lockheed RA-29 *Hudson* ('RELENTLESS')	AF 41-23412		12-43	7-45	Utility; Instrument development, and engine combustion analysis.
Consolidated B-24D *Liberator*	AF 42-40223		5-44	3-46	Improved knock performance with fuel additives and mixture tests.
Boeing B-29 *Superfortress*	AF 42-8357		6-44	8-44	R-3350 cooling problems and engine performance.
Vultee YA-31C *Vengence*	RAF EZ887		7-44	7-45	R-3350 mixture distribution problems.
Vultee YA-31C *Vengence*	RAF AF782		7-44	6-45	R-3350 ground tests and instrumentation development.
North American XB-25E *Mitchell*	41-13029		7-44	2-53	Effects of inflight icing on aircraft performance; all components. wing, tail, engine cowls, nose, props, and antenna. Statistical ice measurements.
Lockheed P-38J *Lightning*	AF 42-103973		10-44	7-45	Induction system icing study.
Federal AT-20	AF 41-35541		10-44	7-46	Exhaust augmented cooling system. (Canadian-built Avro Anson II).
Martin JM-1/B-26C *Marauder*	BuNo.66598		10-44	7-46	Flight tests for exhaust augmented cooling system.
Lockheed P-38J *Lightning*	AF 42-103979		12-44	4-45	Induction system icing study. (No.79).
Fleet	————		-45	——	
Bell P-63A *King Cobra*	AF 42-68868		2-45	7-55	Exhaust flame suppression and associated performance penalties.
Vultee BT-13 'Vibrator'	AF 42-89785		5-45	10-45	Instrument flight trainer.
Republic P-47N *Thunderbolt*	AF 44-88282		6-45	9-45	Fuel cooling tests to reduce boil-off during climb out.
Bell P-59A *Airacomet*	AF 42-2650		9-45	1-49	Jet thrust performance; thrust augmentation, and water injection.
North American AT-6 *Texan*	AF 44-81904		10-45	12-53	Instrument flight trainer. Replaced BT-13.
Northrop P-61 *Black Widow*	AF 42-39754		10-45	10-48	Sub-sonic ram jet test bed. (PK-754).
Fleet	————		-46	——	
Beech AT-11 *Kansan*	AF 42-36941		1-46	7-53	Utility. Development of icing and cloud measurement instruments.
Boeing B-29A *Superfortress*	AF 42-1808		4-46	10-48	Jet engine test bed. Jet engine icing. Ram jet test and launch vehicle.
Douglas R4D-6 *Skytrain*	BuNo.50795		4-10-46	3-65	Metro. R&D, gust survey, and utility transport. From Langley.
Douglas R4D-6 *Skytrain*	BuNo.50791		11-46	3-65	Same as above, plus search and recovery at Wallops Island. Icing research.
Douglas R4D-6/C-47J	BuNo.17268 AF 43-48510 c/n 25771		11-46	3-65	Utility and icing research. Transferred to MSFC as NASA 423, arriving on 6-18-65.
North American P-51H *Mustang*	AF 44-64702		3-47	6-53	Utility. Pace aircraft for air-speed calibration.
General Motors/Allison P-75 *Eagle*	NX 69940		9-47	7-49	Conter-rotating prop performance study.
North American XF-82	AF 44-83887		10-47	7-50	Ram jet test vehicle. Replaced B-29.
Bell L-39	BuNo.90060		12-12-49	——	Modified with swept wings. Transferred from Langley.
Bell L-39	BuNo.90061		12-12-49	——	Same as above.
North American F-82E *Twin Mustang*	AF 46-256	133	1-50	3-54	Hi-altitude icing. Statistical survey of icing encountered above 20,000 feet. Designated EF-82E on 7-6-49.
North American F-82B *Twin Mustang* ('BETTY JO')	AF 44-65168	132	9-50	6-57	Ram jet and aero-dynamic test missile launcher. Replaced XF-82 that was damaged in ground incident. (On 28-2-47 this aircraft set the distance record for fighters by flying from Honolulu to New York in 14-hr 31-min 50-sec, 4,968 miles).
McDonnell F2H-2B	BuNo.124942	209	1-55	12-59	Aero-dynamic missile-ram jet launch vehicle. Replaced F-82B to give higher launch altitude and velocities.
Grumman S2F-1 *Tracker*	BuNo.129140		10-55	2-57	Icing instrumentation development.

191

Manufacturer and Type	Serial No.	NACA No.	Arrived	Departed	Remarks
Lockheed F-94B *Starfire*	AF 51-5329	203	4-56	12-59	Aero-dynamic noise study.
Martin B-57B *Canberra*	AF 52-1576	637	5-56	19-3-73	Use of hydrogen fuel in jet engines. Noise suppressor
		237			evaluation. Solar cell calibration and hi-altitude radiation measurements. To Langley as NASA 516.
Martin B-57A *Canberra*	AF 52-1418	218	19-6-57	9-62	Flight hardware performance tests. Chase for B-57B, LH programme, aero-dynamic tests, rocket launch vehicle. Replaced F2H-2B.
Ryan L-17 *Navion*	48-947	217	3-58	4-65	Utility.
North American AJ-2 *Savage*	BuNo.134069	230	1-60	9-64	Zero gravity tests and facility-behavior of fluids.
North American AJ-2 *Savage*	BuNo.130412 (ZH)				Loaned aircraft.
North American AJ-2 *Savage*	BuNo.130421 (17)				Loaned aircraft.
Douglas C-47A *Skytrain*	AF 43-48086 c/n 13902	636 236	6-61	10-72	Acquired from ARDC Liaison office for additional transport support. Wallops recovery operations, etc.
Convair T-29B	AF 51-7909	5			
		250	5-4-63	24-11-76	Administrative. From Langley as NASA 250. Retired to MASDC-MDA.
Aero Commander Model 680	c/n 577-222	631 31	24-9-64	1-8-68	Small utility transport. To Langley as NASA 503.
Convair NF-106B *Delta Dart*	AF 57-2516	616	10-66	29-1-79	R&D for SST engine inlet design. Modified with two additional jet engines under the wing. Exhaust nozzle studies. To Langley as NASA 816.
Vought F-8A *Crusader*	BuNo.141354	666	5-69	14-7-69	Programme support and chase for NF-106B. Lost in landing accident at Cleveland.
Convair F-102A *Delta Dagger*	AF 56-998	617	6-70	6-74	Chase aircraft for NF-106B. To MASDC-DMA.
Douglas C-47B-51-DK	AF 43-49526	636	28-10-71	14-8-78	Programme support. Water and land quality evaluation. From Langley as NASA 501. To Dryden as NASA 817.
Grumman OV-1B *Mohawk*	64-14244	637	12-72	22-4-78	Project Ice Warn. Transferred to Langley.
Convair NF-106B *Delta Dart*	AF 57-2507	607	26-9-72	12-5-81	Programme support. Solar cell and Ocean colour scanner tests. Water and land quality evaluation. Transferred to Langley.
Vertol CH-46C/YHC-1	58-5514	533	24-3-75	——	From Langley for crash tests.
Convair C-131B	AF 53-7804	635	23-7-76		Acquired from USAF Eglin AFB, Florida. Applications Division Support. Water and land quality evaluation. Donated to University of Georgia.
Convair CV-340	c/n 003	707	3-9-76	19-7-77	Acquired from Ames. inspected, did not keep. To salvage.
Grumman G-159 *Gulfstream I*	c/n 125	5	15-9-76		Administration. Cargo door mod. From MSFC as NASA 10.
Learjet Model 25	c/n 25-035	616	5-79	Current	Water quality and solar cell evaluation. This aircraft was confiscated from Robert Vesco.
Lockheed C-130 *Hercules*	USCG 1351				US Coast Guard — Project Ice Warn.
de Havilland DHC-6 *Twin Otter*	c/n 004	607	-82	Current	From Langley as NASA 508.
Cessna Model M337B/O-2	377-M0236	635	10-83		Rotary engine R&D.
North American OV-10A *Bronco*	BuNo.155390	636	10-83	Current	Noise research programme.
North American YOV-10D *Bronco*	BuNo.155396	6	30-11-83		
Beech T-34 *Mentor*	c/n 144022	614		Current	Educational support services.

Unusual photo of a NACA Douglas R4D-6 Skytrain showing the fuselage markings. It is apparent that what was 'UNITED STATES NAVY' has been adapted to 'UNITED STATES NACA'. Under the port wing can be seen 'NACA' and the NACA winged emblem is on the fin and rudder.
(Peter M. Bowers)

APPENDIX FIVE
Dryden Flight Research Facility Aircraft

Manufacturer and Type	Serial No.	NACA No.	Date Arrived	Date Departed	Remarks etc.
Beechcraft UC-45 *Expeditor*	44-47110	105	May 1947		Transferred from Langley to Muroc FSFS, California in May 1947.
Bell X-1	46-063		7 Oct 1946	23 Oct 1951	1947-1951. Arrived from Pinecastle.
Boeing B-29 *Superfortress*	45-21800		7 Oct 1946		Ex Pinecastle, 'Mother' aircraft for X-1.
Bell X-1	46-062		5 Apr 1947	12 May 1950	Ex Pinecastle.
Douglas D-558-1 *Skystreak*	37971	141	25 Nov 1947	3 May 1948	Crashed on take-off 3 May 1948. Harold C. Lilly killed.
Douglas D-558-1 *Skystreak*	37972	142	22 Apr 1949	10 Jun 1953	Made 78 flights with NACA.
Douglas D-558-1 *Skystreak*	37970	140	14 Apr 1947	29 Sep 1948	Spares support for D-558-1 37972.
Douglas D-558-2 *Skyrocket*	37974	144	24 May 1949	20 Dec 1956	On 20 November 1953 first aircraft to fly twice the speed of sound.
Republic YF-84 *Thunderjet*	45-59490	134	18 Nov 1949		Received ex Langley. Vortex generator research.
Northrop X-4 *Bantam*	46-676		1950		Acquired 1950, but used only for spares support for 46-677.
Northrop X-4 *Bantam*	46-677		18 Aug 1950	Sep 1953	First flight 18 August 1950, last and 82nd flight September 1953.
Douglas D-558-2 *Skyrocket*	37975	145	8 Sep 1950	28 Aug 1956	First NACA flight 22 December 1950. Last and 66th flight 28 August 1956.
North American P-51D-25-NT *Mustang*	44-84958	148	25 Aug 1950		Ex Langley. Dives to 0.8 Mach. Proficiency aircraft.
Republic YF-84A *Thunderjet*	45-59488		1950	1954	Primarily used for proficiency flying.
Boeing EB-50A *Superfortress*	46-006			9 Nov 1951	Destroyed on ground 9 November 1951 with X-1 46-064 on board.
Boeing B-29A *Superfortress*	45-21787	137	1951	1959	'Mother' ship for D-558-1 and 2. 'Fertile Myrtle'.
Bell X-5	50-1838		9 Jan 1952	25 Oct 1955	First flight 9 January 1952. Last and 133rd 25 October 1955.
Bell X-5	50-1839		10 Dec 1951	14 Oct 1953	First flight 10 December 1951. Destroyed 14 October 1953.
Douglas R4D-6 *Skytrain*	50831		7 Jul 1952	1 May 1956	
Boeing B-47A *Stratojet*	49-1900	150	17 Mar 1953	1957	Used by Ames and Langley.
Convair XF-92A *Dart*	46-682		9 Apr 1953	14 Oct 1953	First flight 9 April 1953. Last and 25th 14 October 1953.
Douglas X-3 *Stilleto*	48-2992		23 Aug 1954	23 May 1956	First flight 23 August 1954. Last and 20th 23 May 1956.
Republic YRF-84F *Thunderjet*	51-1828	154	1954	1956	Pitch-up research.
North American F-86F *Sabre*	52-5426		1954		
North American F-100A *Super Sabre*	52-5778		1954	1960	
Convair YF-102 *Delta Dagger*	53-1785		1954	1958	
North American F-86F *Sabre*	52-5426		1954		Pitch-up research for US Air Force.
Bell X-1A	48-1384		14 Feb 1953	8 Aug 1955	First flight 14 February 1953. Destroyed in flight 8 August 1955.
Bell X-1E	46-063		3 Dec 1955	6 Nov 1958	First flight December 1955. Retired from test flying in April 1959.
Bell X-1B	48-1385		24 Sep 1954	27 Jan 1959	First flight 24 September 1954. Retired to USAF Museum 27 January 1959.
Boeing JTB-29A *Superfortress*	45-21800		1955	1958	'Mothership' for X-1 trials.
Douglas C-47H/R4D-5 c/n 12287	17136	817	5 Jan 1956	14 Aug 1978	
Lockheed U-2		320			No details.
Lockheed U-2	56-6700				No dates. Used for gust research and high altitude sampling.
Lockheed U-2A	55741				Fictitious serial — Edwards AFB 6 May 1960.
Douglas D-558-2 *Skyrocket*	37973	143	4 Feb 1948	17 Sep 1956	Returned 15 November 1955. Family flight 17 September 1956. Programme cancelled.
North American F-100C *Super Sabre*	53-1712		1956	1957	
Convair F-102A *Delta Dagger*	54-1374		1956	1959	
Lockheed YF-104A *Starfighter*	55-2961	818	1956	1975	
McDonnell F-101A *Voodoo*	53-2434	219		22 Aug 1956	To Langley 22 August 1956.
North American F-100C *Super Sabre*	53-1717		1957	1961	

Manufacturer and Type	Serial No.	NASA No.	Date Arrived	Date Departed	Remarks etc.
North American YF-107A-1-NA	55-5118	207	6 Nov 1957	3 Jun 1960	Grounded for spares support.
Lockheed F-104A *Starfighter*	56-734		1957	1961	
Lockheed YF-104A *Starfighter*	55-5118		1957	1958	Grounded for spares support.
Boeing KC-135A *Stratotanker*			1957		Retired after damage inflicted in mid-air collision.
Boeing KC-135A *Stratotanker*	55-3124		1958		Replacement aircraft.
North American YF-107A-1-NA	55-5120		10 Feb 1958	1 Sep 1959	Lost in accident.
Lockheed T-33A *Shooting Star*	49-939A		3 Feb 1958		Received from Langley 3 February 1958.
Boeing NB-52A *Stratofortress*	52-003		Nov 1958		Ex North American, Palmdale November 1958. X-15 'Mothership'.
Lockheed F-104A *Starfighter*	56-749		1959	1962	Destroyed.
Lockheed F-104B *Starfighter*	57-1303	819	1959	1978	Two-seat aircraft.
Republic F-105B *Thunderchief*	54-102		1959		Pilot familiarisation.
Boeing NB-52A *Stratofortress*	52-008	DFRC-008	8 Jun 1959	Current	Permanent loan, with effect from 26 April 1976.
North American X-15	56-6670	820	8 Jun 1959	24 Oct 1968	First flight 8 Jun 1959. Last and 199th X-15 flight 24 October 1968.
North American X-15	56-6671	821	17 Sep 1959	3 Oct 1967	First flight 17 September 1959. Last flight 3 October 1967. Became X-15-2.
North American X-15	56-6672	822	25 Nov 1960	15 Nov 1967	First flight 25 November 1960. Destroyed 15 November 1967.
Douglas JC-47D *Skytrain*	43-48273		26 Jun 1960	29 Aug 1960	
North American JF-100C *Super Sabre*	53-1709		1960	1964	Ex Ames. Variable stability studies.
Douglas F5D-1 *Skylancer*	139208a	212	1961	1963	Transferred to Ames for SST studies.
Douglas F5D-1 *Skylancer*	142350	213	Jun 1961	1970	
Cessna O-1A (L-19A *Bird Dog*)	51-2220		1962		
NASA FRC *Parasev*	N9765Z		1962	1964	Rogallo wing. Kite-parachute research aircraft.
Lockheed T-33A *Shooting Star*	55-4351	823	Jan 1963		
Lockheed T-33A *Shooting Star*	57-0721	934	28 Sep 1962	6 Nov 1970	
'Mother'			1967	1968	Radio controlled model flown to launch other models.
Lockheed NF-104N *Starfighter*	L683C-4045	811	19 Jul 1963		Special F-104 version for Dryden FRC.
North American A-5A *Vigilante*	147858		1963		SST Research.
Cessna TO-1A (L-19G *Bird Dog*)	-4128		1963	1964	
Aero Commander 680F ex N-6297X	680F-1288-131	801	May 1963	Current	
NASA FRC M2-F1	N-86652		1963	1964	Lifting body. First ground tow 5 April 1963. Air tow C-47.
Lockheed NF-104N *Starfighter*	L683C-4053	812	Jul 1963	Stored	Special version for Dryden FRC.
Beech *Debonair*	N-4307		1964	1965	
Cessna 310 c/n 8199M16-1			1964	1965	
Bell LLRV No.1			30 Oct 1964	1966	Lunar Landing Research Vehicle. First flew 30 October 1964.
Lockheed NF-104N *Starfighter*	L683C-4058	813	Sep 1963	8 Jun 1966	Collided with XB-70 62-207 8 June 1966. Destroyed.
Lockheed L1329 *Jetstar*	5003	814	May 1963		General purpose airborne simulator.
Beech	N-3849K		1965		
Cessna 210	N-910V		1965		
Piper *Apache*	N-4383P		1965		
Boeing *Stearman*	N-69056				Tow aircraft for Parasev 1A Rogalla kite-parachute vehicle.
Northrop M2-F2, M2-F3	NLB 101	803	Jun 1965	1972	Damaged in landing 10 May 1967. Rebuilt and retired 1972.
Northrop HL-10	NLB 102	804	Jan 1966		First unpowered landing 22 December 1966.
Lockheed F-104A *Starfighter*	56-790	820	1966	30 Oct 1983	Replacement for NASA 813. Withdrawn from use 30 October 1983.
McDonnell F-4A *Phantom* c/n 14	145313	824	3 Dec 1965	25 Jul 1967	Damaged in flight 25 July 1967. 630 hours. Hugh Jackson.
Piper	N-7845Y		1966		
Bell LLRV No.2			1967		Lunar Landing Research Vehicle. First flight January 1967. Retired.
North American XB-70A *Valkyrie*	62-0001		1967	4 Feb 1969	Total of 129 flights. Ferried to USAF Museum 4 February 1969.
General Dynamics F-111A	63-9771		1967	1971	
Piper PA-30-160B *Twin Comanche*	30-1498	808	1967	Current	Later used as RPRV aircraft. General purpose mission support.
Lockheed YF-12A *Blackbird*	60-6935	DFRC 935	10 Dec 1969	7 Nov 1979	First flew Dryden 10 December 1969. To USAF Museum 7 November 1979.
Martin X-24A	66-13551		17 Apr 1969	4 Jun 1971	First flew 17 April 1969. Last flight 4 June 1971. Rebuilt as X-24B.
Hyper III			1969	1969	One flight, launched from helicopter. Langley design.
General Dynamics F-111A	63-9777		1969	1971	
Bell 47G-3BE *Sioux* c/n 6670		948	10 Jan 1968	1 Nov 1973	Acquired 10 Jan 1968. Operated until 1 November 1973.

Manufacturer and Type	Serial No.	NASA No.	Date Arrived	Date Departed	Remarks etc.
Lockheed YF-12A *Blackbird*	60-0936		1970	24 Jun 1971	Lost on 24 June 1971 during a landing approach. USAF pilot.
Vought LTV TF-8A *Crusader*	141353	810	25 May 1969		SCW — Super Critical Wing. Arrived Dryden 25 May 1969.
Cessna U-3A *Blue Canoe*	57-5921A			7 Feb 1971	Transferred to Langley 7 February 1971.
Lockheed YF-12C *Blackbird*	60-0937	DFRC 937	16 Jul 1971	22 Dec 1978	Returned to US Air Force. Last NASA flight 28 September 1978.
Vought LTV F-8C *Crusader*	145546	802	25 May 1972	Current	First flight 25 May 1972. Current. Digital Fly-by-Wire — DFBW.
General Dynamics F-111A	63-9778		1 Nov 1972		First flight 1 November 1972. Transonic Aircraft Technology — TACT.
Northrop T-38A Talon c/n N5772	65-10353	821	28 Sep 1972		Ex Johnson Space Center — JSC.
Martin X-24B	66-13551		1 Aug 1973	26 Nov 1975	First flight 1 August 1973. Last flight 26 November 1975.
McDonnell Douglas F-15 *Eagle* RPRV			4 Dec 1972		1st 4 December 1972. Two vehicles. One is Spin Research Vehicle RSV. B-52.
Bell Model 47G-3BJ *Sioux*	6670	948	10 Jan 1968	1 Nov 1973	Ex Johnson Space Center. Disposed 1 November 1973.
General Dynamics F-111E *Aardvaark*	67-0115		4 Sep 1975		First flew 4 Sep 1975. Integrated Propulsion Control System — IPCS.
Boeing 747-123 c/n 20107 N9668	N905NA	905	22 Jun 1974	Current	Tests with OV-101 Enterprise — shuttle carrier.
Martin B-57B-MA	52-1576	809	8 Feb 1975	Preserved	Ex Langley 8 February 1975. Preserved for AFFTC Museum, Edwards AFB.
General Dynamics YFB-111A c/n 18	63-9783		Apr 1974		Prototype FB-111A. Seen Dryden April 1974.
General Dynamics F-111E c/n 160	67-0115		Oct 1975		Seen Dryden FRC October 1975.
Learjet 23 c/n 23-049	N701NA	701	Aug 1974		Wake vortex research from Boeing 747. Loaned from Ames.
Cessna T-37B-CE	60-84		Aug 1974		Wake vortex research from Boeing 747 NASA 905.
Bell UH-1H *Iroquois* c/n 11519	69-15231		Oct 1975		
Northrop YA-9A	71-11367		Oct 1975		Seen Dryden FRC October 1975.
Northrop YA-9A	71-11368		Oct 1975		Seen Dryden FRC October 1975.
Lockheed TF-104G *Starfighter*		824	Oct 1975		Luftwaffe aircraft 27+37.
Lockheed TF-104G *Starfighter*	66-13628	825	Oct 1975	Current	Luftwaffe aircraft 28+09 c/n 5939.
Lockheed TF-104G *Starfighter*		826	Oct 1975	Current	Luftwaffe aircraft 26+64 c/n 8213.
McDonnell Douglas F-15A *Eagle*	71-0281		1975	1990	Drag research.
McDonnell Douglas F-15A *Eagle*	71-0287		1975	1990	Drag research.
Mini-Sniffer I, II, III			1975		RPRV. Propeller driven used for gathering air samples.
Northrop YF-17 *Cobra*	70-1569		1976		Base drag studies for US Navy.
Space Shuttle Orbiter	OV-101				Enterprise. First launch from B-747 NASA 905 on 18 February 1977.
Northrop T-38A *Talon* c/n N5951	66-8381	901	May 1978	Aug 1970	
Douglas C-47B-51-DK *Skytrain*	43-49526	817	14 Aug 1978	Nov 1984	Ex Lewis Research Center.
Convair NF-106B *Delta Dart*	57-2516	816	29 Jan 1979		Wing leading edge vortex flaps.
Boeing NKC-135A *Stratotanker*	55-3129		1979	1980	Winglets.
Ames-Dryden AD-1	N805NA	805	1979		Oblique wing test-bed. Adjustable wing.
Himat No.1, No.2 (Rockwell-NASA)			1979		Highly Manoeuvrable Aircraft Technology. RPRV.
Grumman F-14A *Tomcat*	157991	991	1979		NASA — US Navy programme.
Martin B-57B-MA	52-1576	516	19 Mar 1973		Ex Langley.
Bell XV-15	N702NA	702	1981		VTOL research aircaft ex Ames.
Bell XV-15	N703NA	703	1981		VTOL research aircraft ex Ames.
Grumman X-29A	82-0003	003	11 Oct 1984	Current	Defense Advanced Research Projects Agency — DARPA. USAF.
PIK 20E Sailplane	N803NA	803	22 Oct 1983	Current	Ex N202NA.
McDonnell Douglas F-18 *Hornet*	160780	840	22 Oct 1984	Current	HARV-TVCS.
McDonnell Douglas F-18 *Hornet*	161213	844			
General Dynamics NF-111A	63-9776		Mar 1983		Seen Dryden FRC March 1983.
Grumman X-29A	82-0049		5 Nov 1988	Current	First flight 23 May 1989. Delivered 7 November 1988.
McDonnell Douglas F-15A *Eagle*		835	5 Jan 1976	Current	HIDEC.
McDonnell Douglas F-18 *Hornet*	160781	845	24 Jul 1986	Current	
General Dynamics F-111	63-9778				Advanced Fighter Technology Integration Flight — AFTI.
General Dynamics F-16 *Fighting Falcon*	75-0749		2 May 1988	Current	
Boeing 720-027 c/n 18066	N833NA	833	30 Oct 1983		Seen Dryden FRC 30 October 1983.
General Dynamics F-16XL *Fighting Falcon*	75-0747		1990	Current	Supersonic laminarflow characteristics.
General Dynamics F-16XL *Fighting Falcon*	75-0750		16 Jun 1982	Current	Supersonic laminarflow characteristics.
Lockheed SR-71A	64-17980		15 Feb 1990	Current	
Lockheed SR-71A	64-17971		19 Mar 1990	Current	
Lockheed SR-71B	64-17956	831	15 Feb 1990	Current	
McDonnell Douglas F-18A *Hornet*	161216	841	1 Oct 1985	Current	

Dryden Flight Research Facility Aircraft — continued

Manufacturer and Type	Serial No.	NASA No.	Date Arrived	Date Departed	Remarks etc.
McDonnell Douglas F-18A *Hornet*	161214	842	24 Aug 1987	Current	
McDonnell Douglas F-18A *Hornet*	161250	843	6 Nov 1987	Current	
McDonnell Douglas F-18A *Hornet*	161520	847	21 Sep 1989	Current	
McDonnell Douglas F-18A *Hornet*	161949	848	28 Dec 1989	Current	
McDonnell Douglas F-18A *Hornet*	161213	844			Crashed 7 October 1988.
Vought F-8A *Crusader*	145385	816	Oct 1975		. Ex NASA 616
Northrop T-38 *Talon* c/n N5772	65-10353	821	28 Sep 1972	Current	
Convair CV.990 *Coronado* N5617NA		810	6 Mar 1989	Current	

Dryden Research Operations Division — NASA Aircraft Inventory, 1 March 1990

NASA No.	Type	c/n	Military Serial No.	Date Acquired	Category/Status
803	PIK 20E *Sailplane*	N202NA		22 Oct 1983	RD NASA
808	PA.30 *Twin Comanche*	c/n 30-1498		5 Jun 1987	PS NASA
810	Convair CV.990 *Coronado*	N5617NA		6 Mar 1989	PS NASA
821	Northrop T.38 *Talon*	c/n N5772	65-10353	28 Sep 1972	PS NASA
825	Lockheed TF-104	c/n 5939	66-13628	3 Jul 1975	RD NASA
826	Lockheed F-104	c/n 8213		3 Jul 1975	RD NASA
835	McDonnell Douglas F-15 *Eagle* HIDEC		71-0287	5 Jan 1976	RD USAF
836	Boeing NB-52 (Permanent loan 26 April 1976)		52-008	8 Jun 1959	RD USAF
840	McDonnell Douglas F-18A *Hornet* HARV-TVCS		160780	22 Oct 1984	RD NAVY
841	McDonnell Douglas F-18A *Hornet*		161216	1 Oct 1985	PS NAVY
842	McDonnell Douglas F-18A *Hornet*		161214	24 Aug 1987	PS NAVY
843	McDonnell Douglas F-18A *Hornet*		161250	6 Nov 1987	PS NAVY
845	McDonnell Douglas TF-18 *Hornet*		160781	24 Jul 1986	PS NAVY
847	McDonnell Douglas F-18A *Hornet*		161520	21 Sep 1989	PS NAVY
848	McDonnell Douglas F-18A *Hornet*		161949	28 Dec 1989	PS NAVY
849	General Dynamics F-16XL *Falcon*		75-0749	2 May 1988	RD USAF
8xx	General Dynamics F-16XL *Falcon*		75-0747	Expect in summer of 1990	RD USAF
750	General Dynamics F-16 *Falcon* AFTI		75-0750	16 Jun 1982	RD USAF
003	Grumman X-29 No.1		82-0003	11 Oct 1984	RD DARPA
049	Grumman X-29 No.2		82-0049	5 Nov 1988	RD DARPA
	Lockheed SR-71A		64-17980	15 Feb 1990	RD USAF
	Lockheed SR-71A		64-17971	19 Mar 1990	RD USAF
	Lockheed SR-71B		64-17956	15 Feb 1990	RD USAF

Some NACA-NASA aircraft carried the name of the centre, an example being this Northrop T-38A Talon ex-65-10353 NASA 821 with 'DRYDEN FLIGHT RESEARCH CENTER' on the tail. Photo was taken at Edwards AFB in February 1977. The Boeing 747 NASA 905 shuttle carrier is in the background showing the attachment arms for the orbiter. (Vic Seeley)

APPENDIX SIX
Johnson Space Flight Centre Aircraft

Manufacturer and Type	Serial No.	NASA No.	Arrived	Departed	Disposition	JSC Time	Remarks
Convair TF-102A *Delta Dagger*	AF 54-1358		9-04-62	7-01-64	Perrin AFB	407.6	Two place.
Convair F-102A *Delta Dagger*	AF 55-3391		9-10-62	7-27-64	Perrin AFB	496.7	
Convair F-102A *Delta Dagger*	AF 55-3405		9-10-62	8-13-64	Perrin AFB	519.6	
Lockheed T-33A	AF 57-722	917/935	9-17-62	3-05-71	Andrews AFB	26661.1	
Lockheed T-33A	AF 57-721	916/934	9-26-70	11-06-70	Maxwell AFB	2582.3	
Lockheed T-33A/TV-2	AF 51-9157	224	10-01-62	11-10-64	MASDC-DMA	149.7	From Langley.
Convair TF-102A *Delta Dagger*	AF 54-1356		11-01-62	7-21-64	Perrin AFB	300.6	
Lockheed T-33A	AF 53-5178	911/944	1-25-63	6-13-67	Pensacola	1460.1	
Lockheed T-33A	AF 53-6009	943	1-25-63	10-19-66	City of Bristol, Tenn.	1417.5	Display.
Grumman G-159 *Gulfstream I*	c/n 98	2	3-28-63			17,979	Administrative.
Lockheed T-33A	AF 58-671	918/936	5-31-63	3-05-71	Hill AFB	2252.6	
Bell Model 47G	N2490B		7-02-63	9-06-63	Houston Helicopter	31.6	Leased.
Lockheed T-33A	AF 52-9225		12-16-63	1-02-64	Robins AFB	1.7	
Lockheed T-33A	AF 53-4972	946	1-02-64	6-02-67	Pensacola	1025.4	
Lockheed T-33A	AF 53-5400	945	1-14-64	1-12-71	NASA Ames	1904.3	
Lockheed T-33A	AF 52-9335	942	2-02-64	10-10-66	MASDC-DMA	783.5	
Lockheed T-33A	AF 52-9360	906/941	2-02-64	8-31--66	MASDC-DMA	801.4	
Bell TH-13N	Bu.145841	930	2-10-64	2-02-68	Pensacola	1597.8	
Sikorsky UH-19D	56-4261		3-04-64	3-04-65	USAF Eagle Mountain Lake	67.0	
Convair CV-240A	c/n 149	26/926	4-07-64	4-02-70	MASDC-DMA	2067.1	Ex-N94272
Northrop T-38A *Talon*	AF 63-8181	901	5-27-64	2-28-66	Accident/Fatal	634.9	See & Bassett, St Louis, Mo.
Northrop T-38A	AF 63-8188		6-10-64	10-31-64	Accident/Fatal	207.8	Ted Freeman, Tallahassee, Fl.
Northrop T-38A	AF 63-8193	902	7-09-64		Current	4361.4	
Northrop T-38A	AF 63-8200	903	7-15-64	12-27-64	Loan to USN Current	4790.4	
		Returned	12-27-76		*Patuxent River*		
Northrop T-38A	AF 63-8204	904	7-22-64	5-04-73	Loan Det.51 Current	4178.7	
		Returned	9-20-78		Palmdale, Ca.		
Sikorsky UH-34G	Bu.138461		11-16-64	8-12-65	Jacksonville NAS	114.2	
Lockheed T-33A	AF 53-6011	940	12-10-64	5-19-70	MASDC-DMA	422.3	Research.
Northrop T-38A *Talon*	AF 61-912	907 *905	5-07-65	6-28-73	Loan to Kelly AFB	2709.3	
		Returned	9-18-78				
Lockheed T-33A	AF 56-3671	919/937	5-24-65	11-20-70	MASDC-DMA	1343.0	
Bell OH-13H *Sioux*	57-6207	931	7-06-65	2-14-68	NASA Langley	1025.3	To NASA 537.
Sikorsky SH-3A *Sea King*	Bu.149723	933	8-26-65	7-20-70	NASA Langley	760.6	To NASA 538
Northrop T-38A	AF 65-10326	906	11-18-65	3-12-73	Loan to Eglin AFB	3636.7	
		Returned	9-14-78		Current		
Northrop	AF 65-10328	908	11-29-65	9-10-72	NASA Langley	2297.5	Loaned. NASA 514.
		Returned	6-20-80		Current		
Northrop T-38A	AF 65-10329		12-07-65	12-15-65	NASA Langley		NASA 511.
Northrop T-38A	AF 65-10327	907	12-17-65	1-05-73	Loan to NAS Patuxtent	2090.5	Lost in accident at NAS Pax.
Lockheed NP-3A *Orion*	Bu.148276	927	12-20-65	9-29-77	NASA Wallops	5276.5	NASA 428
Northrop T-38A	AF 65-10351	909	1-28-66		Current	5771.6	

Johnson Space Flight Center Aircraft — continued

Manufacturer and Type	Serial No.	NASA No.	Arrived	Departed	Disposition	JSC Time	Remarks
Northrop T-38A	AF 65-10352	910	1-31-66	3-12-73	Loan to Eglin AFB	3602.2	
		Returned	8-11-78		Current		
Northrop T-38A	AF 65-10353	911	2-10-66	9-28-72	NASA Dryden	2014.8	NASA 821
Northrop T-38A	AF 65-10354	912	2-09-66	3-12-73	Loan to Eglin AFB	2673.5	
		Returned	8-16-78		Current		
Northrop T-38A	AF 65-10356	914	2-10-66		Current	5546.0	
Northrop T-38A	AF 65-10355	913	2-16-66		Current	6125.4	
Northrop T-38A	AF 65-10357	915	2-16-66	12-20-72	NASA Ames	1936.2	NASA 717
Douglas C-133A *Cargomaster*	AF 54-136	928	4-13-66	8-19-69	Dover AFB	3.5	Prog. support.
Lockheed T-33A	AF 56-1584	947	6-08-66	6-13-67	Pensacola NAS	213.5	
Lockheed T-33A	AF 56-1589	948	6-08-66	6-15-67	Pensacola NAS	230.7	
Lockheed T-33A	AF 56-1720	949	6-08-66	6-02-67	Pensacola NAS	134.9	
Lockheed T-33A	AF 56-1730	950	6-08-66	6-22-67	Dept of Interior, Phoenix, Az.	110.7	
Lockheed T-33A	AF 56-3666	941	9-20-66	9-03-70	MASDC-DMA	805.8	
Lockheed T-33A	AF 57-688	942	10-07-66	2-17-71	Andrews AFB	993.2	
Lockheed T-33A	AF 55-4386	938	10-07-66	3-09-71	Kelly AFB	923.0	
Lockheed T-33A	AF 56-3689	939	10-07-66	9-28-70	NASA Langley	870.6	
Lockheed T-33A	AF 57-720	943	10-13-66	2-25-71	Andrews AFB	1150.3	
Lunar Lander Research Vehicle	#1		12-13-66	5-06-66	Destroyed/ Neil Armstrong	08:21:18	No injuries.
Lunar Lander Research Vehicle (LLRV)	#2		1-26-67	8-27-72	NASA Dryden	00:00:00	
Northrop T-38A *Talon*	AF 66-8381	901	5-04-67		Current	5387.7	
Northrop T-38A	AF 66-8382	916	5-04-67		Current	5343.8	
Northrop T-38A	AF 66-8383	917	5-16-67		Current	5437.9	
Northrop T-38A	AF 66-8384	918	5-16-67		Current	5102.3	
Northrop T-38A	AF 66-8385	919	5-17-67		Current	5162.5	
Northrop T-38A	AF 66-8386	920NS	5-19-67		Current	5170.3	
Northrop T-38A	AF 66-8387	921NS	5-23-67		Current	5038.8	
Northrop T-38A	AF 66-8355	923	7-26-67		Current	4585.6	
Northrop T-38A	AF 66-8354	922	8-04-67	10-05-67	Accident/Fatal	66.5	C. Cliff Williams.
Lunar Landing Training Vehicle	#1	950	10-09-67	12-02-67	Destroyed/J. Algranti	02:08:35	No injuries.
Bell Model 47G-3B1	6663	946	12-05-67	8-02-69	Destroyed/Gibson	916.5	No injuries.
Lunar Lander Training Vehicle	#2	951	12-07-67	1-29-71	Destroyed/Stu Present	24:02:26	No injuries.
Lunar Lander Training Vehicle	#3	952	12-07-67	11-29-72	Termination of flying	33:18:51	
Bell Model 47G-3B1	6665	947	12-11-67	1-23-71	Destroyed/ Gene Cernan	1353.3	No injuries. Florida.
Bell Model 47G-3B1	6670	948	1-10-68	11-01-73	NASA Dryden	2373.8	
Northrop T-38A *Talon*	AF 67-14825	924	4-02-68		Current	5712.6	
Lockheed NC-130B *Hercules*	AF 58-712	929	7-18-68	10-09-81	NASA Ames	5836.8	NASA 707.
Bell 47G-3B2	6754	949	10-29-69	6- -83	Storage	2863.6	Research.
Northrop T-38A	AF 69-7082	955	8-13-70		Current	4949.1	
Northrop T-38A	AF 69-7084	956	9-17-70		Current	4648.1	
Northrop T38A	AF 69-7086	957	10-15-70	5-10-72	Destroyed/ Pete Conrad	610.4	No injuries. Austin, Texas.
Northrop T-38A	AF 69-7088	958	11-13-70	1-20-72	Destroyed/ Present and Heath	333.9	No injuries. Matagorda Is.
Northrop T-38A	AF 70-1550	959	12-10-70		Current	4833.1	
Northrop T-38A	AF 70-1552	960	1-06-71		Current		4610.6
Northrop T-38A	AF 70-1555	961	2-17-71		Current	4669.7	
Northrop T-38A	AF 70-1556	962	3-05-71		Accident/Tom K. Mattingly	4723.0	No injuries. 11-30-82. Repair 9-83.
General Dynamics/Martin WB-57F	AF 63-13501	925	7-19-72	9-15-82	MASDC-DMA	1172.6	Research.
General Dynamics/Martin WB-57F	AF 63-13503	926	7-21-72		Current	3169.0	Research.
Grumman OV-1C *Mohawk*	67-18915	928	11-29-72	4-25-73	Destroyed/Conway Roberts/C. Hayes	64.7	No injuries.
Boeing KC-135A *Stratotanker*	AF 59-1481	930	8-15-73		Current	1755.9	Prog. support, Zero G trainer.
Bell 206B *Jet Ranger*	c/n 508	950NS	10-01-73			1442	Research. From JPL, NASA 12.
Grumman (STA) G-1159 *Gulfstream II*	c/n 146	946	5-20-74		Current	1931.5	
Grumman (STA) G-1159 *Gulfstream II*	c/n 147	947	6-13-74		Current	2005.5	
Boeing (SCA) Model B-747-123	c/n 20107	905	6-22-74		Current	417.0	Prog. support Ex-Am.A/L.
General Dynamics/Martin WB-57F	AF 63-13298	928	6-24-74		Current	2053.0	Prog. support D.O.E./NASA
Boeing YC-97J/SGB377 'Super Guppy'	AF 52-2693	940	7-13-79		Current	320.8	Prog. support.
Grumman (STA) G-1159 *Gulfstream II*	c/n 144	944	5-06-83		Current		
Boeing (SCA) Model B-747-100	c/n 20781	911	4-17-89		Current		ex-Japan A/L.
Grumman (STA) G-1159 *Gulfstream II*	c/n 118	945	7-05-89		Current		ex-NASA 650.
Northrop T-38A	ex 00585	915	5-24-89		Current		ex-Holloman AFB.
Northrop T-38A	59-1603	963	1-27-89		Current		

Aircraft on JSC Records for: National Space Technology Laboratory

Manufacturer and Type	Serial No.	NASA No.	Arrived	Departed	Remarks
Beech D-18S	c/n BA-155	948	9-03-74		
Lear Jet Model 35A	c/n 35-213	935	3-28-79		To Mexico via State Dept. (per telcom with Ken Haugen).
Lear Jet Model 23	c/n 23-049	933	1-11-80		From NASA Ames Research Center as NASA 701.

APPENDIX SEVEN
Wallops Flight Facility

Manufacturer and Type	Serial No.	NASA No.	Arrived	Departed	Remarks
Lockheed EC-121P	BuNo.143201 c/n 4475	671	5-21-65	5- -71	Programme support. To MASDC-DMA.
Piasecki CH-21C *Work Horse*	56-4342	672	7-25-65		Programme support. Curt Allen said 672 and 673 were used in Canada.
Piasecki CH-21C *Work Horse*	56-4344	673	7-25-65		Programme support.
Lockheed EC-121K	BuNo.145937 c/n 5518	670	8-05-65	5- -71	Programme support. To MASDC-DMA.
Beech Model 65-80 *Queen Air*	c/n LD-79	8	7-08-68		Administrative and some R&D. Acquired from NASA JPL. Traded in on new *King Air*.
Douglas C-54G *Skymaster*	AF 45-637 BuNo.36031	432	4-21-69	1- -77	Programme support. Acquired from GSFC. To MASDC-DMA.
Douglas C-54G *Skymaster*	AF 45-578 BuNo.36031 c/n 307	438	4-21-69		Programme support. Acquired from GSFC.
Douglas C-54G *Skymaster*	AF 45-556 BuNo.66009 c/n 291	427	6-01-69	9-05-79	Programme support. Acquired from GSFC. To MASDC-DMA.
Bell UH-1B 'Huey'	64-13907 c/n 1031	424	12-11-72	3-07-74	Programme support. Lost in non-fatal accident, $70K damage.
Bell UH-1B 'Huey'	64-14021	424	3-08-74	Current	Replaced damaged NASA 424. This aircraft is on floats.
Douglas C-118A *Liftmaster*	AF 51-3828 BuNo.43575 c/n 4375	428	7-28-75	12-09-76	Programme support. Acquired from GSFC. To MASDC-DMA.
Beech C-45H *Expediter*	c/n AF-534	650	7-22-76	10-09-79	Programme support. Acquired from MSFC. To MASDC-DMA.
Lockheed NP-3A *Orion*	BuNo.148276	428	9-29-77	Current	Programme support. Acquired from JSC as NASA 927.
Lockheed L-188 *Electra*	c/n 1103	429	9-27-78	Current	Programme support. Acquired from FAA.
Short Brothers & Harland SC-7 *Skyvan*	c/n SH1844	430	7-18-79	Current	Programme support. Ex-N30DA.
Beech Model 200 *Super King Air*	c/n BB-950	8	10-10-81	Current	Administrative.
Bell UH-1B (floats)	63-8695	415		Current	Applications research platform.
NA T-39 *Sabreliner*	BuNo. 150970	431		Current	Platform for small application systems.

BIBLIOGRAPHY

The NASA History Series
The Superintendent of Documents, US Government Printing Office, Washington DC. 20402.
EP-145. *Sixty Years of Aeronautical Research 1917–1977.* David A. Anderton.
SP-468. *Quest for Performance,* The Evolution of Modern Aircraft. Laurence K. Loften Jr.
SP-4103. *Model Research. NACA 1915–1958.* Vol.2. Alex Roland.
SP-4302. *Adventures in Research.* A History of Ames Research Center 1940-1965. Edwin P. Hartman.
SP-4303. *On the Frontier.* Flight Research at Dryden 1946–1981. Dr Richard P. Hallion.
SP-4304. *Searching the Horizon.* A History of Ames Research Center 1940-1976. Elizabeth A. Muenger.
SP-4305. *Engineer in Charge.* A History of the Langley Aeronautical Laboratory 1917–1958. Dr James R. Hansen.
SP-4406. *Orders of Magnitude.* A History of the NACA and NASA 1915–1990. Roger E. Bilstein.
Frontier of Flight. The Story of NACA Research. George W. Gray. (1945).
History of NACA-NASA Rotating Wing Aircraft Research 1915-70. Frederick B. Gustafson.

The X-Planes X-1 to X-31. Jay Miller. Aerofax.
Bell X-1 Variants. Ben Guenther and Jay Miller. Aerofax.
Lockheed U-2 and TR-1. Jay Miller and Chris Pocock. Aerofax.
Dragon Lady. The History of the U-2 Spyplane. Chris Pocock. Airlife.
Black Magic. America's Spyplanes: SR-71 and U-2. Michael O'Leary and Eric Schulzinger. Airlife.
Designers and Test Pilots. Dr Richard P. Hallion. Time-Life.
Northrop. An Aeronautical History. Fred Anderson. Northrop.
U.S. Military Aircraft Designations and Serials since 1909. John M. Andrade. Midland Counties.
Grumman Aircraft since 1929. Rene J. Francillon. Putnam.
Boeing Aircraft since 1916. Peter M. Bowers. Putnam.
Curtiss Aircraft 1907-1947. Peter M. Bowers. Putnam.
McDonnell Douglas Aircraft since 1920. Vol.1. Rene J. Francillon. Putnam.
United States Military Aircraft since 1909. Gordon Swanborough — Peter M. Bowers. Putnam.
The Helicopters. Warren R. Young. Time-Life.